A VARIATIONAL APPROACH TO STRUCTURAL ANALYSIS

A VARIATIONAL APPROACH TO STRUCTURAL ANALYSIS

DAVID V. WALLERSTEIN

A Wiley-Interscience Publication
JOHN WILEY & SONS, INC.

This book is printed on acid-free paper. ♾

Published simultaneously in Canada.

Library of Congress Cataloging-in-Publication Data:

Wallerstein, David V.
A variational approach to structural analysis / David V. Wallerstein
p. cm.
"A Wiley-Interscience publication"
ISBN 0-471-39593-5 (cloth : alk. paper)
1. Structural analysis (Engineering) 2. Calculus of variations. I. Title.
TA646.W35 2002 624.1—dc21 2001026914

Printed in the United States of America.

10 9 8 7 6 5 4 3 2 1

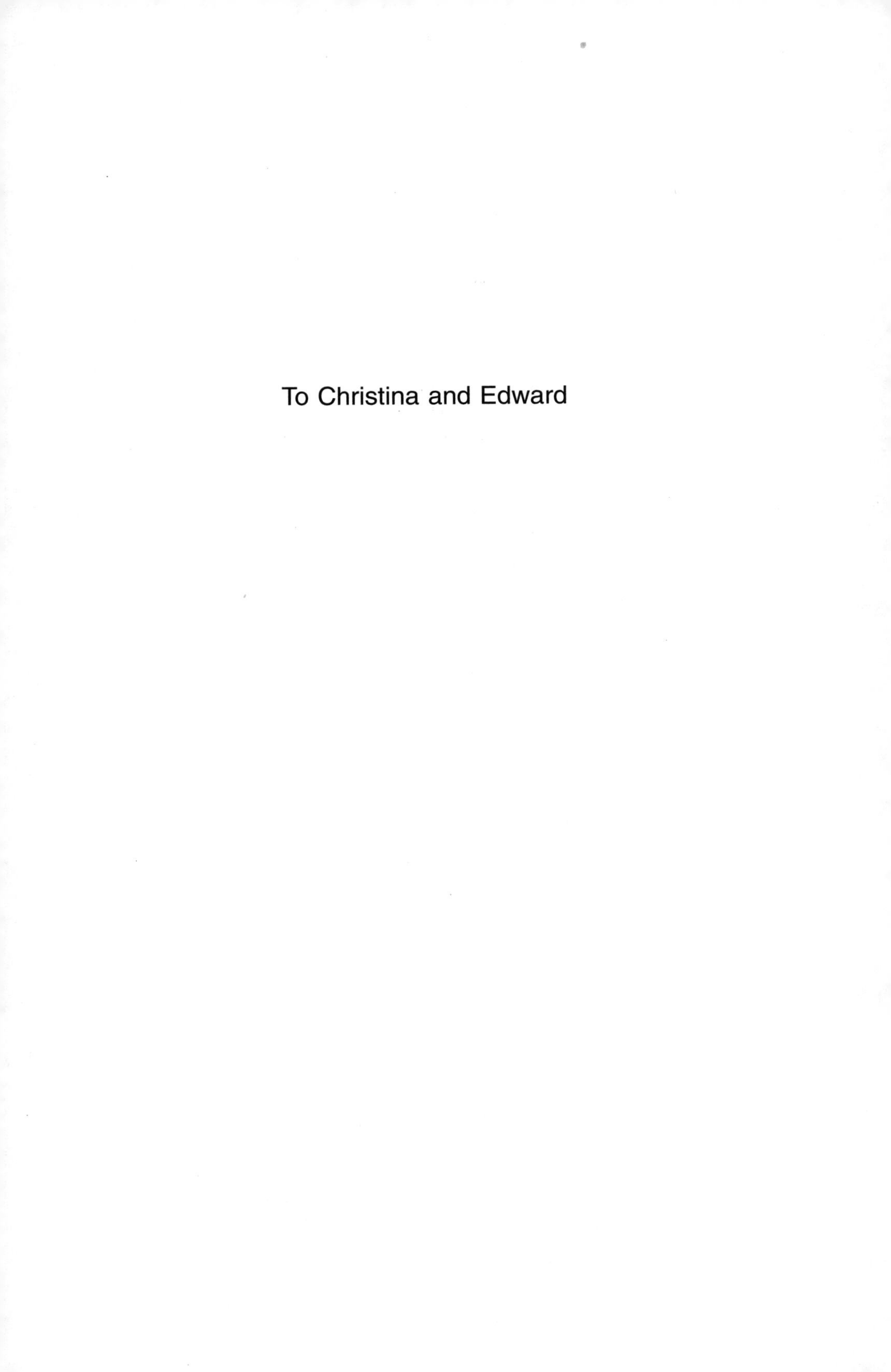

To Christina and Edward

CONTENTS

PREFACE

My objective in writing this book has been to provide a discourse on the treatment of variational formulations in deformable structures for upper-level undergraduate through graduate-level students of aeronautical, civil, and mechanical engineering, as well as engineering mechanics. Its self-containment is also designed to be useful to practicing professional engineers who need to review related topics. Emphasis is placed on showing both the power and the pitfalls of virtual methods.

Today, analysis in structures, heat transfer, acoustics, and electromagnetics currently depends on the finite element method or the boundary element method. These methods, in turn, depend on virtual methods or their generalization represented by techniques such as the Galerkin method.

The notes for this book come from a graduate aerospace course in structures that I have taught at the University of Southern California for the last eighteen years. The students come from such diverse backgrounds as mechanical engineering, structural engineering, physics, fluid dynamics, control theory, and, occasionally, astronomy. At least half of an average class is made up of graduates working in the industry full-time. The majority view the course as a terminal course in structures, but only about half have a firm understanding of structures or plan to make a career in the subject. All, however, need an understanding of virtual methods and structures to survive in the industrial world. The control theory expert must be able to convey his or her needs to the structural-modeling group. It is highly embarrassing to find out three years into a project that there has been a lack of communication between design groups.

Without the necessity of memorizing numerous formulas, virtual methods provide a logical, unified approach to obtaining solutions to problems in mechanics. Thus students have a readily understood method of analysis valid across all areas of mechanics, from structures to heat transfer to electrical–mechanical, that will remain with them throughout their careers.

Students often want to be taught the finite element method without having any basic understanding of mechanics. Although my notes touch in an extremely basic way on finite element methods (FEM), I try to impress on students that FEM codes are nothing more than modern slide rules that are only effective when the underlying mechanics is fully understood. All too often, I have observed in my twenty-odd years as a principal engineer for the world's largest FEM developer that even in industry, there is a blind use and acceptance of FEM results without consideration of whether the model is a good mathematical representation of the physics. That said, however, I should also state that I have encountered clients with a far better understanding of the use of the code and its application to physics than the code developers.

The text is divided into six chapters, with discussion basically limited to springs, rods, straight beams, curved beams, and thin walled open beams (which represent a specialized form of shells). Springs and rods were selected because the reader can easily understand them, and they provide a means in nonlinear discussions to get closed solutions with very general physical interpretations. The three classes of beams were chosen for several reasons. First, my years in aerospace and my current interaction with both auto-industry and aircraft-industry clients indicate that beams are still a very important structural component, as they are used extensively in both automobile and aircraft structures and to model such complex nonlinear problems as blade-out conditions on aircraft engines. Second, beams can be used to clearly demonstrate the ease with which virtual methods can be applied to determine governing equations. Third, beams can be used to clearly demonstrate when and why virtual methods will yield inconsistent results. Fourth, beams yield equations that can generally be solved by the student. And fifth, the physics of the solution can be discussed in relationship to what an FEM code can perform.

The first two chapters are mainly introductory material, introducing the variational notation used and reviewing the equilibrium and compatibility equations of mechanics. Since variational methods rely heavily on integration by parts and on the variational operator functioning in a manner similar to a total differential, these techniques are discussed in great detail. Though the text itself does not extensively use the concept of adjoint operators, a section is included here in an attempt to bridge the gap between basic mechanics courses, which make no mention of the subject, and advanced texts, which assume extensive knowledge of the subject. Legendre transformations are introduced for later use in establishing the duality between variational methods and complementary virtual methods. Lagrange multipliers are discussed, and a possible physical interpretation is given.

The third chapter covers virtual work. It uses kinematical formulations for the determination of the required strain relationships for straight, curved, and thin walled beams. The importance of accounting for all work is emphasized, and it is pointed out that if a particular strain assumption (such as the theory of plane sections) does not contain certain strain components, virtual work will lose information. It is also pointed out, however, that if additional assumptions consistent with the fundamental constraints are included, then virtual work can yield more general results. Examples of this are discussed with the straight beam theory, where the assumption that the tangent to the deflection curve is normal to the plane section is relaxed, yielding a shear deflection theory for straight beams. For curved beams, the requirement that curved beams do not distort in the plane of the cross section under thermal loading is relaxed; the removal of this restriction is required to get correct thermal stress results for curved beams. In thin walled beams of open section, the fundamental relationship is that the shear strain of the median surface is zero. By defining a shear strain consistent with this requirement, the Saint Venant torsion theory for thin walled beams of open section is automatically included via virtual work. Additionally, this chapter shows how easily virtual work concepts can produce the resultant curvature-stress relationships or the differential equations for deflection curves. One observation made in teaching the material of this chapter is that the homework problems that seem to pose the most difficulty when it is asked to have them solved with the use of virtual work are those types usually found early in a strength course involving stress and strain with axial loading. In these problems, the student really has to think in selecting a virtual displacement.

The fourth and fifth chapters cover complementary virtual work and energy methods. These are problem-solving chapters. For complementary virtual work, emphasis is put on the selection of virtual loads that meet the requirement of static equilibrium and answer the question: Does the selected virtual set yield results useful to the solution of the problem at hand? Virtual load diagrams are used extensively for complementary virtual work solutions. In Chapter 5, stationary potential energy, Castigliano's first theorem, and the Engesser-Crotti theorem are derived and used in the solution of problems. Variational statements are discussed and generalized with the introduction of the Galerkin method. Here, the concept of single- and multipoint constraints is introduced. Also introduced are derived variational principles, which are important because even in supposed displacement FEM analysis, mixed methods are often used in element formulations and can also have utility in optimization methods.

The sixth chapter discusses some static and dynamic stability concepts. Various geometric measures of strain are introduced. Straight, curved, and thin walled beams are revisited and studied in a deformed geometry. For curved beams, different forms of curvature arise from different parametric representations of curvature. These are shown to be equivalent. For thin walled beams of open section, there is no conjugate variable for transverse shear. Thus it is shown that a blind use of virtual work yields incorrect stability equations. Classic equilibrium techniques are then used to correct the equations obtained by virtual work. General stability concepts are then introduced. Two of the most important influences on stability are the type of loading and the concept of stiffness. Instationary and stationary loading is discussed, with the latter broken down into gyroscopic, dissipative, circulatory, and noncirculatory categories. Stiffness is broken down into that provided by the material and that provided by the internal loading. Generalized stiffness leading to secant methods of analysis are also introduced. Then, classic rigid-link mass-spring models are used to gain an understanding of the various forms of stability.

I would like to thank Mrs. Julie Suva of Beyond Words for doing the figures for this book. Also, it has been a real pleasure working with Mr. Bob Hilbert of Wiley, who kept the whole process going.

DAVID V. WALLERSTEIN

CHAPTER 1

INTRODUCTION

The object of this course is to provide an introductory treatment of variational formulations in structural analysis. A knowledge of these formulations alone, however, is not adequate in reaching engineering solutions for large or complex structures. Currently, such solutions are most commonly obtained by means of a matrix method classified as the finite element method. Since most modern finite element methods are formulated via variational methods, their inclusion follows in natural fashion.

The power of the variational formulations is that they yield a systematic way of deriving the governing equations and corresponding boundary conditions that relate to the behavior of a structure under loading. Indeed, the governing equations they yield are in general nothing more than the equations of equilibrium or the equations of compatibility.

The boundary conditions are usually classified as two types:

1. Forced boundary conditions, which are mathematical expressions of constraint.
2. Natural boundary conditions, which are necessary conditions for stationary requirements.

The boundary conditions for a free edge of a plate is a classic example in which a variational procedure routinely yields the correct results, while boundary conditions obtained by manipulation of the fourth order governing differential equations yield inconsistent results. Figure 1.1 shows a rectangular plate that is clamped on three edges and free of restraint on the fourth edge. The stress resultants shown on the free edge are M_y, M_{yx}, and Q_y, which repre-

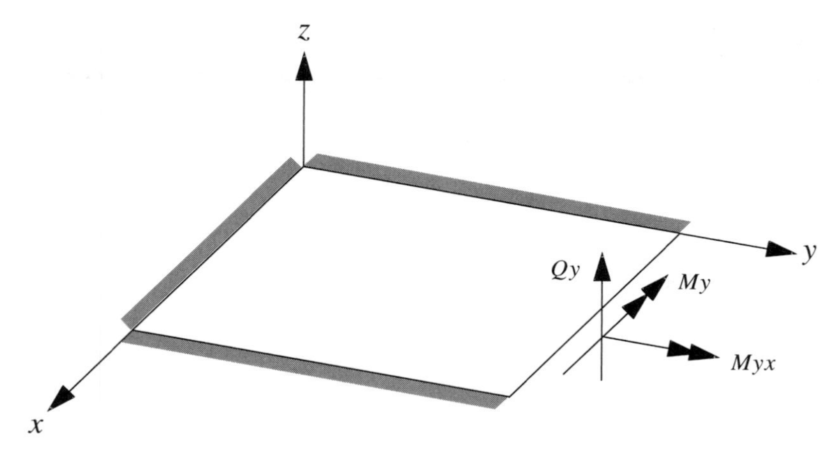

Figure 1.1 Plate with free edge.

sent the bending moment normal to the y axis, the twisting moment, and the transverse shear on the free edge, respectively.

The natural assumption is that on the free edge, $M_y = 0$, $M_{yx} = 0$, and $Q_y = 0$. However, these conditions will not generally satisfy the differential equation of the plate. A variational approach yields the correct boundary conditions.

$$M_y = 0 \tag{1.1}$$

$$Q_y + \frac{\partial M_{yx}}{\partial x} = 0 \tag{1.2}$$

Just as variational formulations yield a systematic procedure in the derivation of the governing equations and their boundary conditions, matrix methods vis-à-vis the finite element method yield a systematic procedure in the solution of these equations. These methods are based on the concept of replacing the actual continuous structure by a mathematically "equivalent" model made up from discrete structural elements having known properties expressible in matrix form.

The range of application of the variational formulations and hence the finite element formulations fall into various categories, such as the following:

- Equilibrium problems—time independent
- Displacement distribution—structures
- Temperature distribution—heat
- Pressure or velocity distribution—fluid
- Eigenvalue problems—steady state
- Natural frequencies

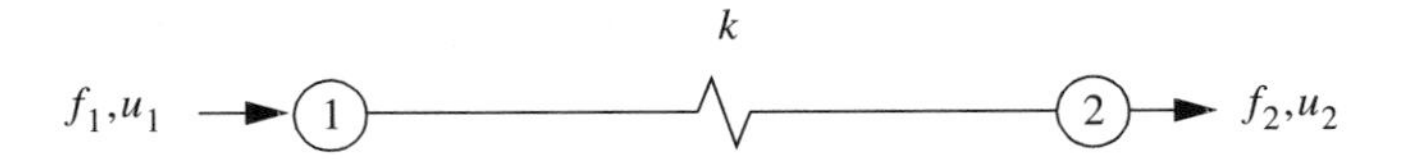

Figure 1.2 Spring element and nodal equilibrium.

- Structural stability
- Lamina flow stability
- Propagation problems—time domain

The finite element method itself yields a major bonus to problem solving.

Example 1.1 Consider first a simple elastic system discretized into a composite of simple spring elements. Figure 1.2 shows one typical such element. By Hooke's law, the force f in the spring is related to the spring elongation u_e by the following relationship:

$$f = ku_e$$

where k is the spring stiffness. The elongation u_e is related to the end displacements at ends 1 and 2 by the following expression:

$$u_e = u_2 - u_1$$

Equilibrium of a free body at end 1 (node 1) requires that the applied force f_1 at end 1 be related to the end displacements by the following relationship:

$$f_1 = ku_1 - ku_2$$

At end 2 (node 2), equilibrium requires the following:

$$f_2 = -ku_1 + ku_2$$

These two expressions for equilibrium may be combined into a single matrix equation of the following form:

$$k\begin{bmatrix} 1 & -1 \\ -1 & 1 \end{bmatrix}\begin{Bmatrix} u_1 \\ u_2 \end{Bmatrix} = \begin{Bmatrix} f_1 \\ f_2 \end{Bmatrix}$$

The 2×2 matrix is called the element-stiffness matrix.

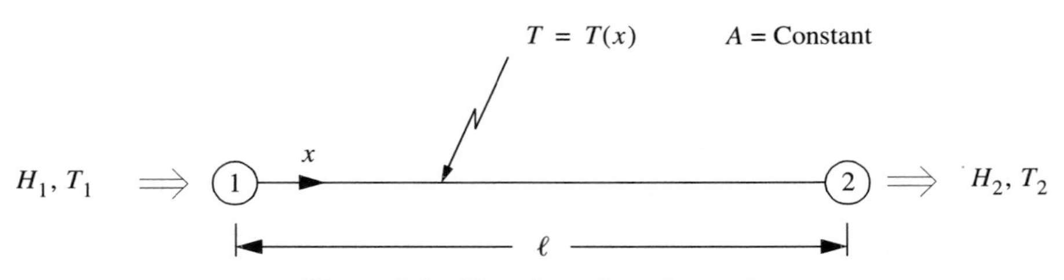

Figure 1.3 Heat flow through a rod.

Example 1.2 As a second example, consider the heat flow through the rod shown in Fig. 1.3. Fourier's law of heat conduction gives the following expression:

$$h = -k(dT/dx)$$

where

h is the steady-state heat input per unit area (joule/m^2)
k is the thermal conductivity (joule/m/°C)
T is the temperature (°C)
A is the area (m^2)
l is the length (m)
H_1, h_1A is the total heat input at node 1 (joule)
H_2, h_2A is the total heat input at node 2 (joule)

Consider a linear temperature distribution of the following form:

$$T = a + bx$$

where a and b are constants. Evaluating the equation for temperature distribution at end 1 ($x = 0$) and end 2 ($x = l$), we obtain the following relations:

$$a = T_1$$

and

$$b = (T_2 - T_1)/l$$

where T_1 and T_2 are the rod-end temperatures. The temperature distribution can then be expressed as follows:

$$T = \lfloor 1 - x/l \quad x/l \rfloor \begin{Bmatrix} T_1 \\ T_2 \end{Bmatrix}$$

with dT/dx expressed as

$$dT/dx = \lfloor -1/l \quad 1/l \rfloor \begin{Bmatrix} T_1 \\ T_2 \end{Bmatrix}$$

Fourier's law of heat conduction then takes the following form:

$$h = (-k/l) \lfloor -1 \quad 1 \rfloor \begin{Bmatrix} T_1 \\ T_2 \end{Bmatrix}$$

where $\lfloor \; \rfloor$ is a row matrix and { } is a column matrix.

The total heat input at node 1 and node 2 can then be expressed as:

$$Ak/l \begin{bmatrix} 1 & -1 \\ -1 & 1 \end{bmatrix} \begin{Bmatrix} T_1 \\ T_2 \end{Bmatrix} = \begin{Bmatrix} H_1 \\ H_2 \end{Bmatrix}$$

The 2×2 matrix is called the element-conductivity matrix.

While the two element matrices formed in these two examples have quite different physical meanings, they are identical in mathematical form. Furthermore, their assembly into a global system is identical. This characteristic assembly procedure is one of the great advantages of the finite element procedure. The equation solver used in the finite element method need never know the physical system being solved.

Variational methods and finite element methods are applicable to general continua with both geometric and material nonlinearities. When solving structural problems with these methods, however, there are additional requirements that must be met. For example, when using these methods to solve the problem of torsion of bars, the strain-displacement assumptions and the ramifications of these assumptions on the nonvanishing components of stress must be known beforehand. For the bar shown in Fig. 1.4, the Saint Venant torsion assumptions [1] are given as follows:

$$\begin{aligned}
\epsilon_x &= \epsilon_y = \epsilon_z = \epsilon_{xy} = 0 \\
\epsilon_{xz} &= \partial w/\partial x - \theta y \\
\epsilon_{yz} &= \partial w/\partial y + \theta x \\
w &= \theta \psi(x, y) \\
\sigma_{xz} &= G\epsilon_{xz} \\
\sigma_{yz} &= G\epsilon_{yz}
\end{aligned}$$

where

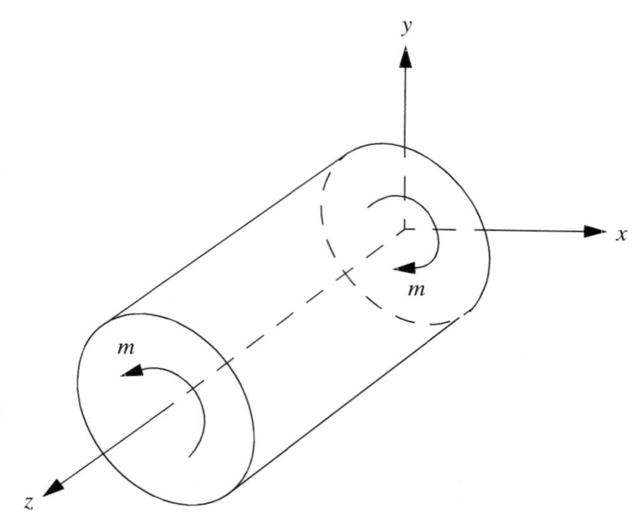

Figure 1.4 Torsion in a circular bar.

ϵ's are the strains
σ's are the shear stresses
G is the torsion modulus
ψ, a function of (x, y), is the warping function
θ, a constant, is the rate of twist produced by constant end twisting moments
w is the component of displacement along the z axis

Similarly, for beams, the Bernoulli-Euler hypothesis, the more general Vlasov hypothesis, or even your own hypothesis must be decided in advance of any variational method.

CHAPTER 2

PRELIMINARIES

2.1 VARIATIONAL NOTATION

Variational principles are in their most general form covered by the methods of the calculus of variations [2]. Their application to mechanics in general is found in its most elegant form in *The Variational Principles of Mechanics* by Lanczos [3].

We will in general discuss pertinent aspects of the calculus of variations as needed. The outstanding difference between calculus and variational calculus when we seek stationary values is that, in the former, we seek the stationary values of a function, whereas in the latter, we seek the stationary value of a functional, which is usually a definite integral of the following form:

$$V = \int_{x_0}^{x_1} F(x, y, y', y'', \ldots, y^{(n)})\, dx$$

To discuss variational principles, it is necessary to discuss variational notation. The most important aspect of this notation is the δ process.

Let f be a function of the independent variable x. Figure 2.1 shows a plot of $y = f(x)$, and df is a change along the curve $f(x)$ from a change in x. Define a new function as follows:

$$\begin{aligned} f^*(x) &\equiv f(x) + \epsilon\, \eta(x) \\ &= f(x) + \delta f(x) \end{aligned}$$

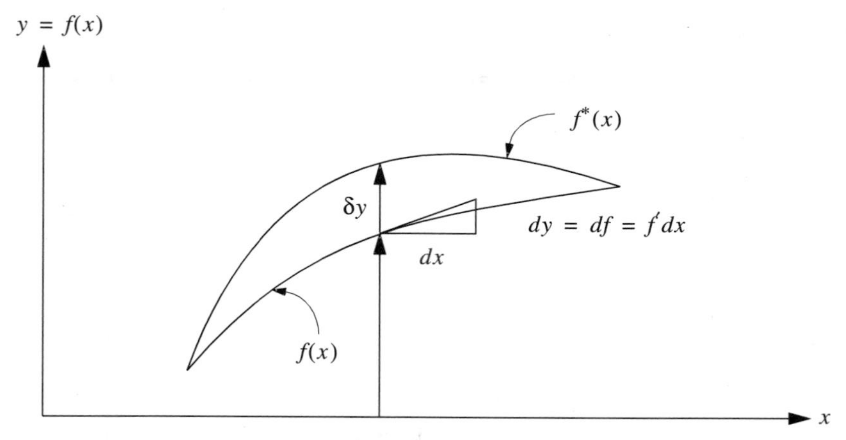

Figure 2.1 Variational notation.

where $\eta(x)$ is a function with at least continuous first derivatives and ϵ is a small parameter. We define the variation in $f(x)$ as follows:

$$\begin{aligned}\delta f(x) &= \delta y \\ &= f^{*}(x) - f(x) \\ &= \epsilon\,\eta(x)\end{aligned}$$

where δf is a change to a different curve with the independent variable x fixed. For the one-dimensional function shown in Fig. 2.1, the variation is the vertical movement δy.

Consider the commutative properties of the δ operator:

$$\begin{aligned}d(\delta y)/dx &= d[f^{*}(x) - f(x)]/dx \\ &= d[\epsilon\,\eta(x)]/dx \\ &= \epsilon\,\eta'(x)\end{aligned}$$

Next, take the derivative of $f^{*}(x)$ and $f(x)$ and define as follows:

$$\begin{aligned}\delta(dy/dx) &= f^{*\prime}(x) - f'(x) \\ &= [y(x) + \epsilon\,\eta(x)]' - y'(x) \\ &= y'(x) + \epsilon\,\eta'(x) - y'(x) \\ &= \epsilon\,\eta'(x)\end{aligned}$$

Thus, on comparing the above two expressions for $\epsilon\,\eta'(x)$, we may write the following:

$$d\,\frac{\delta y}{dx} = \delta\,\frac{dy}{dx} \tag{2.1}$$

This shows that the derivative of the variation is equal to the variation of the derivative. In general, the properties of δ are similar to the properties of the operator d.

Next, consider the behavior of the δ operatorwith integration:

$$\begin{aligned} \delta \int f(x)\,dx &= \int f^*(x)\,dx - \int f(x)\,dx \\ &= \int [f^*(x) - f(x)]\,dx \\ &= \int \delta f(x)\,dx \end{aligned} \tag{2.2}$$

Thus, we see that the δ operator may be taken in or out of the integral at our convenience.

There is often a restriction placed on $\eta(x)$: namely, $\eta(a) = \eta(b) = \eta'(a) = \eta'(b) = 0$. These restrictions impose the forced boundary conditions. Relaxing these restrictions results in the so-called natural boundary conditions.

In our discussion of δ, we treated δ as an operator. This implies that there is a function f that can be operated on by an operator δ to give δf. There are virtual quantities for which we can define, say, $\delta\alpha$ as a symbol, but with the understanding that α has not been defined. In this case, δ does not operate on α, but rather $\delta\alpha$ is a complete, self-contained symbol and the δ cannot be separated from the α. For example, in dynamics we may define a virtual rotation [4] $\delta\alpha$ such that

$$\delta\alpha = \sum_{r=1}^{n} w_{qr}\,\delta q_r$$

In this equation, w_{qr} are the nonholonomic rates of change of orientation and q_r are the generalized coordinates.

While

$$d(\delta q_r)/dt = \delta(dq_r/dt)$$

would be appropriate, no function α is defined; hence

$$d(\delta\alpha/dt) \neq \delta(d\alpha/dt)$$

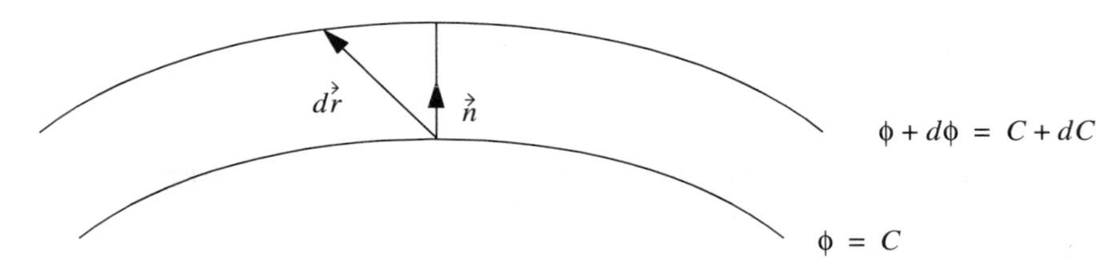

Figure 2.2 Surface contour.

2.2 THE GRADIENT

We may define the gradient by the following expression:

$$\nabla\phi = \text{grad } \phi = (d\phi/dn)\vec{n} \tag{2.3}$$

where $\vec{n}$ is the unit normal to the surface ϕ shown in Fig. 2.2 and $\nabla \equiv$ grad is the del operator.

Equivalently, the gradient [5] may be defined as follows:

$$d\phi = \nabla\phi \cdot d\vec{r} \tag{2.4}$$

To see how Eqs. (2.3) and (2.4) arise, consider again Fig. 2.2, from which we observe that dC is the actual increase of ϕ from the first surface to the second surface. The *rate* of increase depends upon the direction and is expressed as

$$\frac{dC}{\sqrt{d\vec{r} \cdot d\vec{r}}}$$

The expression $\sqrt{d\vec{r} \cdot d\vec{r}}$ has a minimum value for $d\vec{r} \parallel \vec{n}$ and a magnitude of dn. Then, for unit normal $\vec{n}$, $\vec{n}\ dn$ is the least value for $d\vec{r}$, and $(dC/dn)\ \vec{n}$ represents the most rapid rate of increase from the first surface to the second surface. Thus we may write

$$\frac{dC}{dn}\vec{n} = \frac{d\phi}{dn}\vec{n}$$

as the defining equation for the gradient. If we note that $dn = \vec{n} \cdot d\vec{r}$ and then dot Eq. (2.3) with $d\vec{r}$, Eq. (2.4) follows immediately.

From the above definitions for gradient, it should be observed that the gradient is independent of any coordinate system. Also, the del operator is independent of any coordinate system. In a rectangular cartesian coordinate system, the del operator takes the following form:

$$\nabla = \vec{i}\frac{\partial}{\partial x} + \vec{j}\frac{\partial}{\partial y} + \vec{k}\frac{\partial}{\partial z} \tag{2.5}$$

In general, the del operator is defined as

$$\nabla = \lim_{\Delta V \to 0} \frac{1}{\Delta V} \oint_{\partial V} d\vec{S}$$

where ΔV is a simply connected volume enclosing the point P about which the del operator is applied, and $d\vec{S}$ is the bounding surface to the enclosed volume.

A useful mnemonic for the del operator based on the above definition is

$$\nabla \sim \partial/\partial x \sim d\vec{s}/dS \sim d\vec{S}/dV \tag{2.6}$$

In the mnemonic, $d\vec{s}$ is a measure of arc, $dS(d\vec{S})$ is a measure of area, and dV is a measure of volume.

2.3 INTEGRATION OF PARTS

For one-dimensional analysis, if the integrand can be expressed as the product of two functions $u(x)v'(x) = u(x)\, dv(x)/dx$, then

$$\int_{p_1}^{p_2} u(x)v'(x)\, dx = u(x)v(x)\bigg|_{p_1}^{p_2} - \int_{p_1}^{p_2} u'(x)v(x)\, dx \tag{2.7}$$

is the formula for integration by parts.

For two dimensions and three dimensions, we start with the divergence theorem. Consider the two-dimensional case shown in Fig. 2.3. The divergence theorem [by Eq. (2.6)] may be written as

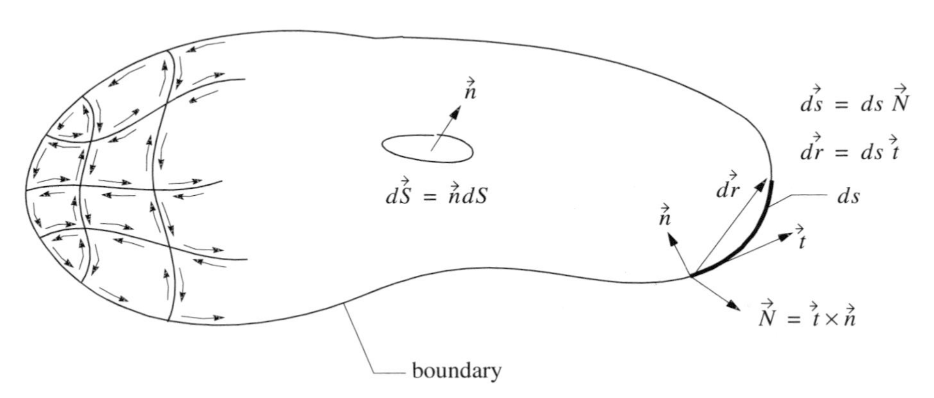

Figure 2.3 Closed surface and boundary.

$$\int_S \nabla \cdot \vec{A}\, dS = \int_{\partial S} \vec{A} \cdot d\vec{s} \tag{2.8}$$

Define $\vec{A}$ as

$$\vec{A} = \phi \nabla \psi$$

Then, $\nabla \cdot \vec{A}$ becomes

$$\nabla \cdot \vec{A} = \nabla \phi \cdot \nabla \psi + \phi \nabla^2 \psi$$

Substitute the above expressions for $\vec{A}$ and $\nabla \cdot \vec{A}$ into Eq. (2.8) with the following result:

$$\int_S \nabla \phi \cdot \nabla \psi \, dS + \int_S \phi \nabla^2 \psi \, dS = \int_{\partial S} \phi \nabla \psi \cdot ds \tag{2.9}$$

Define

$$l = \vec{N} \cdot \vec{i} = \cos(\vec{N}, \vec{i})$$
$$m = \vec{N} \cdot \vec{j} = \cos(\vec{N}, \vec{j})$$

Then Eq. (2.9) becomes

$$\int_S [(\partial \phi / \partial x)(\partial \psi / \partial x) + (\partial \phi / \partial y)(\partial \psi / \partial y)]\, dS$$
$$= -\int_S \phi \nabla^2 \psi \, dS + \int_{\partial S} \phi [l(\partial \psi / \partial x) + m(\partial \psi / \partial y)]\, ds \tag{2.10}$$

For three dimensions, replace dS with dV and $d\vec{s} = \vec{N}\, ds$, with $d\vec{S} = \vec{n}\, dS$, to obtain

$$\int_V [(\partial \phi / \partial x)(\partial \psi / \partial x) + (\partial \phi / \partial y)(\partial \psi / \partial y) + (\partial \phi / \partial z)(\partial \psi / \partial z)]\, dV$$
$$= -\int_V \phi \nabla^2 \psi \, dV + \int_{\partial V} \phi [l(\partial \psi / \partial x) + m(\partial \psi / \partial y) + n(\partial \psi / \partial z)]\, dS \tag{2.11}$$

where

$$l = \vec{n} \cdot \vec{i} = \cos(\vec{n}, \vec{i})$$
$$m = \vec{n} \cdot \vec{j} = \cos(\vec{n}, \vec{j})$$
$$n = \vec{n} \cdot \vec{k} = \cos(\vec{n}, \vec{k})$$

Equation (2.11) is often called Green's first theorem. It can be obtained directly from Eq. (2.7) by term-by-term integration by parts of the left side of Eq. (2.11). For example, consider the first term as follows:

$$\int_V \frac{\partial \phi}{\partial x} \frac{\partial \phi}{\partial x}\, dx\, dy\, dz = \int_{\partial V} \int_x \frac{\partial \phi}{\partial x} \frac{\partial \psi}{\partial x}\, l\, dS\, dx$$

where $l\, dS = dy\ dz$. Let $u = (\partial\psi/\partial x)l\ dS$ and let $dv = (\partial\phi/\partial x)\ dx$; then apply Eq. (2.7) to obtain

$$\int_{\partial V} \phi \frac{\partial \psi}{\partial x}\, l\, dS - \int_V \phi \frac{\partial^2 \psi}{\partial x^2}\, dV$$

which represents the x terms of the right side of Eq. (2.11). The other terms follow in a similar fashion.

In three dimensions, $\int_{\partial V} \vec{A} \cdot \vec{n}\, dS$ is called the flux (or flow) of $\vec{A}$ through the surface and represents

the average normal component of $\vec{A} \times$ the surface area

In addition, $\nabla \cdot \vec{A}$ represents the rate per unit volume per unit time that $\vec{A}$ is leaving a point. Then, $\int_V \nabla \cdot \vec{A}$ is the total flow from the volume and $\int_{\partial V} \vec{A} \cdot d\vec{S}$ is the total flux through the surface.

2.4 STOKES'S THEOREM

Stokes's theorem relates to a closed curve and any surface bounded by this curve. The curve and surface are orientated by positive tangent direction $\vec{t}$ and positive normal direction $\vec{n}$ according to the right-hand rule. In the plane of $d\vec{S}$, an element of surface area

$$d\vec{s} = \vec{N}\, ds$$

represents any element of arc on the boundary of $d\vec{S}$. The unit vectors $(\vec{t}, \vec{n}, \vec{N})$ form a right-handed orthogonal system, with

$$\vec{N} = \vec{t} \times \vec{n}$$
$$d\vec{r} = \vec{t}\, ds$$

representing an element of the closed curve bounding the surface. These relationships are shown in Fig. 2.3.

We are interested in determining a relationship for the line integral of $\vec{A}$. Figure 2.3 shows the surface divided into smaller regions with common boundaries of neighboring regions. Along these boundaries, all line integrals appear twice in both directions. Thus by summing up all smaller line integrals we find that the net result is the line integral of the boundary curve. Therefore we need only concentrate our attention on the triad $(d\vec{r}, d\vec{S}, d\vec{s})$ on the subregions along the boundary curve.

Using the second form for ∇, as shown in Eq. (2.6), and taking the vector cross-product with a vector $\vec{A}$, we may write

$$\nabla \times \vec{A} = \frac{d\vec{s}}{dS} \times \vec{A}$$

Multiplying through the dS and by integrating, we get

$$\int_S \nabla \times \vec{A}\, dS = \int_{\partial S} d\vec{s} \times \vec{A}$$

Next, by dotting both sides with the unit vector $\vec{n}$ normal to the surface shown in Fig. 2.3, we obtain

$$\begin{aligned}\int_S \vec{n} \cdot \nabla \times A\, dS &= \int_{\partial S} \vec{n} \cdot d\vec{s} \times \vec{A} \\ &= \int_{\partial S} \vec{n} \times d\vec{s} \cdot \vec{A} \\ &= \int_{\partial S} \vec{t} \cdot \vec{A}\, ds\end{aligned}$$

Thus we may write

$$\int_S \nabla \times \vec{A} \cdot d\vec{S} = \int_{\partial S} \vec{A} \cdot d\vec{r} \tag{2.12}$$

The $\int_{\partial S} \vec{A} \cdot d\vec{r}$ in Eq. (2.12) is called the circulation and represents

the average tangential component of $\vec{A} \times$ the distance around the contour

2.5 GREEN'S THEOREM IN THE PLANE

An important result of Stokes's theorem is when we consider the following case:

$$\nabla \times \vec{A} = 0$$

By using the vector identity

$$\nabla \times \nabla\phi = 0$$

we then define the vector $\vec{A}$ as

$$\vec{A} = \nabla\phi$$

Then, the right side of Eq. (2.12) becomes

$$\oint_{\partial S} \vec{A} \cdot d\vec{r} = \oint_{\partial S} \nabla\phi \cdot d\vec{r} = \oint_{\partial S} d\phi = 0 \tag{2.13}$$

Equation (2.13) is the requirement that $d\phi$ be an exact differential. From this result, we can get Green's theorem in the plane. Define $\vec{A}(x, y)$, a function of independent variables x and y, as follows:

$$\vec{A} = P(x, y)\vec{i} + Q(x, y)\vec{j}$$

If the integral $\int_{\partial S} \vec{A} \cdot d\vec{r}$ is to represent an exact differential, then the left side of Eq. (2.12) must vanish by virtue of Eq. (2.13). Thus, by substituting the above relation for $\vec{A}$ into the left side of Eq. (2.12), we get

$$\int_S \left(\vec{i}\,\frac{\partial}{\partial x} + \vec{j}\,\frac{\partial}{\partial y}\right) \times (P\vec{i} + Q\vec{j}) \cdot \vec{n}\; dx\; dy$$

$$= \int_S \left(\frac{\partial Q}{\partial x} - \frac{\partial P}{\partial y}\right) dx\; dy = 0$$

The above equation must hold for any arbitrary surface; hence, the necessary and sufficient condition for $\vec{A} \cdot d\vec{r} = d\phi$ to be an exact differential is

$$\frac{\partial P}{\partial y} - \frac{\partial Q}{\partial x} = 0 \tag{2.14}$$

Some very important mathematical quantities are not exact differentials. For example, consider a differential line element ds. If ds were integrable, it would be impossible to find the shortest distance between two points, because the length of any curve would be the same. Riemann geometry is based on this single differential quantity. Often, nonintegrable differentials are written as $\overline{d}s$ to emphasize that the symbol is not the d of some function s, but rather that it is a self-contained symbol.

2.6 ADJOINT EQUATIONS

The purpose of this section is to introduce the concept of adjoint equations and how they relate to the solution of differential equations. This concept provides a powerful tool in advanced studies of mechanics, and detailed descriptions are given by Stakgold [6] and by Readdy and Rasmussen [7]. The idea of an adjoint equation starts with the concept of an integrating factor. Consider a linear first-order differential equation of the following form:

$$u' + p(x)u = q(x), \qquad u(0) = u_0 \tag{2.15}$$

We would like to find, if possible, a function $v(x)$ called an integrating factor so that if Eq. (2.15) is multiplied by $v(x)$, then the left side of Eq. (2.15) can be written as the derivative of the single function $v(x)u$ or as

$$\begin{aligned} v(x)\,[u' + p(x)u] &= [v(x)u]' \\ &= v(x)u' + v'(x)u \end{aligned}$$

implying that

$$v'(x) - v(x)p(x) = 0 \tag{2.16}$$

The homogeneous Eq. (2.16) is known as the *adjoint* of Eq. (2.15), and its solution is

$$v(x) = \exp\left[\int^x p(t)\,dt\right]$$

With $v(x)$ so determined, we can write Eq. (2.15) as follows:

$$\frac{d}{dx}\,[v(x)u] = v(x)q(x)$$

with the following solution:

$$u = \frac{1}{v(x)}\left[\int^x v(s)q(s)\,ds + u_0\right]$$

The above ideas can be extended to problems of the following form:

$$P(x,y)\,dx + Q(x,y)\,dy = 0 \tag{2.17}$$

If Eq. (2.17) is not exact, it often can be made so by finding a function $v(x,y)$ so that the equation

$$v[P\,dx + Q\,dy] = 0 \tag{2.18}$$

is exact. Equation (2.18) is exact if and only if

$$\frac{\partial}{\partial y}[vP] = \frac{\partial}{\partial x}[vQ]$$

or if the following adjoint equation for v is satisfied:

$$P\frac{\partial v}{\partial y} - Q\frac{\partial v}{\partial x} + \left(\frac{\partial P}{\partial y} - \frac{\partial Q}{\partial x}\right)v = 0 \tag{2.19}$$

Equation (2.19) is usually at least as difficult as Eq. (2.17) to solve; however, whenever a nontrivial solution of the adjoint is known, every solution of the original differential equation can be found by quadratures (or indicated integrations). The solution can be written in terms of integrals that may or may not be easy to evaluate.

These concepts can also be extended to higher-order equations. A second-order equation

$$P(x)u'' + Q(x)u' + R(x)u = 0$$

is called exact if it can be written as

$$[P(x)u']' + [f(x)u]' = 0$$

where $f(x)$ is $f(P,Q,R)$. By equating the coefficients of the above two equations and by eliminating f, it is seen that the requirement for exactness is

$$P'' - Q' + R = 0$$

If the above second-order equation is not exact, it can be made so by an appropriate integrating factor $v(x)$. Thus we would require

$$vP(x)u'' + vQ(x)u' + vR(x)u = 0$$

or

$$[vP(x)u']' + [f(x)u]' = 0$$

Equating coefficients of the above equation with the expression for the exact form and by eliminating f results in the adjoint equation written as follows:

$$Pv'' + (2P' - Q)v' + (P'' - Q' + R)v = 0$$

If $P' = Q$, the second-order equation and its adjoint are identical in coefficients and in form, and the equations are said to be self-adjoint.

In more advanced work, it is no longer convenient to consider the differential equation but rather the differential operator. Thus, rather than the second-order equation above, consider the following second-order operator:

$$L = P(x) \frac{d^2}{dx^2} + Q(x) \frac{d}{dx} + R(x)$$

Next, we introduce the concept of Hilbert space, which is a special vector space defined by a Cauchy sequence of vectors.[1] For our discussion, the most important property of the Hilbert space is that it allows for the definition of a scalar or inner product defined as

$$\langle u, v \rangle = \int_V uv \, dV$$

over the domain V. Thus, rather than studying equations per se, we can study the properties of the operators. Consider the twice-differentiable functions $u(x)$, $v(x)$ and form the inner product

$$\int_a^b vLu \; dx = \int_a^b (vPu'' + vQu' + vRu) \, dx$$

Integrate by parts to get

$$\int_a^b vLu \, dx = \int_a^b u[(Pv)'' - (Qv)' + Rv] \, dx$$
$$+ [P(vu' - uv') + uv(Q - P')]_a^b$$

The bracketed term in the integrand is the adjoint of L, denoted as L^*, and may be expressed as

$$L^* = P \frac{d^2}{dx^2} + (2P' - Q) \frac{d}{dx} + (P'' - Q' + R)$$

The other term may be shortened by defining

[1]See Section 3.20.

$$J(u, v) = P(vu' - uv') + uv(Q - P')$$

Some manipulation would show that

$$\int_a^b [(vLu - uL^* v)\, dx = J(u, v)]_a^b$$

which is called Green's formula. In differential form, it is expressed as

$$vLu - uL^* v = \frac{d}{dx} J(u, v)$$

which is called Lagrange's identity. If $L = L^*$, the operator L is called the self-adjoint. A study of those functions v that make $J(u, v)_a^b$ vanish yields the conditions on the adjoint boundary conditions.

This whole concept of operators is then applicable to higher-order equations.

2.7 MEANING OF ∇^2

To see the meaning of $\nabla^2 \phi$, consider the following example:

$$\phi = T(x, y, z)$$

the temperature at a point. Surround the point P by a cube $2a$ on a side and consider the average of T throughout the volume $8a^3$. This average may be defined as

$$T_{\text{ave}} = \frac{1}{8a^3} \int_{-a}^{a} \int_{-a}^{a} \int_{-a}^{a} T(x, y, z)\, dx\, dy\, dz$$

Since we are considering the average value of T in the neighborhood of P, we may expand $T(x, y, z)$ in a Taylor's series to obtain the following expression:

$$T(x, y, z) = T_P + \nabla T \cdot \vec{r} + \frac{1}{2} \left(\left. \frac{\partial^2 T}{\partial x^2} \right|_P x^2 + \cdots + \left. \frac{\partial^2 T}{\partial z^2} \right|_P z^2 \right) + O^3$$

By symmetry, odd powers of x, y, and z are zero. Then, a typical term of the above integral becomes

$$\int_{-a}^{a} \int_{-a}^{a} \int_{-a}^{a} x^2\, dx\, dy\, dz = \frac{8}{3} a^3 a^2$$

Collecting terms and taking the limit, we get the following result:

$$\nabla^2 T = 2 \lim_{a \to 0} \frac{T_{\text{ave}} - T_P}{a^2}$$

We see, therefore, that $\nabla^2 T$ is proportional to $T_{\text{ave}} - T_P$ and represents the degree to which T differs from T_{ave} in the neighborhood of point P. A negative value of $\nabla^2 T$ means that at point P, T_P is greater than T_{ave} in the neighborhood of the point (a source). If $\nabla^2 T = 0$, then we have a steady condition.

2.8 TOTAL DIFFERENTIALS

Let $f = f(x, y, z)$ be a function of three independent variables: x, y, z. The expression

$$df = \frac{\partial f}{\partial x}\, dx + \frac{\partial f}{\partial y}\, dy + \frac{\partial f}{\partial z}\, dz$$

is called the total differential of f, where dx, dy, dz are three fixed increments that are otherwise arbitrary and are assigned to the three independent variables.

The total differential of the second-order $d^2 f$ is the total differential of the differential of the first order, the increments of dx, dy, dz remaining the same as we pass from one differential to the next higher one. Thus

$$d^2 f = \frac{\partial\, df}{\partial x}\, dx + \frac{\partial\, df}{\partial y}\, dy + \frac{\partial\, df}{\partial z}\, dz$$

We can obtain the expansion of the above expression mnemonically if we agree that ∂f^2 is really $\partial^2 f$ when we write

$$d^2 f = \left(\frac{\partial f}{\partial x}\, dx + \frac{\partial f}{\partial y}\, dy + \frac{\partial f}{\partial z}\, dz \right)^2$$

In general, we may write

$$d^n f = \left(\frac{\partial f}{\partial x}\, dx + \frac{\partial f}{\partial y}\, dy + \frac{\partial f}{\partial z}\, dz \right)^n$$

Let $g = G(u, v, w)$ be a composite function of u, v, w, with these themselves serving as functions of the independent variables x, y, z, t. Then, the chain rule of partial differentiation yields

$$\frac{\partial g}{\partial x} = \frac{\partial G}{\partial u} \frac{\partial u}{\partial x} + \frac{\partial G}{\partial v} \frac{\partial v}{\partial x} + \frac{\partial G}{\partial w} \frac{\partial w}{\partial x}$$

with three similar expressions for $\partial g/\partial y$, $\partial g/\partial z$, and $\partial g/\partial t$.

If the four resulting equations are multiplied by dx, dy, dz, dt, respectively, and then added, the left side becomes

$$\frac{\partial g}{\partial x}\,dx + \frac{\partial g}{\partial y}\,dy + \frac{\partial g}{\partial z}\,dz + \frac{\partial g}{\partial t}\,dt$$

That is, it becomes dg, whereas the coefficients of

$$\frac{\partial G}{\partial u}, \qquad \frac{\partial G}{\partial v}, \qquad \frac{\partial G}{\partial w}$$

on the right side become du, dv, dw, respectively. Thus we have

$$dg = \frac{\partial G}{\partial u}\,du + \frac{\partial G}{\partial v}\,dv + \frac{\partial G}{\partial w}\,dw$$

We see, therefore, that the expression for the total differential of the first order of a composite function is the same as if the auxiliary functions were the independent variables. Using the above rule, we can then calculate d^2g by first noting that the right side of the expression for dg involves the six auxiliary functions u, v, w, du, dv, dw. By using our mnemonic notation, we find

$$d^2g = \left(\frac{\partial G}{\partial u}\,du + \frac{\partial G}{\partial v}\,dv + \frac{\partial G}{\partial w}\,dw\right)^2 + \frac{\partial G}{\partial u}\,d^2u + \frac{\partial G}{\partial v}\,d^2v + \frac{\partial G}{\partial w}\,d^2w$$

The terms in d^2u and so on only drop out if u, v, w are the independent variables.

2.9 LEGENDRE TRANSFORMATION

The Legendre transformation is in essence a change of variables. It is one of many transformations developed by Legendre and Ampère in their geometric studies of contact transformations [8]. We shall use it in our study of complementary energy. The transformation arises in many other applications. For example, in thermodynamics, there is a class of parameters that have values in a composite system equal to the sum of their values in each of the subsystems; these parameters are called extensive parameters. In thermodynamics, the fundamental energy equation is a function of the extensive parameters entropy (S), volume (V), and mole numbers (N_r) [9] and can be expressed as

$$U = U(S, V, N_1, \ldots, N_r)$$

However, since S cannot be measured, it would be more convenient to express the energy equation in terms of entropy's conjugate (or related) variable T, an intensive (or derived) parameter obtained by the following relationship:

$$T = \left.\frac{\partial U}{\partial S}\right|_{V, N_1, \ldots, N_r}$$

This change of variable is accomplished via the Legendre transformation without loss of informational content, thereby arriving at the following expression of Helmholtz free energy:

$$F = F(T, V, N_1, \ldots, N_r)$$

For simplicity, consider a function $f(x, y)$ of two independent variables x, y. Let y be considered a passive variable that is to remain unchanged and let x be an active variable we wish to replace with a new independent variable ξ. Write the total derivative of f to get

$$\begin{aligned} df &= \frac{\partial f}{\partial x}\,dx + \frac{\partial f}{\partial y}\,dy \\ &= \xi\,dx + \eta\,dy \end{aligned}$$

where we have defined

$$\xi = \frac{\partial f}{\partial x}$$

$$\eta = \frac{\partial f}{\partial y}$$

Observe that $\xi\,dx$ satisfies the identity

$$\xi\,dx = d(\xi x) - x\,d\xi$$

Substituting this identity into the expression for df, we get

$$df = d(\xi x) - x\,d\xi + \eta\,dy$$

By rearranging terms, we can define a new function $dg(\xi, y)$ as

$$dg(\xi, y) = d(\xi x) - df = x\,d\xi - \eta\,dy$$

The new function $g(\xi, y)$ defines the Legendre transformation as follows:

$$g(\xi, y) = \xi x - f(x, y) \tag{2.20}$$

Then,

$$dg(\xi, y) = \frac{\partial g}{\partial \xi}\, d\xi + \frac{\partial g}{\partial y}\, dy$$
$$= x\, d\xi - \eta\, dy$$

where we have defined

$$x = \frac{\partial g}{\partial \xi}$$

$$\eta = -\frac{\partial g}{\partial y}$$

Equation (2.20) has transformed a function f to a function g.

The inverse transformation is

$$f(x, y) = x\xi - g(\xi, y)$$
$$df = \xi\, dx + x\, d\xi - dg$$
$$= \xi\, dx + \eta\, dy$$

Also, for the passive variable, we have the following on comparing values for η:

$$\frac{\partial f}{\partial y} = -\frac{\partial g}{\partial y}$$

To get a feel for the meaning of the Legendre transformation, consider the function

$$y = f(x) = (x - 5)^2 + 2; \qquad x \geq 5$$

Then,

$$\xi = \frac{df}{dx} = 2x - 10$$

or

$$x = \frac{\xi}{2} + 5$$

Let the transformation be

$$-p = g(\xi) = \xi x - (x - 5)^2 - 2$$

or

$$p = -\frac{\xi^2}{4} - 5\xi + 2$$

The independent variable ξ represents the tangent to the curve, and the dependent variable p represents the y-intercept point of the line drawn from the curve tangent point to the y axis. Thus the curve is now represented by its envelope of tangent points. Then (ξ, p) determine the homogeneous line coordinates [10] $(\vec{i} + \xi\vec{j}, -p\vec{k})$ of the envelope of the tangents to the curve.

Such a curve could now be sketched from the family of straight lines defined as

$$y = \xi x - \frac{\xi^2}{4} - 5\xi + 2; \qquad \xi \geq 0$$

The reader should now note that the transformation has changed from x as a variable to ξ as a variable without loss of informational content. The exact same curve is sketched by either the original function or the homogeneous coordinate representation. If only the relationship

$$\xi = \frac{df}{dx} = 2x - 10$$

had been used to define a new function $f(\xi)$, an attempt to get back to the original function would have yielded

$$y = (x - 5)^2 + C$$

Thus the function would have been obtained only within an arbitrary constant.

2.10 LAGRANGE MULTIPLIERS

Chapter 5 discusses the conditions under which functions of several variables are stationary—that is, under what conditions does the function have a minimum, maximum, or saddle-point value? Consider a function of two independent variables (x, y). Let (x_0, y_0) be a stationary point and expand the function by a Taylor's series about the point to get

$$f(x, y) = f(x_0, y_0) + (x - x_0)\left.\frac{\partial f}{\partial x}\right|_0 + (y - y_0)\left.\frac{\partial f}{\partial y}\right|_0 + \text{higher-order terms}$$

Chapter 5 shows that the condition that the point (x_0, y_0) be stationary is that the first-order derivatives vanish. If constraint conditions exist, the condition for stationarity is more difficult to determine.

Let $g(x, y) = 0$ be a constraint condition applied to the function $f(x, y)$, and let us compute dy/dx in two different ways. Considering the constraint condition first, we have

$$\frac{dg}{dx} = \frac{\partial g}{\partial x} + \frac{\partial g}{\partial y}\frac{\partial y}{dx} = 0$$

or

$$\frac{dy}{dx} = \frac{\partial g/\partial x}{\partial g/\partial y}$$

Next, consider the function $f[x, y(x)] = C$, and use the fact that its implicit derivative is obtained as $df/dx = 0$. Thus we have

$$\frac{df}{dx} = \frac{\partial f}{\partial x} + \frac{\partial f}{\partial y}\frac{dy}{dx} = 0$$

or

$$\frac{dy}{dx} = -\frac{\partial f/\partial x}{\partial f/\partial y}$$

By equating the two expressions for dy/dx and arranging the result so that functions of x are on one side of the equal sign and functions of y are on the other, we get

$$\frac{\partial f/\partial x}{\partial g/\partial x} = \frac{\partial f/\partial y}{\partial g/\partial y} = -\lambda$$

Therefore, $-\lambda$ represents the point where the level curve $f[x, y(x)] = C$ and the constraint $g(x, y)$ becomes tangent. Rearranging, we have

$$\frac{\partial f}{\partial x} + \lambda\frac{\partial g}{\partial x} = 0; \qquad \frac{\partial f}{\partial y} + \lambda\frac{\partial g}{\partial y} = 0$$

These are the conditions that a function of three independent variables defined as

$$\Phi(x, y, \lambda) = f(x, y) + \lambda g(x, y)$$

must satisfy to be stationary. λ is called a Lagrange multiplier.

Physically, we can interpret λ as follows: Let $M(x, y) = f(x, y)$ represent $f(x, y)$ at the stationary point; then, compute

$$dM = \frac{\partial f}{\partial x}\, dx + \frac{\partial f}{\partial y}\, dy$$

$$= -\lambda \left(\frac{\partial g}{\partial x}\, dx + \frac{\partial g}{\partial y}\, dy \right)$$

$$= -\lambda\, dg$$

or

$$\lambda = -\frac{dM}{dg}$$

Thus λ represents the rate of change of the stationary point with respect to the constraint. In Chapter 5, we introduce the concept of strain energy as a function U to be made stationary. Often, there are displacement constraints Δ, in which case

$$-\lambda = \frac{dU}{d\Delta}$$

physically represents the forces of constraints required to impose the displacement constraints.

In general, the stationary point $(x_0, y_0, \ldots)$ of a function of n variables $f(x, y, \ldots)$ with r constraints $g_i(x, y, \ldots) = 0$ can be obtained by forming the function of $n + r$ variables as follows:

$$\Phi(x, y, \ldots, \lambda_1, \ldots, \lambda_r) = f(x, y, \ldots) + \lambda_1 g_1(x, y, \ldots) + \cdots + \lambda_r g_r(x, y, \ldots)$$

and solve the $n + r$ equations:

$$\frac{\partial f}{\partial x} + \sum_{i=1}^{r} \lambda_i \frac{\partial g_i}{\partial x} = 0$$

$$\frac{\partial f}{\partial y} + \sum_{i=1}^{r} \lambda_i \frac{\partial g_i}{\partial y} = 0$$

$$\vdots$$

$$g_1 = 0$$

$$\vdots$$

$$g_r = 0$$

by eliminating λ_i.

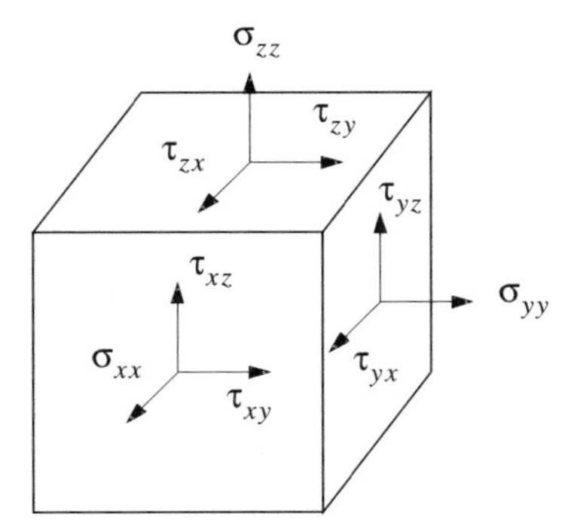

Figure 2.4 Differential rectangular parallelepiped.

2.11 DIFFERENTIAL EQUATIONS OF EQUILIBRIUM

Figure 2.4 represents a vanishingly small rectangular parallelepiped in a deformed continuum. The edges of the rectangular parallelepiped are parallel to the orthogonal reference axes x, y, z. Its sides are dx, dy, dz.

Using the notation σ_{xx}, τ_{xy}, τ_{xz} for the components of stress acting on the surface whose normal is in the x direction with similar notation for the other two directions and the notation X_b, Y_b, Z_b for the components of body force per unit volume, we can write the following expression for equilibrium in the x direction:

$$
\begin{aligned}
&\left(\sigma_{xx} + \frac{\partial \sigma_{xx}}{\partial x}\, dx\right) dy\ dz - \sigma_{xx}\, dy\ dz \\
&\quad + \left(\tau_{yx} + \frac{\partial \tau_{yx}}{\partial y}\, dy\right) dx\ dz - \tau_{yx}\, dx\ dz \\
&\quad + \left(\tau_{zx} + \frac{\partial \tau_{zx}}{\partial z}\, dz\right) dx\ dy - \tau_{zx}\, dx\ dy \\
&\quad + X_b\, dx\ dy\ dz = 0
\end{aligned}
$$

In the above, we have considered the normal stress on the rear face as σ_{xx} and the normal stress on the front face as a small variation $\sigma_{xx} + (\partial \sigma_{xx}/\partial x)\ dx$. Similar variations have been considered for τ_{zx} between the bottom and top faces and for τ_{yx} between the left and right faces.

The above relation must hold for any arbitrary volume $dx\ dy\ dz$; hence, we write

$$\frac{\partial \sigma_{xx}}{\partial x} + \frac{\partial \tau_{yx}}{\partial y} + \frac{\partial \tau_{zx}}{\partial z} + X_b = 0$$

with similar results for the other two directions.

If we take moments about an axis through the center of the element parallel to the z axis, we can show that

$$\tau_{yx} = \tau_{xy}$$

with similar results for moments about the other two axes. Note, however, that these results hold only when the planes of τ_{yx}, τ_{xy}, and so forth are orthogonal.

Thus, the equations of equilibrium may be written as follows:

$$\frac{\partial \sigma_{xx}}{\partial x} + \frac{\partial \tau_{xy}}{\partial y} + \frac{\partial \tau_{xz}}{\partial z} + X_b = 0 \tag{2.21}$$

$$\frac{\partial \sigma_{yy}}{\partial y} + \frac{\partial \tau_{yx}}{\partial x} + \frac{\partial \tau_{yz}}{\partial z} + Y_b = 0 \tag{2.22}$$

$$\frac{\partial \sigma_{zz}}{\partial z} + \frac{\partial \tau_{zx}}{\partial x} + \frac{\partial \tau_{zy}}{\partial y} + Z_b = 0 \tag{2.23}$$

These equations must be satisfied throughout the deformed volume of the body, and they are completely general in nature in that no assumptions have been made regarding the size of the deformation involved.

The stresses vary over the volume of the body and must be so distributed that, at the surface of the body, they are in equilibrium with the external forces on the surface of the body. The conditions of equilibrium at the surface in two dimensions can be obtained by considering a small triangular element, as shown in Fig. 2.5. The figure represents a vanishingly small triangular element, of unit thickness in the z direction, at the boundary of a two-dimensional body that is in equilibrium under the action of surface forces X_s and Y_s on the element boundary AB. These forces are balanced by the internal forces σ_{xx} and $\tau_{xy} = \tau_{yx}$ on the internal faces of the triangular element, along with the body forces X_b and Y_b. Summation of forces in the x direction gives

$$X_s \Delta s - \sigma_{xx} \Delta y - \tau_{yx} \Delta x + X_b \frac{1}{2} \Delta x \Delta y = 0$$

On taking the limit as $\Delta x \Delta y$ goes to zero, and on noting that dy/ds and dx/ds are the direction cosines l and m of the angle that a normal to AB makes with the x and y axes, respectively, we obtain

$$X_s = l\sigma_{xx} + m\tau_{xy}$$

Extending to three dimensions by considering a small tetrahedron, we obtain the following three-dimensional boundary conditions:

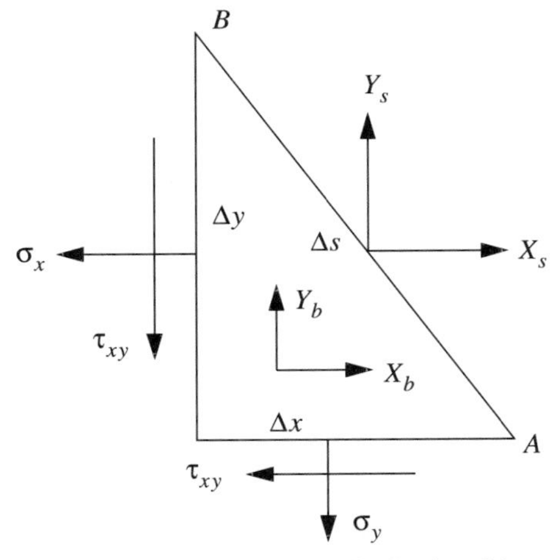

Figure 2.5 Stresses on a two-dimensional boundary.

$$X_s = l\sigma_{xx} + m\tau_{xy} + n\tau_{xz} \tag{2.24}$$

$$Y_s = l\tau_{yx} + m\sigma_{yy} + n\tau_{yz} \tag{2.25}$$

$$Z_s = l\tau_{zx} + m\tau_{zy} + n\sigma_{zz} \tag{2.26}$$

in which l, m, n are the direction cosines of the external normal to the surface of the body at the point under consideration, and X_s, Y_s, Z_s are the components of surface forces per unit area at this point.

2.12 STRAIN-DISPLACEMENT RELATIONS

While stresses and forces are used to describe the equilibrium of a structure, strains and displacements are used to describe the kinematics of deformation in a structure. There are at least ten different definitions of direct strain in use [11].

Some, such as natural (or logarithmic) strain, have the important characteristic that, if the so-called principal axes of strain and stress coincide during plastic deformation, the increments of natural strain are additive. Also, the natural strain is important in the understanding of the simple tension test. Natural strain is expressed as

$$\epsilon_n = \int_{l_0}^{l} \frac{dl}{l} = \ln \frac{l}{l_0}$$

where l_0 is the initial length and l is the final length of the tensile specimen.

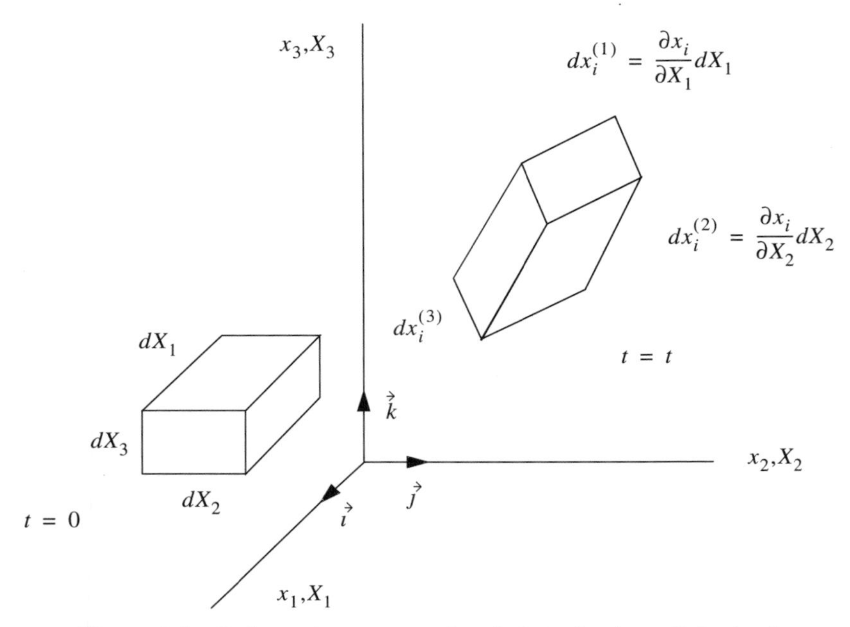

Figure 2.6 Deformed geometry of an infinitesimal parallelepiped.

Others, such as Lagrangian and Eulerian strain, are mathematically convenient. They arise from examining the mathematical relationships involving the square of the elements of length between two points before and after deformation. The square of length is used to avoid the awkwardness of the square root symbol.

In what follows [12], quantities associated with the undeformed structure are denoted by majuscules (capital letters) and those associated with the deformed structure by minuscules (small letters). At times, expanding equations out in full form or using vector notation becomes an impediment rather than a help. When this occurs, we will use Cartesian tensor notation. As an example, consider the divergence Eq. (2.8). In Cartesian tensor notation, we would write the divergence as $\int_S A_{i,i}\, dS = \int_{\partial S} A_i\, ds_i$. In this expression, summations over any repeated subscript is understood, and the range of the subscript is 1, 2 for a two-dimensional space, 1, 2, 3 for a three-dimensional space, and so on. $A_{i,i}$ is equivalent to

$$A_{i,i} = \frac{\partial A_i}{\partial x_i} = \frac{\partial A_1}{\partial x_1} + \frac{\partial A_2}{\partial x_2}$$

for a two-dimensional space, $x_1 = x$, and $x_2 = y$.

Figure 2.6 shows the geometry of an infinitesimal parallelepiped before and after deformation.

To describe the position occupied at time t by a particle that occupied position $\vec{X}$ in the reference configuration at time $t = 0$ (Lagrangian description), we write

$$\vec{x} = \vec{x}(\vec{X}, t) \tag{2.27}$$

with the following rectangular Cartesian components:

$$\begin{aligned} x_1 &= x_1(X_1, X_2, X_3, t) \\ x_2 &= x_2(X_1, X_2, X_3, t) \\ x_3 &= x_3(X_1, X_2, X_3, t) \end{aligned}$$

Consider a given material vector $d\vec{X}$. The squared length $(ds)^2$ of this vector in its deformed position is

$$(ds)^2 = dx_k \, dx_k \tag{2.28}$$

Its squared length $(dS)^2$ in the initial position is

$$(dS)^2 = dX_k \, dX_k \tag{2.29}$$

A useful measure of strain is defined as

$$(ds)^2 - (dS)^2 = dx_k \, dx_k - dX_k \, dX_k \tag{2.30}$$

Using Eq. (2.27), we can by use of the chain rule write

$$dx_k = \frac{\partial x_k}{\partial X_j} \, dX_j \tag{2.31}$$

Equation (2.30) then becomes

$$(ds)^2 - (dS)^2 = \left(\frac{\partial x_k}{\partial X_i} \frac{\partial x_k}{\partial X_j} - \delta_{ij} \right) dX_i \, dX_j \tag{2.32}$$

where δ_{ij} is called the Kronecker delta and is defined as follows:

$$\delta_{ij} = \begin{cases} 1 & \text{if } i = j \\ 0 & \text{if } i \neq j \end{cases}$$

Equation (2.32) may then be expressed as follows:

$$(ds)^2 - (dS)^2 = 2E_{ij} \, dX_i \, dX_j \tag{2.33}$$

where E_{ij} is called the Lagrangian strain tensor. If we define the displacement u_k of a particle form its initial position $\vec{X}$ to the current position $\vec{x}$ as

$$u_k = x_k - X_k$$

we may write

$$x_k = u_k + X_k$$

Then, E_{ij} can be expressed as

$$E_{ij} = \frac{1}{2}\left[\left(\frac{\partial u_k}{\partial X_i} + \delta_{ki}\right)\left(\frac{\partial u_k}{\partial X_j} + \delta_{kj}\right) - \delta_{ij}\right] \tag{2.34}$$

Or, on expansion, it can be expressed as

$$E_{ij} = \frac{1}{2}\left(\frac{\partial u_i}{\partial X_j} + \frac{\partial u_j}{\partial X_i} + \frac{\partial u_k}{\partial X_i}\frac{\partial u_k}{\partial X_j}\right) \tag{2.35}$$

Typical terms of E_{ij} are

$$E_{11} = \frac{\partial u_1}{\partial X_1} + \frac{1}{2}\left[\left(\frac{\partial u_1}{\partial X_1}\right)^2 + \left(\frac{\partial u_2}{\partial X_1}\right)^2 + \left(\frac{\partial u_3}{\partial X_1}\right)^2\right] \tag{2.36}$$

$$E_{12} = \frac{1}{2}\left(\frac{\partial u_1}{\partial X_2} + \frac{\partial u_2}{\partial X_1}\right) + \frac{1}{2}\left(\frac{\partial u_1}{\partial X_1}\frac{\partial u_1}{\partial X_2} + \frac{\partial u_2}{\partial X_1}\frac{\partial u_2}{\partial X_2} + \frac{\partial u_3}{\partial X_1}\frac{\partial u_3}{\partial X_2}\right) \tag{2.37}$$

The large strain tensor may also be expressed in terms of spatial coordinates. It is then called the Eulerian strain tensor and takes the following form:

$$e_{ij} = \frac{1}{2}\left(\frac{\partial u_i}{\partial x_j} + \frac{\partial u_j}{\partial x_i} - \frac{\partial u_k}{\partial x_i}\frac{\partial u_k}{\partial x_j}\right) \tag{2.38}$$

The finite strain components involve only linear and quadratic terms in components of the displacement gradient. This comes about from the defining equation; thus the finite strain is complete, not a second-order approximation.

For classical strain theory (small displacements compared to the linear dimensions of the body; small rotations, elongations, and shears compared to unity; and second-degree terms in angles of small rotation compared to the corresponding strain components [13]), the distinction between $\vec{X}$ and $\vec{x}$ vanishes totally and the strains become

$$\epsilon_x = \frac{\partial u}{\partial x} \tag{2.39}$$

$$\epsilon_y = \frac{\partial v}{\partial y} \tag{2.40}$$

$$\epsilon_z = \frac{\partial w}{\partial z} \tag{2.41}$$

$$\epsilon_{xy} = \gamma_{xy} = \frac{\partial v}{\partial x} + \frac{\partial u}{\partial y} \tag{2.42}$$

$$\epsilon_{xz} = \gamma_{xz} = \frac{\partial w}{\partial x} + \frac{\partial u}{\partial z} \tag{2.43}$$

$$\epsilon_{yz} = \gamma_{yz} = \frac{\partial w}{\partial y} + \frac{\partial v}{\partial z} \tag{2.44}$$

Notice the absence in the shear strain term of $1/2$. These strains are often called engineering strains.

2.13 COMPATIBILITY CONDITIONS OF STRAIN

If the three components of displacement are explicitly given (as, for example, in classical beam theory), then the components of strain are known functions. If, however, the six components of strain are given, six partial-differential equations for strain determine three unknown functions of displacement. Hence the system is overdetermined. Equations are needed to provide the necessary and sufficient conditions to uniquely determine the components of displacement u, v, w.

Finite strain compatibility conditions are complex in form and require a knowledge of general tensor concepts [14].

For small classical elasticity, the compatibility equations are written as

$$\epsilon_{ik,lm} + \epsilon_{lm,ik} + \epsilon_{il,km} + \epsilon_{lm,il} = 0 \tag{2.45}$$

and were first obtained by Saint Venant in 1860. Note that the indices i, k, l, m $= 1, 2, 3$ represent $3^4 = 81$ equations. However, some equations are identically satisfied, and because of symmetry in ik and lm, some are repetitious. As a result, only six independent equations remain.

These equations are the necessary conditions for compatibility and occur by simple elimination of u, v, w from the strain-displacement equations. For example, consider the two-dimensional case. Eliminate u and v by taking the second derivative of Eq. (2.39) with respect to y and the second derivative of Eq. (2.40) with respect to x; then add the result. Next, take the derivative of Eq.

(2.42) with respect to both x and y and subtract this result from the previously obtained sum. The result is as follows:

$$\frac{\partial^2 \epsilon_y}{\partial x^2} + \frac{\partial^2 \epsilon_x}{\partial y^2} - \frac{\partial^2 \gamma_{xy}}{\partial x \partial y} = 0 \tag{2.46}$$

To prove sufficiency, it is necessary to integrate along a path from point P_1, about which it is assumed that everything is known, to point P_2 in a deformed body and show that the value of the displacement at P_2 is independent of the path chosen [15]. Briefly, consider the following example: We require the u component of displacement to be expressed as

$$u_{P_2} = u_{P_1} + \int_{P_1}^{P_2} du = 0 \tag{2.47}$$

For this to hold, du must be an exact differential. In two dimensions, u may be expressed as $u = f(x, y)$. Then we may write

$$du = \frac{\partial u}{\partial x}\, dx + \frac{\partial u}{\partial y}\, dy$$

Or, on noting that $\partial u/\partial y$ can be written as

$$\frac{\partial u}{\partial y} = \gamma_{xy} - \frac{\partial v}{\partial x}$$

we get for du:

$$du = \epsilon_x\, dx + \left(\gamma_{xy} - \frac{\partial v}{\partial x}\right) dy$$

We are now going to substitute this expression for du into Eq. (2.47) and integrate by parts. We assume that ϵ_x, ϵ_y, ϵ_{xy}, u, v are known at point P_1 but are unknown at point P_2. If, however, we perform the integration by parts on the current differentials, terms such as the following appear:

$$x\epsilon_x\Big|_{P_1}^{P_2} \cdots$$

Unfortunately, the upper limit in the above expression is not known and hence cannot be evaluated. However, since P_2 is a fixed point, terms such as $dx_{P_2} = 0$, and we may use the following identities:

$$dx = d(x - x_{P_2})$$
$$dy = d(y - y_{P_2})$$

With these identities, integration by parts yields terms such as the following:

$$(x - x_{P_2})\epsilon_x\Big|_{P_1}^{P_2} \cdots$$

In such an expression, the upper limit is identically zero.

Integration by parts on Eq. (2.47) now yields

$$u_{P_2} = u_{P_1} + (x - x_{P_2})\epsilon_x\Big|_{P_1}^{P_2} + (y - y_{P_2})\left(\gamma_{xy} - \frac{\partial v}{\partial y}\right)\Bigg|_{P_1}^{P_2}$$

$$- \left[\int_{P_1}^{P_2} P(x,y)\,dx + \int_{P_1}^{P_2} Q(x,y)\,dy\right]$$

where

$$P(x,y) = (x - x_{P_2})\frac{\partial \epsilon_x}{\partial x} + (y - y_{P_2})\frac{\partial \epsilon_x}{\partial y}$$

$$Q(x,y) = (x - x_{P_2})\frac{\partial \epsilon_x}{\partial y} + (y - y_{P_2})\left(\frac{\partial \gamma_{xy}}{\partial y} - \frac{\partial \epsilon_y}{\partial x}\right)$$

Remembering that the integrated terms represent known constants evaluated at point P_1, we can see that they represent rigid-body translations and rotations of P_2 relative to P_1. The integral terms, on the other hand, represent relative displacements from strain and rotation in material elements located along the path between P_1 and P_2. If u_{P_2} is to be single-valued, these integrals must be independent of path. Substituting $P(x,y)$ and $Q(x,y)$ into Eq. (2.14), this requirement takes the following form:

$$(y - y_{P_2})\left(\frac{\partial^2 \epsilon_y}{\partial x^2} + \frac{\partial^2 \epsilon_y}{\partial y^2} - \frac{\partial^2 \gamma_{xy}}{\partial x \partial y}\right)$$

This expression must hold for all choices of $(y - y_{P_2})$. Hence Eq. (2.46) follows.

2.14 THERMODYNAMIC CONSIDERATIONS

In later chapters, we will discuss some thermodynamic restrictions in the definitions of such quantities as potential energy. Thus, we find it useful at this point to review some basic principles of physics. To begin, recall the first law of thermodynamics. It states that the total change in mean energy of a system is expressed as

$$\Delta K + \Delta U = W_e + H \tag{2.48}$$

In this expression:

W_e is the work performed *on* the system by mechanical means
H is the heat that flows *into* the system
ΔK is the increase of kinetic energy
ΔU is the increase in internal potential energy

Equation (2.48) simply splits the total energy change of a system into two parts: the part W_e from mechanical interaction and the part H from thermal interaction. If we are dealing with an infinitesimal change in state, the first law is written as follows:

$$dK + dU = \bar{d}W_e + \bar{d}H$$

In the above expression, dK and dU are well-defined functions associated with an initial state and a final state; hence they are exact differentials. $\bar{d}W_e$ and $\bar{d}H$ are inexact differentials; they represent a symbol to indicate a very small quantity. The reason $\bar{d}W_e$ and $\bar{d}H$ are inexact or path-dependent quantities is that while we can add a certain amount of each to a system by some process, after the completion of the process all we can measure is the change in energy of the system. We cannot discern how much of the change in energy is due to $\bar{d}W_e$ and how much is due to $\bar{d}H$. If the system is thermally isolated (adiabatic) so that $H = 0$, Eq. (2.48) implies that

$$\Delta K + \Delta U = W_e$$

Then, the work depends only on the energy difference and constitutes an exact differential.

Next, consider the work-energy theorem for a particle. Let the particle have a mass m and a velocity $\vec{v}$. Newton's law then states that the force $\vec{F}$ acting on a particle is

$$\vec{F} = m\,\frac{d\vec{v}}{dt} \tag{2.49}$$

The work done on the particle in moving it from an initial position to a final position can be expressed as

$$W = \int \vec{F} \cdot d\vec{r} \tag{2.50}$$

Next, consider the useful identities

$$d\vec{r} = \vec{v}\,dt$$

and

$$\frac{1}{2}\frac{d}{dt}(\vec{v}\cdot\vec{v}) = \vec{v}\cdot\frac{d\vec{v}}{dt}$$

Thus, with use of these identities and Eq. (2.49), Eq. (2.50) becomes

$$W = \int \vec{F}\cdot\vec{v}\,dt = \int m\vec{v}\cdot\frac{d\vec{v}}{dt}\,dt$$

$$= \int d\left(\frac{1}{2}mv^2\right) = \int dK$$

or

$$W = \Delta K \tag{2.51}$$

In this expression, K, the kinetic energy, has been defined as $1/2mv^2$. Extending to a system of particles, we have

$$W_e + W_i = \Delta K \tag{2.52}$$

where W_i is the work performed by the internal forces in the system. Equation (2.52) represents the work-energy theorem for a system of particles and is valid for both a system of particles and a continuum. For a continuum, however, it is arrived at through momentum balance and the resulting Cauchy's [16] equations of motion. These take the following form:

$$\nabla\cdot\mathbf{T} + \rho\vec{b} = \rho\frac{d\vec{v}}{dt}$$

where

$\mathbf{T}$ is the stress tensor
$\rho\vec{b}$ are the body forces
ρ is the mass density

Combining Eqs. (2.48) and (2.52), we arrive at the following relationship:

$$W_i = H - \Delta U$$

For a rigid body, the internal forces do no work, and we may write

$$H = \Delta U$$

For an adiabatic process, $H = 0$, and we get

$$W_i = -\Delta U \tag{2.53}$$

In dealing with elastic materials, assumptions between adiabatic and isothermal[2]

[2]Isothermal deformation is described by the Helmholtz free-energy function $F_0 = U - TS$, where T is the temperature and S is the entropy (see Section 2.9) [17].

deformations appear only as differences between adiabatic and isothermal constants. In general, these differences are experimentally negligible. Therefore, for our purposes we consider Eq. (2.53) to define the relationship between internal work and internal energy.

Notice that in going from Eq. (2.50) to Eq. (2.51), we arrived at the following intermediate results:

$$W = \int \vec{F} \cdot \vec{v}\, dt$$

where $\vec{F} \cdot \vec{v}$ is defined as the power or activity. The advantage of this definition of work is that it depends only on knowledge of the force and the instantaneous velocity of the particle, whereas Eq. (2.50) has the restriction that the force needs to always act on the same particle.

In closing this section, it is useful to show the tie between the mechanics of a deformable solid and the thermodynamics. As discussed in Section 2.9, the fundamental energy equation of thermodynamics is

$$U = U(S, V, N_1, \ldots, N_r)$$

For a solid, the strain components (say, the Lagrangian strain) provide a complete description of the internal configuration of the solid [18]. Then we may replace the extensive parameter V in the above equation with the nine extensive parameters $V_0E_{11}, \ldots, V_0E_{33}$, where V_0 is the volume in the initial state. Then, the fundamental equation becomes

$$U = U(S, V_0, E_{11}, \ldots, V_0E_{33}, N_1, \ldots, N_r)$$

The related intensive parameters or thermodynamic conjugate [19] are

$$\tilde{\mathbf{T}}_{ij} = \frac{1}{V_0}\left(\frac{\partial U}{\partial E_{ij}}\right)$$

where $\tilde{\mathbf{T}}_{ij}$ is the so called Piola-Kirchhoff stress tensor of the second kind.

PROBLEMS

2.1 If $\vec{i}$ and $\vec{j}$ are mutually perpendicular unit vectors fixed in a plane and the position vector of a point P constrained to move in the plane is given by the expression

$$\vec{r} = b \cos \phi \vec{i} + b \sin \phi \vec{j}$$

where b, a constant radius, and ϕ, a variable angle measured from the $\vec{i}$ axis, are the polar coordinates of P. Then give the expression for $\delta \vec{r}$.

2.2 Derive Eqs. (2.21), (2.22), and (2.23).

2.3 Derive Eqs. (2.24), (2.25), and (2.26).

2.4 Starting with Eq. (2.47), fill in the steps leading to Eq. (2.46).

2.5 A uniform, rigid disk of radius r and mass M rolls without slipping on a horizontal surface. A load P of constant magnitude is applied to the rim of the disk in the direction of motion at a distance $h > r$ above the horizontal surface. Show that the center C of the disk will travel a distance $3Mv^2r/4Ph$ when it acquires the speed v after starting from rest.

2.6 For two-dimensional small-strain theory, to any displacement u we may always add the term $\tilde{u} = a - by$, and to any v displacement we may add the term $\tilde{v} = c + bx$ without affecting the components of strain. Show that b represents a small, rigid-body rotation.

CHAPTER 3

PRINCIPLE OF VIRTUAL WORK

3.1 VIRTUAL WORK DEFINTION

Real work is normally defined as a real force moving through a real displacement. Symbolically, we may write this as

$$dW = \vec{P} \cdot d\vec{u}$$

Virtual work is defined as a real force moving through a virtual displacement, or

$$\delta W = \vec{P} \cdot \delta\vec{u}$$

Virtual means having the effect but not the actual form of what is specified. A virtual displacement $\delta\vec{u}$ can be defined in terms of virtual velocities $\delta\vec{u} = \delta\vec{v}\, dt$. It is essential to the whole development to realize that the force $\vec{P}$ remains unchanged in all its vector attributes during its virtual displacement. Consider the load-deflection curve shown in Fig. 3.1.

In Fig. 3.1 and the related material in this section, δ stands for a small, real change in quantity. The shaded area represents the work done in taking the real force δP through the real displacement δu. Assuming that δu is small enough that $\widehat{AB}$ can be considered linear, for ΔW we can write the following expression:

$$\Delta W = \frac{1}{2}\,[P + (P + \delta P)]\;\delta u = P\,\delta u + \frac{\delta P\,\delta u}{2}$$

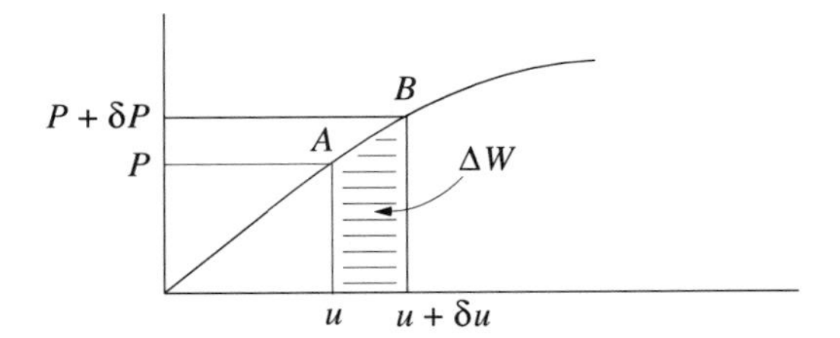

Figure 3.1 Load-deflection curve.

In the above equation, the first-order term $P\,\delta u$ is defined as the virtual work. The increment of force δP causing the actual incremental change δu does not enter into the definition of virtual work.

3.2 GENERALIZED COORDINATES

If a mechanical system consists of a finite number of material points or rigid bodies, its configuration (or position) can be specified by a finite number of real vector quantities (say, N) or, equivalently, $3N$ scalar quantities. For example, the position of a rigid sphere that is free to move in space can be described by six variables: namely, the translational Cartesian coordinates of the center of the sphere and the three rotational coordinates usually represented by some form of Euler angle system [20]. Generally, the coordinates meet the following four conditions:

- Their values determine the configuration of the system.
- They are independent of each other.
- They may be varied in an arbitrary fashion.
- Their variation does not violate the constraints of the system.

In general, a constraint is defined as a requirement that imposes a restriction on the positions of the particles or rigid bodies comprising the system under consideration. As an example, consider a rigid sphere constrained to roll on a smooth, fixed horizontal plane without slipping. It may be described by the coordinates represented by the two horizontal Cartesian coordinates of the center of the sphere and three Euler angles. It would be possible to give these five coordinates any arbitrary independent variation without violating the constraint. If R is the radius of the sphere, the constraint may be defined by the following relationship:

$$z - R = f(x, y, z, t) = 0$$

Constraints that can be expressed in the form

$$f(x_1, y_1, z_1, \ldots, x_N, y_N, z_N, t) = 0 \tag{3.1}$$

are called holonomic. When a set of N particles is subject to M-independent holonomic constraints, only

$$n = 3N - M$$

of the $3N$ quantities x_1, y_1, $z_1, \ldots, x_N$, y_N, z_N are independent of each other. Thus, it may be possible to express each of the x_1, y_1, $z_1, \ldots, x_N$, y_N, z_N as a single valued function of n functions of t in such a way that the constraint equations are satisfied for all values of the new functions $q_1(t), \ldots, q_n(t)$ and t. The $q_1(t), \ldots, q_n(t)$ are called the *generalized coordinates* of the system possessing *n degrees of freedom*. Holonomic constraints are further classified as rheonomic and scleronomic according to whether the function f does or does not contain t explicitly.

Constraints that are not holonomic are called nonholonomic or nonintegrable and are represented by nonintegrable differential equations. For example, the differential equation

$$dx - z\,dy = 0$$

relating the rectangular coordinates (x, y, z) does not possess an integrating factor. In general, nonholonomic constraints are represented by nonintegrable differential equations of the form

$$\sum_{r=1}^{n} A_{rs}\dot{q}_r + B_s = 0, \qquad s = 1, \ldots, m \tag{3.2}$$

where m is the number of linearly independent nonholonomic constraints. As an example, consider a rigid sphere constrained to roll on a rough, fixed horizontal plane without slipping. It may be described by the coordinates represented by the two horizontal Cartesian coordinates of the center of the sphere and by the three Euler angles. Since the plane is not smooth, we must impose two conditions of rolling by equating to zero the horizontal velocity of the particle of the sphere at the point of contact with the plane. Since the equations of rolling are nonintegrable, we cannot use them to express two of the generalized coordinates in terms of the other three. Thus an arbitrary variation in any of the five generalized coordinates will in general violate the constraint of rolling, unless the variation should happen to satisfy the nonholonomic constraint.

3.3 VIRTUAL WORK OF A DEFORMABLE BODY

Consider the equilibrium of a deformable body represented by Eq. (2.21)–Eq. (2.23). Each of these equations represents a force balance. If we now multiply each of these equations by a virtual displacement corresponding to the direction of the force balance, integrate over the volume of the body, and sum the results, we obtain the following relationship:

$$\delta W \stackrel{\text{def}}{=} \int_V \left(\frac{\partial \sigma_{xx}}{\partial x} + \frac{\partial \tau_{xy}}{\partial y} + \frac{\partial \tau_{xz}}{\partial z} + X_b \right) \delta u \, dV$$
$$+ \int_V \left(\frac{\partial \sigma_{yy}}{\partial y} + \frac{\partial \tau_{yx}}{\partial x} + \frac{\partial \tau_{yz}}{\partial z} + Y_b \right) \delta v \, dV$$
$$+ \int_V \left(\frac{\partial \sigma_{zz}}{\partial z} + \frac{\partial \tau_{zx}}{\partial x} + \frac{\partial \tau_{zy}}{\partial y} + Z_b \right) \delta w \, dV = 0 \qquad (3.3)$$

Equation (3.3) defines the virtual work for a deformable body. It is important to understand the following fine points of the definition.

- The deformable body is in equilibrium.
- Associated with the deformed body in its equilibrium state is a unique set of values for stress in equilibrium with the external forces.
- Associated with the deformed body in its equilibrium state is a unique set of values for strain and displacement that is consistent with the given boundary conditions.
- The virtual displacements of δu, δv, and δw are superimposed on the true existing displacements of the deformable body.
- Since the virtual displacements δu, δv, and δw describes a new configuration for the deformable body, they may be considered generalized coordinates; hence
  - they are independent of each other,
  - they may be varied in an arbitrary fashion, and
  - their variation does not violate the constraints of the system.
- Since it is virtual work that is being defined, it is important to remember that the internal and external force distributions remain unchanged by the virtual displacements.

While Eq. (3.3) represents the definition of virtual work for a deformable body, its present form is not very convenient for the solution of actual problems. Thus we recast the equation into a more useful form by use of Green's theorem [see Eq. (2.11)]. To apply Green's theorem, consider the first term of Eq. (3.3):

$$\int_V \frac{\partial \sigma_{xx}}{\partial x} \delta u \, dV$$

Let $\phi = \sigma_{xx}$ and $\partial\psi/\partial x = \delta u$, and note that since δu may be considered a function of x only, $\partial\psi/\partial y = 0$ and the $\partial\psi/\partial z = 0$. Hence we may write the following expression:

$$\int_V \frac{\partial \sigma_{xx}}{\partial x} \delta u \, dV = -\int \sigma_{xx} \frac{\partial \delta u}{\partial x} \, dV + \oint_{\partial V} \sigma_{xx} l \, \delta u \, dS$$

Expanding each of the terms in this manner, we obtain

$$\begin{aligned}\delta W = -\int_V \bigg[& \sigma_{xx} \frac{\partial \delta u}{\partial x} + \sigma_{yy} \frac{\partial \delta v}{\partial y} + \sigma_{zz} \frac{\partial \delta w}{\partial z} \\ & + \tau_{xy} \left(\frac{\partial \delta u}{\partial y} + \frac{\partial \delta v}{\partial x} \right) + \tau_{yz} \left(\frac{\partial \delta v}{\partial z} + \frac{\partial \delta w}{\partial y} \right) + \tau_{zx} \left(\frac{\partial \delta w}{\partial x} + \frac{\partial \delta u}{\partial z} \right) \bigg] \, dV \\ & + \int_V (X_b \, \delta u + Y_b \, \delta v + Z_b \, \delta w) \, dV + \oint_{\partial V} (X_s \, \delta u + Y_s \, \delta v + Z_s \, \delta w) \, dS \end{aligned} \tag{3.4}$$

where X_s, Y_s, Z_s are defined by Eqs. (2.24), (2.25), and (2.26). Thus the last two integrals of Eq. (3.4) are identified as the virtual work done by the prescribed body and surface forces. These forces are in general considered known.[1]

The first integral of Eq. (3.4) still needs to be clarified. To this end, we must now explicitly define δu, δv, δw. Our aim is to define a useful analytic tool, for which reason we choose to define these quantities as small variations in real displacements u, v, w. Then δ takes on operator significance, and we may write the expression for virtual work as follows:

$$\begin{aligned}\delta W = -\int_V & (\sigma_{xx} \, \delta \epsilon_{xx} + \sigma_{yy} \, \delta \epsilon_{yy} + \sigma_{zz} \, \delta \epsilon_{zz} \\ & + \tau_{xy} \, \delta \gamma_{xy} + \tau_{yz} \, \delta \gamma_{yz} + \tau_{zx} \, \delta \gamma_{zx}) \, dV\end{aligned}$$

[1]This is not always the case, however. For instance, consider a deformable elastic beam of magnetic material in the presence of a magnetic field. The electromechanical coupling can be modeled via Amperian current densities in the form of surface tractions on the boundary of the beam. In general, this problem is nonlinear, with the resulting beam deformations causing changes in the magnetic field and vice versa. Thus the surface tractions are dependent on the unknown displacements of the beam and on the unknown magnetic field at the surface of the beam [21].

$$+ \int_V (X_b\, \delta u + Y_b\, \delta v + Z_b\, \delta w)\, dV$$

$$+ \oint_{\partial V} (X_s\, \delta u + Y_s\, \delta v + Z_s\, \delta w)\, dS \tag{3.5}$$

The first integral in Eq. (3.5) therefore represents the total internal virtual work δU. The remaining two integrals represent the virtual work of the applied body forces and surface forces denoted by δW_e. Then, Eq. (3.5) may be expressed as

$$\delta W = -\delta U + \delta W_e = 0$$

or

$$\delta W_e = \delta U \tag{3.6}$$

Thus we may state the following:

A deformable system is in equilibrium if the total external virtual work is equal to the total internal virtual work for every virtual displacement consistent with the constraints.

Starting with the equations of equilibrium, we have proved the necessary conditions for the principle of virtual work. To prove that the principle of virtual work is a sufficient condition of equilibrium, we simply reverse the process, starting with Eq. (3.5) and applying Green's theorem to obtain Eq. (3.3).

There are several points about the principle of virtual work that should be noted:

- The principle holds irrespective of any material stress-strain relationships.
- The principle is an alternative way of expressing equilibrium conditions. Thus, once a strain-displacement relationship compatible with the constraints has been assumed, a set of equilibrium equations will result.
- The principle has nothing to do with the conservation of energy. Hence it is valid for nonconservative systems such as follower forces and plastic deformation.
- The fact that the virtual displacements and the associated strains (and rotations)[2] are infinitesimal places no restrictions on the actual displace-

[2]An expansion of terms in Eq. (3.4) actually contains, on the assumption of small displacements, terms representing small rigid-body rotations that are representable by skew-symmetric matrices—say, $\delta\omega_{ij}$. The stress terms are representable by symmetric matrices—say, σ_{ij}. These two types of matrices always enter as the product $\sigma_{ij}\delta\omega_{ij} = 0$; thus, only virtual strains occur in Eq. (3.5).

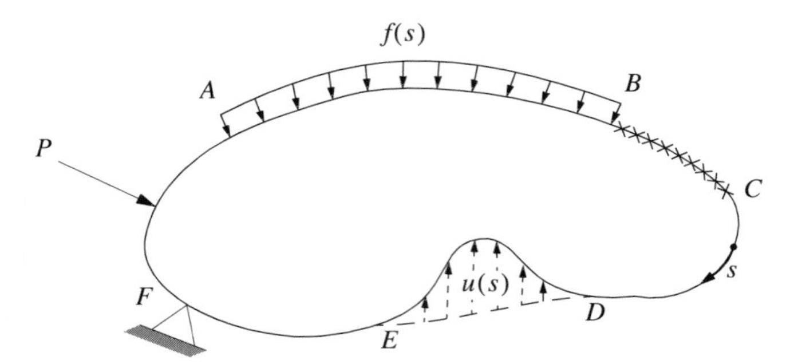

Figure 3.2 Prescribed boundary forces and displacements.

ments. Thus the principle can be used for finite-displacement large-strain problems.

At times, it is convenient to discuss a special continuum called a rigid body. A rigid body is characterized by the requirement that the distance between any two points is a constant after each displacement. Therefore, the internal virtual work of a rigid body is zero, from which we have the following lemma:

A rigid body or a system of rigid bodies is in equilibrium if the total (external) virtual work is zero for every virtual displacement consistent with the constraints.

Before proceeding further, it is important to clearly understand the meaning of the prescribed force and kinematically admissible displacements. Figure 3.2 shows a deformable body with both prescribed forces and kinematically admissible displacements. Forces are prescribed (known) along $\widehat{AB}$, $\widehat{CD}$, $\widehat{EF}$, and $\widehat{FA}$. Displacements are prescribed along $\widehat{BC}$, $\widehat{DE}$, and F.

Generally, on the boundary where forces are known, the displacements are unknown, whereas on the boundary where displacements are known, the forces are unknown.

A kinematically admissible displacement distribution is one that satisfies any prescribed displacement boundary condition and possesses continuous first partial derivatives in the interior of the deformable body. Since virtual displacements are considered additional displacements from the equilibrium configuration, a virtual displacement component must be zero at a boundary point whenever the actual displacement is prescribed by the boundary conditions at that boundary point.

If ∂V represents the total boundary, we may write

$$\partial V = S_\sigma + S_u$$

in which S_σ is the portion of the boundary where the forces are prescribed or are statically admissible, and S_u is the portion where the displacements are prescribed. Then, the last integral in Eq. (3.5) may be written as

$$\int_{S_\sigma} (X_s\,\delta u + Y_s\,\delta v + Z_s\,\delta w)\,dS$$

If on a portion of the boundary where the displacement is prescribed we release the prescribed displacements and replace them by the corresponding unknown reaction forces, we may then consider these reaction forces as "prescribed" and add to the above integral the following terms:

$$\int_{\hat{S}_\sigma} (\hat{X}_s\,\delta u + \hat{Y}_s\,\delta v + \hat{Z}_s\,\delta w)\,dS$$

where $\hat{X}_s$, $\hat{Y}_s$, $\hat{Z}_s$ are the reaction forces at the boundary and $\hat{S}_\sigma$ represents the portion of the boundary S_u where the prescribed displacements have been replaced by their reaction forces.

3.4 THERMAL STRESS, INITIAL STRAIN, AND INITIAL STRESS

In general, because of the weak coupling between elastic deformation and heat transfer, the Cauchy stress σ_{ij} can be considered functionally related to the Lagrangian strain and the change in temperature distribution $\Delta T(X_1, X_2, X_3)$ by the following relationship (see Washizu [22]):

$$\sigma_{ij} = \sigma_{ij}(E_{ij}, \Delta T)$$

With this assumption, the variational principles developed thus far remain unchanged.

Reference [22] also demonstrates that initial strain–incremental strain relationships can be developed and written in the following form:

$$\epsilon_{ij} = \epsilon_{ij}^m + \epsilon_{ij}^0$$

where

ϵ_{ij} is the strain required to satisfy compatibility (actually, the expression is more generally valid for the Lagrangian strain E_{ij})

ϵ_{ij}^m is the incremental or mechanical strain that causes internal stress

ϵ_{ij}^0 is the initial strain

With this assumption, the variational principles developed thus far remain unchanged.

Three important points should be noted:

- Thermal strains may be considered initial strains
- Thermal strains are considered prescribed—$\delta T = 0$
- Neither ϵ_{ij}^m nor ϵ_{ij}^0 by itself satisfies compatibility, but its sum must

While thermal stress and initial strains do not modify the form of the variational principles developed thus far, initial stresses do cause a modification. Initial stresses are those stresses that exist in the deformable body before the start of the deformation of interest. This is the reference state for the initial stress problem. The initial stress problem causes modifications to the variational principles because the initial stresses represent a self-equilibrating system with a corresponding set of equilibrium equations that must be included into the formulation represented by Eq. (3.3). A brief derivation is given in ref. [22], and a detailed exposition is given in ref. [23].[3]

3.5 SOME CONSTITUTIVE RELATIONSHIPS

The principle of virtual work is independent of any stress-strain relationships; however, for practical solutions to problems, it becomes necessary at some point to consider equations characterizing the individual material or materials that comprise the deformable body and their reaction to applied loads. Reference [24] gives an excellent introduction to the modern theory of constitutive equations.

In general, nonlinear material behavior can be expressed as

$$\boldsymbol{\sigma} = \boldsymbol{\sigma}(\boldsymbol{\epsilon})$$

where $\boldsymbol{\sigma}$ and $\boldsymbol{\epsilon}$ are column matrices of stress and strain components, respectively. For classical linear-elastic material, this relationship may be written as

$$\boldsymbol{\sigma} = \mathbf{D}(\boldsymbol{\epsilon} - \boldsymbol{\epsilon}_0) + \boldsymbol{\sigma}_0 \tag{3.7}$$

where $\mathbf{D}$ is the elasticity matrix containing the appropriate material properties, and $\boldsymbol{\sigma}_0$ and $\boldsymbol{\epsilon}_0$ are the initial incremental stress and initial strains (which may be caused by temperature), respectively.

Some examples of linear-elastic stress-strain relationships are discussed in the following paragraphs.

A linear-isotropic material with Young's modulus E, Poisson's ratio ν, and coefficient of thermal expansion α:

[3]Reference [23] presents nonlinear theories of deformation, with applications to problems of finite strains, by the extension of the concept of incremental deformation of deformable bodies under initial stress using a "strength of materials" approach.

$$
\begin{Bmatrix} \epsilon_{xx} \\ \epsilon_{yy} \\ \epsilon_{zz} \\ \gamma_{xy} \\ \gamma_{yz} \\ \gamma_{zx} \end{Bmatrix} = \frac{1}{E} \begin{bmatrix} 1 & -\nu & -\nu & 0 & 0 & 0 \\ -\nu & 1 & -\nu & 0 & 0 & 0 \\ -\nu & -\nu & 1 & 0 & 0 & 0 \\ 0 & 0 & 0 & 2(1+\nu) & 0 & 0 \\ 0 & 0 & 0 & 0 & 2(1+\nu) & 0 \\ 0 & 0 & 0 & 0 & 0 & 2(1+\nu) \end{bmatrix}
\cdot \begin{Bmatrix} \sigma_{xx} \\ \sigma_{yy} \\ \sigma_{zz} \\ \tau_{xy} \\ \tau_{yz} \\ \tau_{zx} \end{Bmatrix} + \alpha \Delta T \begin{Bmatrix} 1 \\ 1 \\ 1 \\ 0 \\ 0 \\ 0 \end{Bmatrix} \tag{3.8}
$$

or

$$
\begin{Bmatrix} \sigma_{xx} \\ \sigma_{yy} \\ \sigma_{zz} \\ \tau_{xy} \\ \tau_{yz} \\ \tau_{zx} \end{Bmatrix} = \frac{E}{(1+\nu)(1-2\nu)}
\cdot \begin{bmatrix} 1-\nu & \nu & \nu & 0 & 0 & 0 \\ \nu & 1-\nu & \nu & 0 & 0 & 0 \\ \nu & \nu & 1-\nu & 0 & 0 & 0 \\ 0 & 0 & 0 & \dfrac{1-2\nu}{2} & 0 & 0 \\ 0 & 0 & 0 & 0 & \dfrac{1-2\nu}{2} & 0 \\ 0 & 0 & 0 & 0 & 0 & \dfrac{1-2\nu}{2} \end{bmatrix}
\cdot \begin{Bmatrix} \epsilon_{xx} \\ \epsilon_{yy} \\ \epsilon_{zz} \\ \gamma_{xy} \\ \gamma_{yz} \\ \gamma_{zx} \end{Bmatrix} - \frac{\alpha E \Delta T}{1-2\nu} \begin{Bmatrix} 1 \\ 1 \\ 1 \\ 0 \\ 0 \\ 0 \end{Bmatrix} \tag{3.9}
$$

For two-dimensional problems, the above constitutive equation can take two forms. Let the z direction be normal to the x–y plane of interest; then, consider the case where $\sigma_{zz} = \tau_{zx} = \tau_{zy} = 0$, which is called plane stress. Thus Eq. (3.8) becomes

$$\begin{Bmatrix} \epsilon_{xx} \\ \epsilon_{yy} \\ \gamma_{xy} \end{Bmatrix} = \frac{1}{E} \begin{bmatrix} 1 & -\nu & 0 \\ -\nu & 1 & 0 \\ 0 & 0 & 2(1+\nu) \end{bmatrix} \begin{Bmatrix} \sigma_{xx} \\ \sigma_{yy} \\ \tau_{xy} \end{Bmatrix} + \alpha \Delta T \begin{Bmatrix} 1 \\ 1 \\ 0 \end{Bmatrix} \tag{3.10}$$

or

$$\begin{Bmatrix} \sigma_{xx} \\ \sigma_{yy} \\ \tau_{xy} \end{Bmatrix} = \frac{E}{1-\nu^2} \begin{bmatrix} 1 & \nu & 0 \\ \nu & 0 & 0 \\ 0 & 0 & \dfrac{1-\nu}{2} \end{bmatrix} \begin{Bmatrix} \epsilon_{xx} \\ \epsilon_{yy} \\ \gamma_{xy} \end{Bmatrix} - \frac{\alpha E \Delta T}{1-\nu} \begin{Bmatrix} 1 \\ 1 \\ 0 \end{Bmatrix} \tag{3.11}$$

Next, consider the case where $\partial u_z / \partial z = 0$ or $\epsilon_{zz} = \gamma_{zx} = \gamma_{zy} = 0$, which is called plane strain. Thus Eq. (3.8) becomes

$$\begin{Bmatrix} \epsilon_{xx} \\ \epsilon_{yy} \\ \gamma_{xy} \end{Bmatrix} = \frac{1+\nu}{E} \begin{bmatrix} 1-\nu & -\nu & 0 \\ -\nu & 1-\nu & 0 \\ 0 & 0 & 2 \end{bmatrix} \begin{Bmatrix} \sigma_{xx} \\ \sigma_{yy} \\ \tau_{xy} \end{Bmatrix} + (1+\nu) \alpha \Delta T \begin{Bmatrix} 1 \\ 1 \\ 0 \end{Bmatrix} \tag{3.12}$$

or

$$\begin{Bmatrix} \sigma_{xx} \\ \sigma_{yy} \\ \tau_{xy} \end{Bmatrix} = \frac{E}{(1+\nu)(1-2\nu)} \begin{bmatrix} 1-\nu & -\nu & 0 \\ -\nu & 1-\nu & 0 \\ 0 & 0 & \dfrac{1-2\nu}{2} \end{bmatrix} \begin{Bmatrix} \epsilon_{xx} \\ \epsilon_{yy} \\ \gamma_{xy} \end{Bmatrix} - \frac{\alpha E \Delta T}{1-2\nu} \begin{Bmatrix} 1 \\ 1 \\ 0 \end{Bmatrix} \tag{3.13}$$

For the one-dimensional case, we have

$$\epsilon = \frac{\sigma}{E} + \alpha \Delta T \tag{3.14}$$

There are, of course, many other forms for constitutive equations. For example, in the case of a linear two-dimensional orthotropic material, D takes the following form:

$$D = \frac{1}{1 - \nu_1 \nu_2} \begin{bmatrix} E_1 & \nu_2 E_1 & 0 \\ \nu_1 E_2 & E_2 & 0 \\ 0 & 0 & G(1 - \nu_1 \nu_2) \end{bmatrix} \tag{3.15}$$

where $\nu_1 E_2 = \nu_2 E_1$ and the subscripts refer to principal directions. For this case, $\boldsymbol{\epsilon}_0$ take the following form:

$$\boldsymbol{\epsilon}_0 = \begin{Bmatrix} \alpha_1 \\ \alpha_2 \\ \alpha_{12} \end{Bmatrix} \Delta T$$

3.6 ACCOUNTING FOR ALL WORK

The principle of virtual work is an elegant, powerful tool, but it still yields correct results only when all the work is accounted for. An example of this is given by the computation of the force on a parallel plate capacitor, as shown in Fig. 3.3. The defining relationships are as follows:

$$V = \frac{Q}{C}$$

and

$$\frac{1}{C} = \frac{z}{\epsilon_0 A}$$

where

Q is the charge (coulombs)
V is the potential difference (volts = newton-meters/Q = joules/Q)

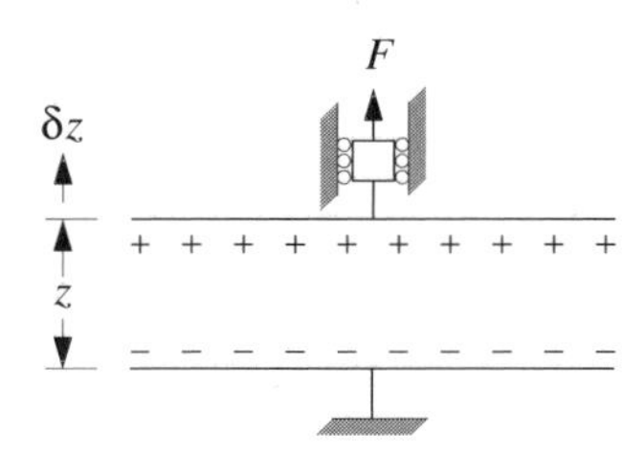

Figure 3.3 Parallel plate capacitor.

C is the capacitance (farad = coulombs/volt)
A is the area (in meters squared, or m^2)
ϵ_0 is the dielectric constant in a vacuum (Q^2/m/joule)
z is the distance between plates (m)

The work to transfer charge is

$$dU = V\,dQ = \frac{Q\,dQ}{C}$$

Then, the total work to charge the plates is

$$U = \int_0^Q \frac{Q\,dQ}{C}$$

$$= \frac{1}{2}\frac{Q^2}{C} \tag{3.16}$$

$$= \frac{1}{2}V^2C \tag{3.17}$$

There are two ways to view this example: case I, in which we can consider the capacitor to be held at constant charge (an isolated system), and case II, in which we can consider holding the plates at constant potential.

Considering case I first, the mechanical work necessary to provide a virtual increase δz in the plate distance is $\delta W_e = F\,\delta z$, where F is the mechanical force on the plates. Then, by using Eq. (3.16) we can compute the internal virtual work caused by the separation of the plates as $(1/2)Q^2\,\delta(1/C)$, since C is changed by plate separation and Q is held fixed. From the above definition of $1/C$, we have

$$\delta\,\frac{1}{C} = \frac{\delta z}{\epsilon_0 A}$$

The principle of virtual work may be stated as

$$\left(F - \frac{1}{2}\,\frac{Q^2}{\epsilon_0 A}\right)\,\delta z = 0$$

Since δz is arbitrary, the above expression can hold only if the terms in the bracket are identically zero. Hence we have

$$F = \frac{Q^2}{2\epsilon_0 A}$$

Considering case II next, Eq. (3.17) can then be used to compute the internal virtual work caused by the separation of the plates while the potential V is held constant. Then, the internal virtual work caused by the separation of the plates may be computed as $(1/2)V^2\,\delta C$. This expression may also be written as

$$\frac{1}{2}\,V^2\,\delta C = \frac{1}{2}\,\frac{Q^2}{C^2}\,\delta C = -\frac{1}{2}\,Q^2\,\delta\left(\frac{1}{C}\right)$$

The internal virtual work can now be equated to the external virtual work. However, the external virtual work is no longer just the term $F\,\delta z$—we must account for all the external virtual work that, in this case, requires the addition of the term $V(V\,\delta C)$. This term represents the charge added to the system by an external source to keep the potential V a constant as the plates are separated. The principle of virtual work may now be stated as follows:

$$\left(F - \frac{Q^2}{\epsilon_0 A} + \frac{1}{2}\,\frac{Q^2}{\epsilon_0 A}\right)\,\delta z = 0$$

The expression for F is, as before,

$$F = \frac{Q^2}{2\epsilon_0 A}$$

This example brings out two important points: First, that all work must be accounted for, and second, that virtual work carries across all fields of mechanics.

3.7 AXIALLY LOADED MEMBERS

Axially loaded members (usually referred to as bars or rods) are characterized by the requirement that the axial stress for any arbitrary cross section is uniform. Any stress risers from a sudden change in the cross section (stress concentrations) can be neglected in the study of the gross behavior of the member (Saint Venant's principle). The requirement of uniform stress has immediate consequences for the location of the resultant load. Figure 3.4 shows the geometry of an arbitrary cross section. By using the principle of virtual work, it is possible to determine the location of the resultant forces.

Consider first the virtual work of the internal stress. This may be determined as

$$\delta U = \int \sigma\; dA\overline{u}_0 + \int \sigma y\; dA\overline{\theta}_z + \int \sigma z\; dA\overline{\theta}_y$$

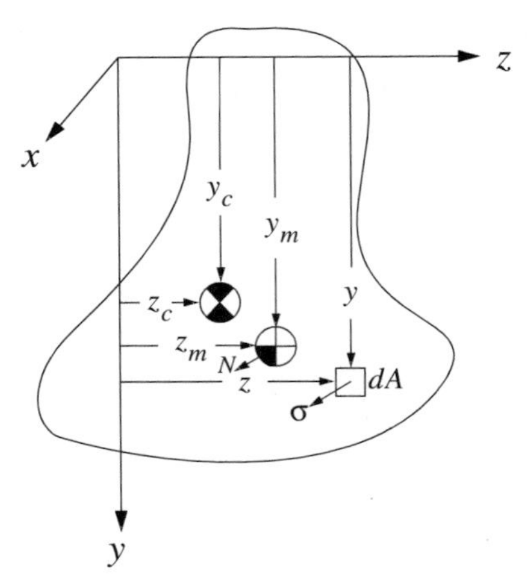

Figure 3.4 Axial member of an arbitrary cross section.

where

- dA is an infinitesimal area
- y, z are the locations of the small area
- σ is the uniform stress acting on this small area

Also in the above equation, $\overline{u}_0$ represents a virtual rigid-body motion of the cross section along the bar axis, whereas $\overline{\theta}_y$ and $\overline{\theta}_z$ represent virtual rigid-body rotations of the cross section about the y and z axis, respectively. Since the stress σ is required to be uniform, it may be removed from the integrals, and the above expression becomes

$$\delta U = \sigma A \overline{u}_0 + \sigma A y_c \overline{\theta}_z + \sigma A z_c \overline{\theta}_y \tag{3.18}$$

where y_c and z_c locate the centroid of the cross section.

The virtual work of the external resultant load acting on the cross section can be expressed as

$$\delta W_e = N \overline{u}_0 + y_m N \overline{\theta}_z + z_m N \overline{\theta}_y \tag{3.19}$$

where N represents the resultant axial load on the bar cross section, and y_m and z_m represent the load's location in the cross section.

Equating Eqs. (3.18) and (3.19) and collecting terms, we get

$$(N - \sigma A)\overline{u}_0 + (y_m N - \sigma A y_c)\overline{\theta}_z + (z_m N - \sigma A z_c)\overline{\theta}_y = 0 \tag{3.20}$$

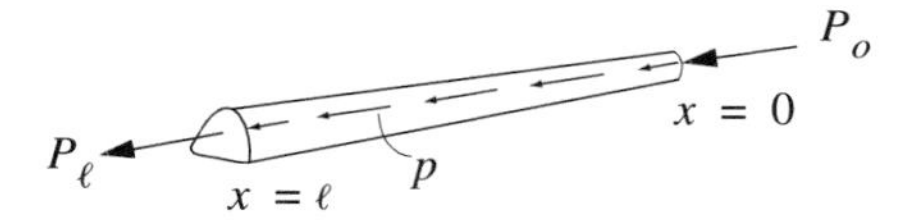

Figure 3.5 Equilibrium of a bar.

Since the virtual displacements represent generalized coordinates, each of the terms in parentheses must in turn be equated to zero, with the following result:

$$N = \sigma A$$
$$y_m = y_c$$
$$z_m = z_c$$

Thus we see that the resultant load must pass through the centroid of the cross section.

At this juncture, a comment on sign convention is in order. Normally, a positive normal stress produces tension. The reader should, however, be careful. In finite element method codes, this is the normal convention except for contact elements, in which compression is often considered positive. When computing work terms, we normally select a coordinate direction as positive; then, all loads and displacement moving in the selected positive direction produce positive work. In computing work, the selected direction is not of particular importance. What is important is consistency. For a given problem, once a convention has been established it must be maintained throughout the analysis.

Figure 3.5 represents a bar of length ℓ with a general loading condition: namely, an axially distributed load $p(x)$, point loads P_ℓ and P_0 at the ends, and some axial temperature distribution $\Delta T = T(x) - T_0$. Additionally, let the cross-sectional area be some function $A(x)$. For simplicity of discussion (but with no implied limits on the applicability of virtual work), consider the bar to remain geometrically and materially linear with a modulus of elasticity $E(x)$. The stress-strain relationship can then be written as

$$\epsilon = u' = \frac{\sigma}{E} + \alpha \Delta T$$

where $u' = du/dx$ and α is the coefficient of thermal expansion. The equation of virtual work becomes

$$\int (Eu' - \alpha E \Delta T)\overline{u}' \, dA \, dx = \int p\overline{u} \, dx + P_\ell \overline{u}(\ell) + P_0 \overline{u}(0) \tag{3.21}$$

In Eq. (3.21), $\overline{u}$ represents a virtual change in the bar's axial displacement and can be considered a generalized coordinate. However, $\overline{u}'$ also occurs and is not independent of $\overline{u}$. Thus we need to integrate the left side of Eq. (3.21) by parts to eliminate $\overline{u}'$. The result is

$$(AEu' - \alpha AE\Delta T)\overline{u}\,|_0^\ell - \int (AEu' - \alpha AE\Delta T)'\overline{u}\, dx$$
$$= \int p\overline{u}\, dx + P_\ell \overline{u}(\ell) + P_0 \overline{u}(0)$$

or

$$\begin{aligned} -\int & [(AEu'(x) - \alpha AE\Delta T)' + p]\overline{u}(x)\, dx \\ &+ [(AEu'(\ell) - \alpha AE\Delta T) - P_\ell]\overline{u}(\ell) \\ &+ [-(AEu'(0) - \alpha AE\Delta T) - P_0]\overline{u}(0) = 0 \end{aligned} \tag{3.22}$$

In Eq. (3.22), we have arrived at a single set of generalized coordinates given by $\overline{u}$. The next step in a variational method is to collect terms into the following form:

$$\begin{aligned} &(\text{expression}_1)(\text{generalized coord}_1) + \cdots + \\ &(\text{expression}_i)(\text{generalized coord}_i) + \cdots = 0 \end{aligned} \tag{3.23}$$

Then, using the property of the arbitrariness of generalized coordinates, set all but one of the generalized coordinates to zero (say, generalized coord$_i$) to get the following:

$$(\text{expression}_i)(\text{generalized coord}_i) = 0$$

Since generalized coord$_i$ is arbitrary, (expression$_i$) = 0. This is done for each generalized coordinate in turn to generate a sequence of equations that may then be solved for the appropriate parameters.

To acquire any useful information from Eq. (3.22), we must now put it into the form of Eq. (3.23) by considering the set of generalized coordinates $\overline{u}$ to consist of the subsets of admissible generalized coordinates that lie in the interior (inside the integral) of the bar and those that lie on the boundary (end points) of the bar. Then, any attribute that holds for the set of generalized functions must also hold for any of the subsets. Thus, in Eq. (3.22) we may consider $\overline{u}(x)$, $\overline{u}(0)$, and $\overline{u}(\ell)$ separately as subsets of $\overline{u}$ while setting the remaining two equal to zero. Thus, letting $\overline{u}(0) = \overline{u}(\ell) = 0$ and $\overline{u}(x) \neq 0$, we arrive at the following expression:

$$-\int [(AEu'(x) - \alpha AE\Delta T)' + p]\overline{u}(x)\, dx = 0$$

If $[(AEu'(x) - \alpha AE\Delta T)' + p]$ is a continuous function in the interval of integration, if for all functions $\overline{u}(x)$ is continuous through the $n - 1$ derivative of u (where n is the highest derivative of u under the integral), and, finally, if $\overline{u}$ satisfies the forced boundary conditions, then a fundamental lemma of calculus [25] states that $[(AEu'(x) - \alpha AE\Delta T)' + p]$ is identically zero. We arrive now at the following differential equation of equilibrium for the bar:

$$-\frac{d}{dx}\left(AE\,\frac{du(x)}{dx} - \alpha AE\Delta T\right) = p \tag{3.24}$$

Depending on the actual boundary conditions, the other two sets of expressions in Eq. (3.22) yield either the forced or the natural boundary conditions. Then, for the end $x = \ell$, we have

$$u(\ell) = 0 \quad \text{or} \quad \left(AE\,\frac{du(\ell)}{dx} - \alpha AE\Delta T\right) = P_\ell \tag{3.25}$$

For the end $x = 0$:

$$u(0) = 0 \quad \text{or} \quad -\left(AE\,\frac{du(0)}{dx} - \alpha AE\Delta T\right) = P_0 \tag{3.26}$$

Example 3.1 Figure 3.6 represents a pillar of circular cross section that is to be designed as an equal-strength structure so that the compressive stress at any cross section x is a constant value σ_W. Using the principle of virtual work, we wish to determine for a pillar of unit weight γ the radius r of the pillar to meet the design requirement.

Equating the internal and external virtual work, we obtain the following relationship:

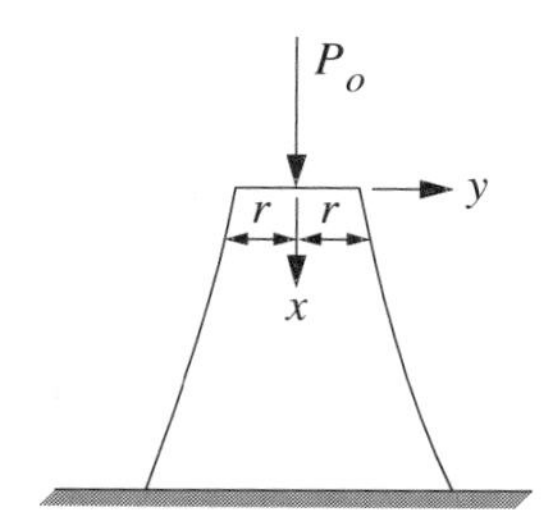

Figure 3.6 Constant stress pillar.

$$\int_V \sigma\bar{\epsilon}\, dV = \int_V \gamma A(x)\bar{u}(x)\, dx + P_0\bar{u}(0)$$

However, in $\sigma = -\sigma_W$ the compressive stress is a constant. By using the relationship $\bar{u}' = \bar{\epsilon}$ and replacing σ with the required value σ_W on the left side of the above expression, we obtain

$$-\sigma_W \int_x A(x)\bar{u}'\, dx = \int_x \gamma A(x)\bar{u}(x)\, dx + P_0\bar{u}(0)$$

Integration by parts then yields

$$-(\sigma_W A\bar{u})_0^\ell + \int_x \sigma_W \frac{dA}{dx}\,\bar{u}(x)\, dx = \int_x \gamma A(x)\bar{u}(x)\, dx + P_0\bar{u}(0)$$

Considering the subsets of $\bar{u}(x)$ on the boundaries, we obtain the boundary conditions

$$u(\ell) = 0 \quad \Rightarrow \quad \bar{u}(\ell) = 0$$

and

$$\sigma_W A(0) = P_0$$

Considering the interior subset $\bar{u}(x)_{\text{interior}}$ of $\bar{u}(x)$, we obtain the differential equation

$$\frac{dA}{A} = \frac{\gamma}{\sigma_W}\, dx$$

with the solution

$$\ln A = \frac{\gamma}{\sigma_W}\, x + C; \qquad C = \ln A(0) = \ln\left(\frac{P_0}{\sigma_W}\right)$$

or

$$A = \frac{P_0}{\sigma_W}\, \exp\left(\frac{\gamma x}{\sigma_W}\right)$$

Using the relationship that $A = \pi r^2$, we arrive at the result

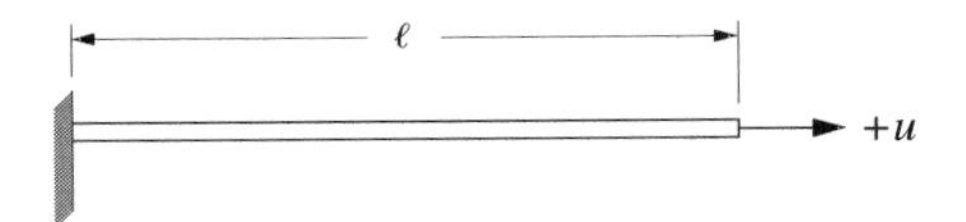

Figure 3.7 Clamped free rod under a uniform temperature increase.

$$r = \left(\frac{P_0}{\pi \sigma_W} \right)^{1/2} \exp\left(\frac{\gamma x}{2\sigma_W} \right)$$

To arrive at this result, we used Eq. (3.6) directly because it is a fundamental principle and is easy to remember. However, we could have used the less easily remembered Eq. (3.24) directly to obtain

$$-\frac{d}{dx} \left[A(x)E \, \frac{du(x)}{dx} \right] = p(x) = \gamma A(x)$$

or

$$\frac{d}{dx} \, [A(x)\sigma_W] = \gamma A(x)$$

which yields the same result.

Example 3.2 As a further example, consider the rod shown in Fig. 3.7 to undergo a uniform temperature increase ΔT. Since there are no external forces, $\delta W_e = 0$, and we may write

$$\delta U = \int_V \sigma \overline{\epsilon} \, dV = 0$$

Consider the compatible strain $\overline{\epsilon} = 1/\ell$. Then, the above expression becomes

$$\delta U = \sigma \left(\frac{1}{\ell} \right) (A\ell) = 0$$

or $\sigma = 0$. Then, by using, for example, Eq. (3.14), we get

$$\epsilon = \alpha \Delta T$$

or

$$u(x) = \alpha \Delta T x \quad \text{and} \quad u(\ell) = \alpha \Delta T \ell$$

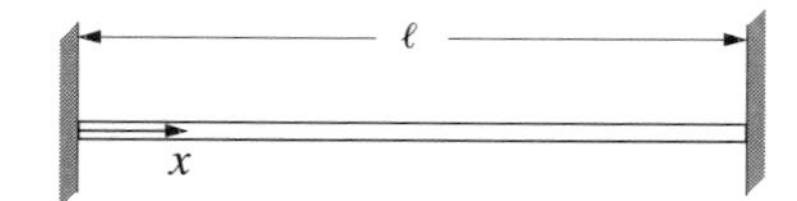

Figure 3.8 Clamped rod under a uniform temperature increase.

Example 3.3 If the rod shown in Fig. 3.7 is clamped as shown in Fig. 3.8, the external virtual work is $\delta W_e = 0$, and the internal virtual work is given as

$$\delta U = \int_V \sigma \bar{\epsilon}\, dV = 0$$

Intuition precludes the condition $\sigma = 0$ in this case. Rather, the expression is satisfied because the only compatible virtual strain is $\bar{\epsilon} = 0$. Therefore, we conclude that the rod undergoes no deformation. From Eq. (3.14), the internal loads are computed as

$$\sigma = -\alpha E \Delta T$$

3.8 THE UNIT-DISPLACEMENT METHOD

Figure 3.9 shows a coplanar linearly elastic truss of n members. Each member is pinned at the ground, and all are joined at a common pin O, as shown in the figure. A typical member has length ℓ_m and area A_m, and it is inclined at an angle β_m with respect to the horizontal. A load P is applied at O, making an angle θ with respect to the horizontal. Assuming a strain-displacement relationship compatible with the constraints, as well as applying the principle of virtual work to this structure, yields the necessary equations of equilibrium. The equations are obtained in a form similar to Eq. (3.23).

For complex structures, it can often be difficult to factor the resulting composite expression for virtual work into the form of Eq. (3.23). However, since virtual displacements are considered to be generalized coordinates, we know that as long as we do not violate any constraints we can arbitrarily vary each virtual displacement independent of any other virtual displacement. A logical choice of variation is to set all but one virtual displacement to zero and to vary the remaining virtual displacement. By repeating this procedure for each of the virtual displacements, the total set of equilibrium equations can be generated one at a time. The process can be further simplified by using a value of one for each varied virtual displacement. This process is often called the *unit-displacement method* or the *unit-dummy-displacement method*. Equation (3.6) then takes the following symbolic form:

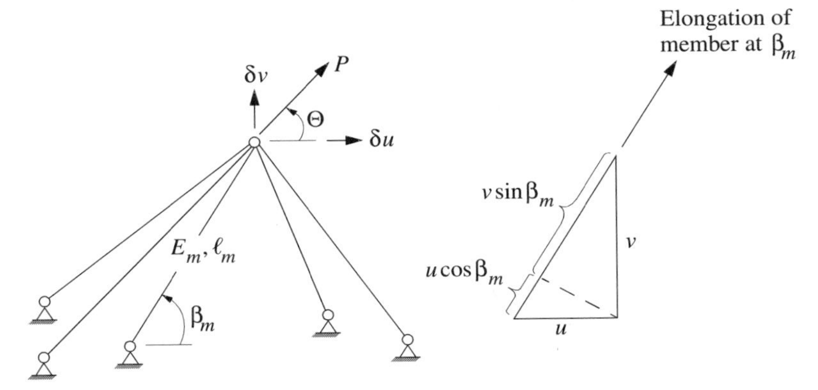

Figure 3.9 A simple indeterminate truss.

(Unit displacement applied in direction of actual external force)
· (Actual external force)
= (Deformation compatible with unit displacement)
· (Actual internal forces)

For the truss shown in Fig. 3.9, u and v—the horizontal and vertical displacements of point O—serve as generalized coordinates. The virtual displacements are $\overline{u}$ and $\overline{v}$. For a small deformation, the elongation of any rod can be expressed as

$$\epsilon_m = u \cos \beta_m + v \sin \beta_m \tag{3.27}$$

and a compatible corresponding strain is

$$\epsilon_m = \frac{u \cos \beta_m + v \sin \beta_m}{\ell_m} \tag{3.28}$$

The stress in a typical member is

$$\sigma_m = \frac{E_m}{\ell_m} (u \cos \beta_m + v \sin \beta_m) \tag{3.29}$$

Virtual work requires the selection of a virtual strain compatible with the constraints. The most logical choice in this case is the true strain, as given by Eq. (3.28), with true displacements u and v replaced by the virtual displacements $\overline{u}$ and $\overline{v}$. For virtual strain, this yields

$$\bar{\epsilon}_m = \frac{\bar{u} \cos \beta_m + \bar{v} \sin \beta_m}{\ell_m} \tag{3.30}$$

Consider first a virtual displacement $\bar{u} = 1$ and $\bar{v} = 0$; then, the expression for virtual work becomes

$$\sum_{m=1}^{n} \underbrace{\frac{E_m}{\ell_m} (u \cos \beta_m + v \sin \beta_m)}_{\text{true stress}} \underbrace{\frac{\cos \beta_m}{\ell_m}}_{\text{virtual strain}} \underbrace{(A_m \ell_m)}_{\text{volume}} = \underbrace{P \cos \theta}_{\text{true load}} \cdot \underbrace{1}_{\text{virtual displacement}}$$

Collecting terms in u and v, we get the following equation for equilibrium of forces in the u direction:

$$u \sum_{m=1}^{n} \frac{E_m A_m}{\ell_m} \cos^2 \beta_m + v \sum_{m=1}^{n} \frac{E_m A_m}{\ell_m} \sin \beta_m \cos \beta_m = P \cos \theta \tag{3.31}$$

Next, consider a virtual displacement $\bar{u} = 0$ and $\bar{v} = 1$; then, the expression for virtual work becomes

$$\sum_{m=1}^{n} \frac{E_m}{\ell_m} (u \cos \beta_m + v \sin \beta_m) \frac{\sin \beta_m}{\ell_m} (A_m \ell_m) = P \sin \theta \cdot 1$$

Collecting terms in u and v, we get the following equation for equilibrium of forces in the v direction:

$$u \sum_{m=1}^{n} \frac{E_m A_m}{\ell_m} \cos \beta_m \sin \beta_m + v \sum_{m=1}^{n} \frac{E_m A_m}{\ell_m} \sin^2 \beta_m = P \sin \theta \tag{3.32}$$

Equations (3.31) and (3.32) can now be solved for the unknown displacements u and v. Before doing so, however, it is convenient to put these equations into the following matrix form:

$$[\mathbf{K}]\{\mathbf{u}\} = \{\mathbf{P}\}$$

or

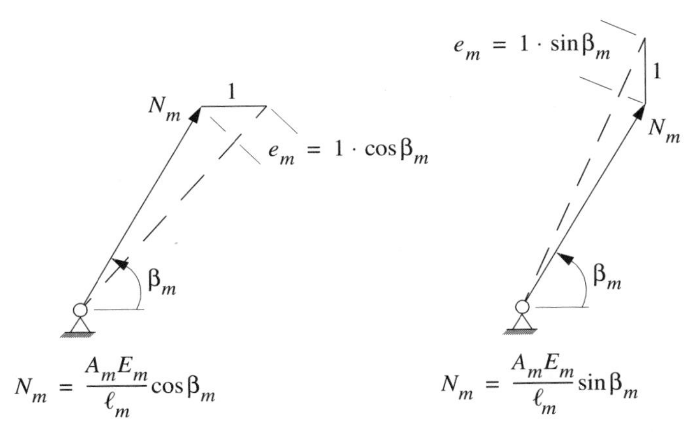

Figure 3.10 Loads on a typical member of a simple indeterminate truss.

$$\begin{bmatrix} \sum_{m=1}^{n} \frac{A_m E_m}{\ell_m} \cos^2 \beta_m & \sum_{m=1}^{n} \frac{A_m E_m}{\ell_m} \sin \beta_m \cos \beta_m \\ \sum_{m=1}^{n} \frac{A_m E_m}{\ell_m} \cos \beta_m \sin \beta_m & \sum_{m=1}^{n} \frac{A_m E_m}{\ell_m} \sin^2 \beta_m \end{bmatrix} \cdot \begin{Bmatrix} u \\ v \end{Bmatrix} = \begin{Bmatrix} P \cos \theta \\ P \sin \theta \end{Bmatrix} \tag{3.33}$$

The matrix [**K**] is called the stiffness matrix, and the unit-displacement method is a statement of the stiffness method of analysis. To obtain a better feeling for the physical significance of the stiffness matrix, we draw our attention to Fig. 3.10, which shows the loading on a typical member of the truss in Fig. 3.9 for a unit displacement in both the u direction and v direction.

Looking at the resolution of forces in the u and v direction for the case $\overline{u} = 1$, we recognize the u component as the first term of the first column of the stiffness matrix in Eq. (3.33). The v component is identified as the second term of the first column of the stiffness matrix in Eq. (3.33). Similarly, for the case of $\overline{v} = 1$, we identify the u and v components of load with the elements of the second column of the stiffness matrix in Eq. (3.33). Thus the columns of the stiffness matrix may be generated one at a time by setting all of the displacements of the structure to zero except the displacement corresponding to the desired column, which is set equal to unity. For example, the forces necessary to maintain the displacement configuration $u = 0$, $v = 1$ are given as

$$\begin{bmatrix} 0 & \sum_{m=1}^{n} \dfrac{A_m E_m}{\ell_m} \sin\beta_m \cos\beta_m \\ 0 & \sum_{m=1}^{n} \dfrac{A_m E_m}{\ell_m} \sin^2\beta_m \end{bmatrix} \begin{Bmatrix} 0 \\ v=1 \end{Bmatrix} = \begin{Bmatrix} P_u \\ P_v \end{Bmatrix}$$

A typical term of [**K**] expressed as k_{uv} then represents the force in the direction of u necessary to sustain a unit displacement at v. (Another point of view is that the term is the force required in the direction u to cause a unit displacement at v.)

A term of the stiffness matrix can be directly related to the integral for virtual work. To do this, consider the virtual displacement $\overline{u} = 1$ and its compatible virtual strain $\{\overline{\epsilon}\}_u$. The subscript used is to indicate that the strain must be compatible with the virtual u displacement. Next, consider a true displacement $v = 1$. This will generate a real internal stress system $\lfloor\sigma\rfloor_v$, where the subscript is used to indicate that the stress is a true stress from the real displacement v. The forces required to sustain the unit v displacement are k_{uv}, as was discussed. Applying the principle of virtual work, we have

$$(\overline{u} = 1) \cdot k_{uv} = \int_V \lfloor\sigma\rfloor_v \{\overline{\epsilon}\}_u \, dV \tag{3.34}$$

For example, to compute k_{12} for Fig. 3.9, Eq. (3.30) gives $\overline{\epsilon} = \cos\beta_m/\ell_m$ and Eq. (3.29) gives $\sigma_m = E_m \sin\beta_m/\ell_m$. By substituting into Eq. (3.34), we get

$$k_{12} = \sum_{m=1}^{n} \frac{E_m}{\ell_m} \sin\beta_m \frac{\cos\beta_m}{\ell_m} \ell_m A_m$$

Equation (3.34) gives the interpretation of k_{uv} as the magnitude of the virtual work done by internal loads in state v on the corresponding internal virtual displacements in state u.

Through the interpretation of k_{uv} as a virtual-work term via Eq. (3.34), we are provided with a method for assembling a stiffness matrix. For a given term in the stiffness matrix, we simply add the contribution of all possible elements to obtain the term. For linear-elastic systems in the absence of initial stress and thermal strains, Eq. (3.34) yields directly the Maxwell-Betti reciprocity law, which may be expressed as

$$k_{uv} = k_{vu}$$

To see this, consider an elastic system with the following stress-strain relationship:

$$\sigma = \mathbf{D}\epsilon$$

where **D** is the symmetric elasticity matrix [see, for example, Eq. (3.9), with $\Delta T = 0$]. Also, note that the transpose of a symmetric matrix is itself; then define two separate systems and compute their virtual work as follows.

For system I, a unit u displacement yields

$$(u = 1) \cdot k_{uv} = \int_V \lfloor \sigma \rfloor_v \{\epsilon\}_u \, dV$$

or

$$k_{uv} = \int_V \lfloor \epsilon \rfloor_v [D] \{\epsilon\}_u \, dV$$

For system II, a unit v displacement yields

$$(v = 1) \cdot k_{vu} = \int_V \lfloor \sigma \rfloor_u \{\epsilon\}_v \, dV$$

or

$$k_{vu} = \int_V \lfloor \epsilon \rfloor_u [D] \{\epsilon\}_v \, dV$$

Since k_{vu} is a scalar number, its transpose is unchanged, and we may write

$$k_{vu} = \int_V \lfloor \epsilon \rfloor_v [D] \{\epsilon\}_u \, dV$$

from which it follows that $k_{uv} = k_{vu}$.

3.9 FINITE ELEMENTS FOR AXIAL MEMBERS

Typical two- and three-dimensional truss structures are composed of axial members (or elements) joined at pin connections (or nodes) to form an assemblage of two force members. The external loading of these structures is applied at the pin connections or nodal points. Thus a truss structure represents not only one of the simplest discretizations of a continuum but also one of the simplest forms of a finite element model.

In this section, we examine some of the basic concepts of the assembly procedure that is common to all finite element analyses. Additionally, we will study some of the aspects in defining an axial finite element. Figure 3.11(a) represents

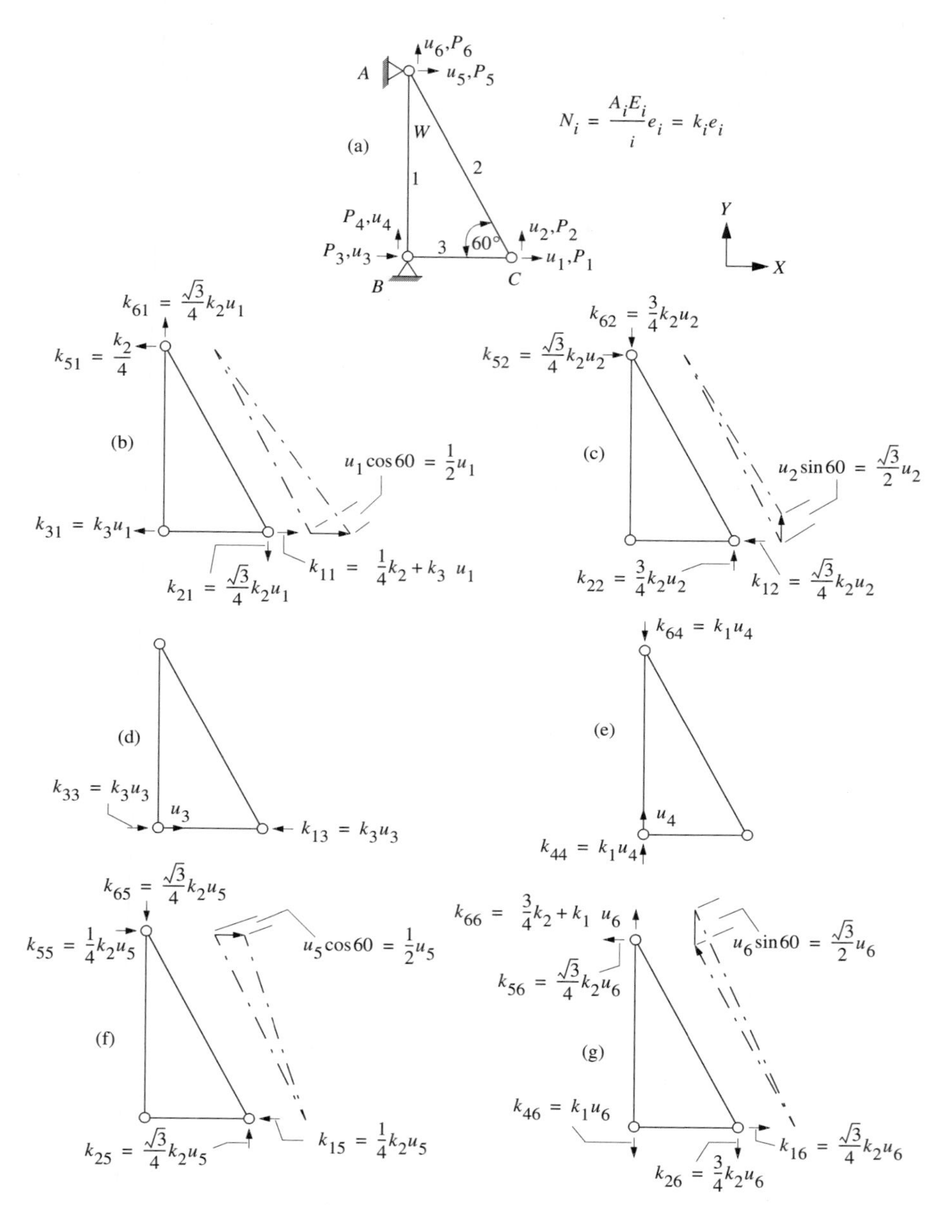

Figure 3.11 Three-rod planar truss.

a simple truss structure supported as shown with degrees of freedom u_1–u_6 indicated at each connection (node) point. Figures 3.11(b)–(g) represent free bodies of the same structure, with the supports temporarily removed and a displacement u_i at a specified degree of freedom while all other u_j, $j \neq i$ are held fixed.

The stiffness of the ith rod is denoted by $k_i = A_i E_i / \ell_i$, and the k_{ij} are defined by Eq. (3.34).

From the collection of the six free-body diagrams, we may determine the equilibrating force P_i required at any node for a specific direction defined by u_i. For example, the equilibrium equation corresponding to the degree of freedom u_1 for any arbitrary displacement of the structure is

$$\left(\frac{1}{4}\,k_2 + k_3\right) u_1 - \frac{\sqrt{3}}{4}\,k_2 u_2 - k_3 u_3 + 0 u_4 - \frac{1}{4}\,k_2 u_5 + \frac{\sqrt{3}}{4}\,k_2 u_6 = P_1$$

Collecting all six equations, we get the following system of equations:

$$\begin{bmatrix} \frac{1}{4}\,k_2 + k_3 & -\frac{\sqrt{3}}{4}\,k_2 & -k_3 & 0 & -\frac{1}{4}\,k_2 & \frac{\sqrt{3}}{4}\,k_2 \\ -\frac{\sqrt{3}}{4}\,k_2 & \frac{3}{4}\,k_2 & 0 & 0 & \frac{\sqrt{3}}{4}\,k_2 & -\frac{3}{4}\,k_2 \\ -k_3 & 0 & k_3 & 0 & 0 & 0 \\ 0 & 0 & 0 & k_1 & 0 & -k_1 \\ -\frac{1}{4}\,k_2 & \frac{\sqrt{3}}{4}\,k_2 & 0 & 0 & \frac{1}{4}\,k_2 & -\frac{\sqrt{3}}{4}\,k_2 \\ \frac{\sqrt{3}}{4}\,k_2 & -\frac{3}{4}\,k_2 & 0 & -k_1 & -\frac{\sqrt{3}}{4}\,k_2 & k_1 + \frac{3}{4}\,k_2 \end{bmatrix} \cdot \begin{Bmatrix} u_1 \\ u_2 \\ u_3 \\ u_4 \\ u_5 \\ u_6 \end{Bmatrix} = \begin{Bmatrix} P_1 \\ P_2 \\ P_3 \\ P_4 \\ P_5 \\ P_6 \end{Bmatrix} \tag{3.35}$$

The following observations can be made concerning Equation 3.35:

- Each row represents the contribution of the elements of the structure to the overall equilibrium of a specific nodal degree of freedom in response to an external load applied at the node corresponding to the row number.
- Each column of the matrix represents the equilibrium of one of the individual Figs. 3.11(b)–(g). Therefore, each column represents an equilibrium state for a specific unit displacement of a particular nodal degree of freedom of the structure, with all other nodal degrees of freedom fixed to zero displacement.

- Each column sums to zero. (For a general stiffness matrix, the sum and moments of each column about an arbitrary point in space must be zero.)
- Each diagonal term is positive.

The following global observations about the matrix can also be made:

- The matrix is symmetric.
- The matrix is singular because the structure is as yet unconstrained and contains, in the plane, three rigid-body displacements.

To see the rigid-body displacements, let $\{u\}$ have any of the following three displacements and observe that the right side of Eq. (3.35) is identically zero.

$$\{u\} = \begin{Bmatrix} 1 \\ 0 \\ 1 \\ 0 \\ 1 \\ 0 \end{Bmatrix}_{X\ \text{Disp}} ; \quad \begin{Bmatrix} 0 \\ 1 \\ 0 \\ 1 \\ 0 \\ 1 \end{Bmatrix}_{Y\ \text{Disp}} ; \quad \begin{Bmatrix} 0 \\ 0 \\ 0 \\ 1 \\ \sqrt{3} \\ 1 \end{Bmatrix}_{\text{Rot about } C}$$

If the structure is constrained, as in Fig. 3.11(a), then the matrix equations to be solved take the following form:

$$\begin{bmatrix} \frac{1}{4}k_2 + k_3 & -\frac{\sqrt{3}}{4}k_2 & \frac{\sqrt{3}}{4}k_2 \\ -\frac{\sqrt{3}}{4}k_2 & \frac{3}{4}k_2 & -\frac{3}{4}k_2 \\ \frac{\sqrt{3}}{4}k_2 & -\frac{3}{4}k_2 & k_1 + \frac{3}{4}k_2 \end{bmatrix} \begin{Bmatrix} u_1 \\ u_2 \\ u_6 \end{Bmatrix} = \begin{Bmatrix} P_1 \\ P_2 \\ P_6 \end{Bmatrix} \tag{3.36}$$

This relationship was arrived at by simply striking out the rows and columns that correspond to the constrained nodal degrees of freedom. The columns no longer sum to zero because by striking out a row and a column, we introduce physical reactive forces necessary to constrain the motion to zero for the support represented by the removed row and column. Consider Eq. (3.35) in symbolic form, where the subscript g stands for global (or fully assembled) but unrestrained stiffness.

$$[K_{gg}]\{u_g\} = \{P_g\}$$

Partition this matrix into degrees of freedom restrained or supported (s) and degrees of freedom that are left free (f), as follows:

$$\begin{bmatrix} K_{ff} & K_{fs} \\ K_{sf} & K_{ss} \end{bmatrix} \begin{Bmatrix} u_f \\ u_s = 0 \end{Bmatrix} = \begin{Bmatrix} P_f \\ P_s + Q_s \end{Bmatrix} \tag{3.37}$$

where Q_s represents the restraint force on the boundary needed to restrain the structure and P_s represents any applied loading to the boundary. Equation (3.36) then takes the following symbolic form from the first of Eq. (3.37):

$$[K_{ff}]\{u_f\} = \{P_f\}$$

and using the second of Eq. (3.37), the boundary reaction is

$$\{Q_s\} = [K_{sf}]\{u_f\} - \{P_s\}$$

For the three-bar truss of Fig. 3.11(a), the matrix $[K_{sf}]$ is of the form

$$[K_{sf}] = \begin{bmatrix} k_{31} & k_{32} & k_{36} \\ k_{41} & k_{42} & k_{46} \\ k_{51} & k_{52} & k_{56} \end{bmatrix} = \begin{bmatrix} -k_3 & 0 & 0 \\ 0 & 0 & -k_1 \\ -\dfrac{1}{4}k_2 & \dfrac{\sqrt{3}}{4}k_2 & -\dfrac{\sqrt{3}}{4}k_2 \end{bmatrix}$$

The first column of $[K_{sf}]$ corresponds to the support reactions of Fig. 3.11(b); the second column, to the support reactions of Fig. 3.11(c); and the third column, to the support reactions of Fig. 3.11(g). We conclude from these observations that $[K_{sf}]$ represents the support reactions in the s direction required to sustain a unit displacement at the free degree of freedom f.

To mechanize the above procedure for assembling a structure, first consider a typical axial-rod finite element as shown in Fig. 3.12. The element is described in terms of a "natural" coordinate [26] ξ, which represents a mapping of the element's physical coordinate x_e into a nondimensionalized system where ξ takes on the value of one or zero at the node points. The origin of the ξ and

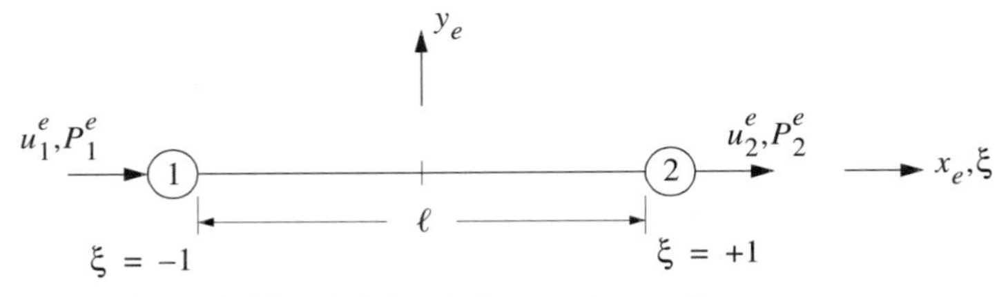

Figure 3.12 Axial-rod element in an element system.

x_e coordinates is at the midpoint of the element. The x_e and ξ coordinates are related by

$$x_e = \frac{\ell}{2}\,\xi \quad \text{and} \quad dx_e = \frac{\ell}{2}\,d\xi = |J|\,d\xi$$

where $|J| = \ell/2$ is the determinant of the Jacobian matrix.

The displacement $\tilde{u}$ can be expressed as a function of ξ and two unknown parameters u_1^e and u_2^e, which represents the axial displacements of node 1 and node 2, respectively. Thus we have in matrix form:

$$\begin{aligned}\tilde{u} &= \frac{1}{2}\,\lfloor (1-\xi) \quad (1+\xi) \rfloor \begin{Bmatrix} u_1^e \\ u_2^e \end{Bmatrix} \\ &= \lfloor N_1 \quad N_2 \rfloor \begin{Bmatrix} u_1^e \\ u_2^e \end{Bmatrix} \\ &= [N]\{u^e\}\end{aligned} \tag{3.38}$$

The matrix $[N]$ is the matrix of shape functions and, in general, contains as many rows as the dimensionality of the element.

The strain in the element is defined as

$$\epsilon = \frac{d\tilde{u}}{dx_e} = \frac{d\tilde{u}}{d\xi}\,\frac{d\xi}{dx_e}$$

or in terms of the as-yet-to-be-determined parameters u_1^e and u_2^e:

$$\begin{aligned}\epsilon &= \frac{1}{\ell}\,\lfloor (-1) \quad (1) \rfloor \begin{Bmatrix} u_1^e \\ u_2^e \end{Bmatrix} \\ &= \lfloor B_1 \quad B_2 \rfloor \begin{Bmatrix} u_1^e \\ u_2^e \end{Bmatrix} \\ &= [B]\{u^e\}\end{aligned} \tag{3.39}$$

where $[B]$ in general relates the strains ϵ to the parameters u_1^e and u_2^e and is completely defined when the shape functions $[N]$ are chosen.

The stress σ for a linear-elastic rod is determined as

$$\begin{aligned}\sigma &= \frac{E}{\ell}\,\lfloor (-1) \quad (1) \rfloor \begin{Bmatrix} u_1^e \\ u_2^e \end{Bmatrix} \\ &= [D][B]\{u^e\}\end{aligned} \tag{3.40}$$

Then, applying the principle of virtual work,

$$\delta U = \delta W_e$$

we get

$$\begin{aligned}
\int \lfloor \bar{\epsilon} \rfloor \{\sigma\}\, dV &= \frac{AE}{\ell^2} \int \lfloor \bar{u}_1^e \quad \bar{u}_2^e \rfloor \begin{Bmatrix} -1 \\ 1 \end{Bmatrix} \lfloor -1 \quad 1 \rfloor \begin{Bmatrix} u_1^e \\ u_2^e \end{Bmatrix} dx_e \\
&= \lfloor \bar{u}_1^e \quad \bar{u}_2^e \rfloor \frac{AE}{\ell^2} \begin{bmatrix} 1 & -1 \\ -1 & 1 \end{bmatrix} \begin{Bmatrix} u_1^e \\ u_2^e \end{Bmatrix} \int_{-1}^{1} \frac{\ell}{2}\, d\xi \\
&= \lfloor \bar{u}_1^e \quad \bar{u}_2^e \rfloor \frac{AE}{\ell} \begin{bmatrix} 1 & -1 \\ -1 & 1 \end{bmatrix} \begin{Bmatrix} u_1^e \\ u_2^e \end{Bmatrix} \\
&= \lfloor \bar{u}_1^e \quad \bar{u}_2^e \rfloor \begin{Bmatrix} P_1^e \\ P_2^e \end{Bmatrix}
\end{aligned} \tag{3.41}$$

Since the virtual displacement $\bar{u}_1^e$ and $\bar{u}_2^e$ are arbitrary, we arrive at the following relationship:

$$\frac{AE}{\ell} \begin{bmatrix} 1 & -1 \\ -1 & 1 \end{bmatrix} \begin{Bmatrix} u_1^e \\ u_2^e \end{Bmatrix} = \begin{Bmatrix} P_1^e \\ P_2^e \end{Bmatrix} \tag{3.42}$$

where

$$[K^e] = \frac{AE}{\ell} \begin{bmatrix} 1 & -1 \\ -1 & 1 \end{bmatrix} \tag{3.43a}$$

$$= \int [B]^{\mathrm{T}} [D][B] |J|\, dV_\xi \tag{3.43b}$$

is the element-stiffness matrix in element coordinates and dV_ξ is the volume of the element in natural coordinates.

3.10 COORDINATE TRANSFORMATIONS

Equations (3.42) and (3.43a) represent the element-equilibrium and -stiffness relationships for the rod degrees of freedom in the element coordinates. To place a particular element into its proper position and orientation in the actual structure, it is necessary to transform the element coordinates' degrees of freedom into the corresponding degrees of freedom of the "structures." In a general case, some of the degrees of freedom to be transformed may not have any physical meaning and may not equal the number of degrees of freedom of the system to which they are to be transformed.

Consider the following transformation between two systems of coordinates:

$$\{u^e\} = [T_{eg}]\{u_g\} \tag{3.44}$$

where $\{u^e\}$ represents the original coordinates, $\{u_g\}$ represents the new coordinates, and $[T_{eg}]$ represents the coordinate transformation; the $\{u_g\}$ are required to represent a set of generalized independent coordinates. Next, give each of these generalized coordinates $\{u_g\}$ in turn a virtual displacement while holding the others fixed. Under this virtual-displacement pattern, Eq. (3.44) takes the following form:

$$[\overline{u}^e] = [T_{eg}]\lceil \overline{u}_g \rfloor \tag{3.45}$$

where $\lceil\ \rfloor$ represents a diagonal matrix, and each row of $[\overline{u}^e]$ contains a column entry for the corresponding column in $\lceil \overline{u}_g \rfloor$. Associate with the original coordinates $\{u^e\}$ the set of forces $\{P^e\}$; with the generalized displacements $\{u_g\}$, the generalized forces $\{P_g\}$. We wish to define a transformation between these two sets of forces so that the virtual work is invariant under the transformation of coordinates. This requirement may be expressed as

$$\lceil \overline{u}_g \rfloor \{P_g\} = [\overline{u}^e]^{\mathrm{T}}\{P^e\} \tag{3.46}$$

Substituting the transpose of Eq. (3.45) for $[\overline{u}^e]^{\mathrm{T}}$ and the collecting terms, we arrive at the expression

$$\lceil \overline{u}_g \rfloor (\{P_g\} - [T_{eg}]^{\mathrm{T}}\{P^e\}) = 0 \tag{3.47}$$

Since the $\lceil \overline{u}_g \rfloor$ are arbitrary, we have

$$\{P_g\} = [T_{eg}]^{\mathrm{T}}\{P^e\} \tag{3.48}$$

Force and displacement vectors related in this manner are called associated (or conjugate) vectors. The respective transformations are called contragradient transformations, since they are contrary to each other. Transformations of this kind were first studied in detail by Clebsch, Maxwell, and Krohn [27]. To get a clearer picture of the relationship between the displacement variables and the force variables, consider Eq. (3.46) for a coordinate transformation of the following form:

$$\begin{Bmatrix} u_1^e \\ u_2^e \end{Bmatrix} = \begin{bmatrix} T_{11} & T_{12} & T_{13} \\ T_{21} & T_{22} & T_{23} \end{bmatrix} \begin{Bmatrix} u_{g1} \\ u_{g2} \\ u_{g3} \end{Bmatrix}$$

For the virtual work, we get

$$\begin{Bmatrix} \overline{u}_{g1} P_{g1} \\ \overline{u}_{g2} P_{g2} \\ \overline{u}_{g3} P_{g3} \end{Bmatrix} = \begin{Bmatrix} \overline{u}_{11}^e P_1^e + \overline{u}_{12}^e P_2^e \\ \overline{u}_{21}^e P_1^e + \overline{u}_{22}^e P_2^e \\ \overline{u}_{31}^e P_1^e + \overline{u}_{32}^e P_2^e \end{Bmatrix}$$

In the expression for virtual work, it should be observed that each work term only involves a force term and its corresponding displacement component.

If the force vector is related to the displacement vector in the original coordinate system by the relationship

$$[K^e]\{u^e\} = \{P^e\}$$

we may use Eqs. (3.44) and (3.48) to get

$$[T_{eg}]^{\mathrm{T}}[K^e][T_{eg}]\{u_g\} = [T_{eg}]^{\mathrm{T}}\{P^e\} = \{P_g\}$$

from which we conclude that

$$[K_{gg}] = [T_{eg}]^{\mathrm{T}}[K^e][T_{eg}] \tag{3.49}$$

We may now use Eq. (3.49) to transform the element-stiffness matrix given by Eq. (3.43a) into its corresponding degrees of freedom in the global stiffness. To do so, however, the transformation expressed by Eq. (3.44) must first be determined. To do this, first consider the following relationships from Fig. 3.13, which shows the rod element of Fig. 3.12 at an arbitrary orientation in a two-dimensional structural system. In Fig. 3.13, $\vec{n}$ is a unit vector orientated along the axis of the element; u_1^e is to be transformed into u_1, u_2 of the global system; and u_2^e is to be transformed into u_3, u_4 of the global system.

From Fig. 3.13, we may obtain the following relationships:

$$u_1^e \vec{n} = u_1 \vec{i} + u_2 \vec{j}$$
$$u_2^e \vec{n} = u_3 \vec{i} + u_4 \vec{j}$$

Taking the vector-dot product with $\vec{n}$ of each of these expressions, we obtain the following result expressed in vector form:

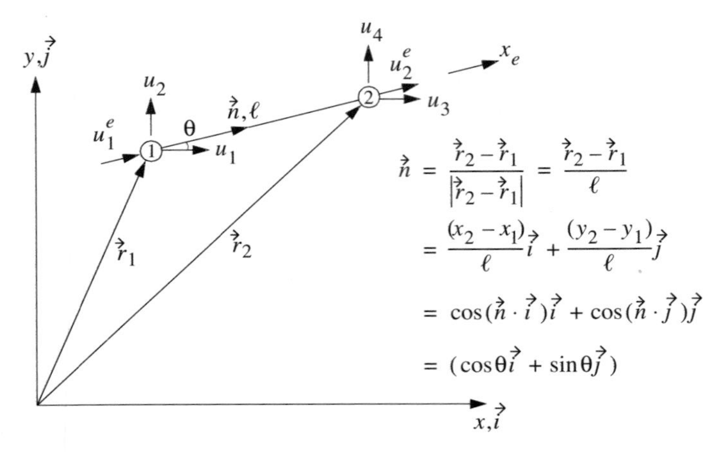

Figure 3.13 Element system to global system—two-dimensional.

$$\begin{Bmatrix} u_1^e \\ u_2^e \end{Bmatrix} = \begin{bmatrix} \cos\theta & \sin\theta & 0 & 0 \\ 0 & 0 & \cos\theta & \sin\theta \end{bmatrix} \begin{Bmatrix} u_1 \\ u_2 \\ u_3 \\ u_4 \end{Bmatrix}$$

From this expression, we identify $[T_{eg}]$ as

$$[T_{eg}] = \begin{bmatrix} \cos\theta & \sin\theta & 0 & 0 \\ 0 & 0 & \cos\theta & \sin\theta \end{bmatrix}$$

With this transformation, we may apply Eq. (3.49) directly to Eq. (3.43a) to obtain the element-stiffness matrix in the structural system with the following result:

$$[K] = \frac{AE}{\ell} \begin{bmatrix} \cos^2\theta & \cos\theta\sin\theta & -\cos^2\theta & -\cos\theta\sin\theta \\ \cos\theta\sin\theta & \sin^2\theta & -\cos\theta\sin\theta & -\sin^2\theta \\ -\cos^2\theta & -\cos\theta\sin\theta & \cos^2\theta & \cos\theta\sin\theta \\ -\cos\theta\sin\theta & -\sin^2\theta & \cos\theta\sin\theta & \sin^2\theta \end{bmatrix} \tag{3.50}$$

3.11 REVIEW OF THE SIMPLE BEAM THEORY

Beams are characterized by their capability to carry axial loads, bending loads, torsion loads, and often cross-sectional warping loads. Most beam theories are based on some adaptation of the strength-of-materials approach.

Classical beam theory, often referred to as the Bernoulli-Euler (sometimes, the Bernoulli-Navier) hypothesis, deals with slender beams and the assumption that the cross sections perpendicular to the centroid axis before bending remain undistorted in their plane and perpendicular to the deformed centroid axis. This theory has several limitations. For example, in vibration analysis, such a beam under impact predicts an infinite wave speed. This difficulty is removed by incorporating into the analysis the shear deformation and the effect of rotary inertia. The beam is then often referred to as a Timoshenko beam. Other considerations arise if the beam is not slender but curved. These types of beams are often analyzed by the Winkler-Bach theory.

If the cross-sectional dimensions of the beam are such that the thickness of the cross section is small in relation to other cross-sectional dimensions, such as the overall height and width, the beam is classified as thin walled. Beams of this nature are further classified as thin walled beams of open (e.g., a channel section) or closed (e.g., a box section) cross section, and they may be even further characterized by warping normal to the plane of the cross section and their torsional stiffness.

In the following sections, simple beam theory will be reviewed first; then, a brief introduction to curved and thin walled beams of open section will be presented. Through the concepts of virtual work, we will get an appreciation of how easily thermal effects can be included into beam theory. However, thermal effects included through a strength-of-materials approach have restrictions, of which some of the details may be found in ref. [28]. We will also get an appreciation of how engineering assumptions on strain can introduce problems in the application of virtual work.

To begin the review of simple beam theory, consider the segment Δx of the beam shown in Fig. 3.14. The cross section of the beam is considered symmetrical with the xy plane defining the plane of symmetry. All loading lies in the plane of symmetry. By convention, the bending moment is considered positive if it bends the beam concave upwards, as shown in Fig. 3.14. The coordinated system is a right-hand system with its origin at the centroid of the cross section. For the given symmetry and loading conditions, the y axis coincides with the neutral axis, which is the axis of the cross section on which the strain and hence the stress from bending is zero.

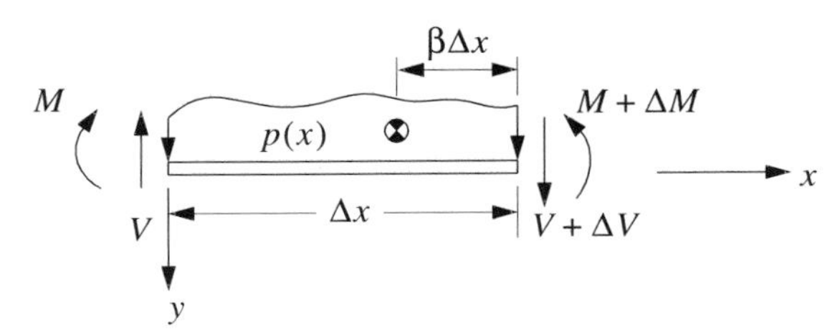

Figure 3.14 Equilibrium of Δx segment of a beam.

From the sum of forces in the y direction and moments about the right edge of the segment, the following two equilibrium equations are obtained:

$$-V + (V + \Delta V) + p(x)\Delta x = 0$$

and

$$-M + (M + \Delta M) + p(x)\Delta x(\beta \Delta x) - V\Delta x = 0$$

where V and M are, respectively, the shear and moment acting on a beam section, and ΔV and ΔM are their respective variations over Δx. The spanwise distributed load is $p(x)$ and $\beta \Delta x$, with $\beta < 1$ defining the location of its centroid over the segment Δx.

In the limit, as $\Delta x \rightarrow 0$, we obtain the following two equations:

$$\frac{dV}{dx} = -p(x) \tag{3.51}$$

and

$$\frac{dM}{dx} = V \tag{3.52}$$

These two equations can be combined into the following single equation of equilibrium:

$$\frac{d^2M}{dx^2} + p(x) = 0 \tag{3.53}$$

Next, consider the elastic curve of the beam. Figure 3.15 represents a typical elastic curve where x is measured positive to the right and y is measured positive downward. Since we are interested in the curvature of the elastic curve, we note in Fig. 3.15 that as the curve is traversed in the positive s direction, the angle ϕ measured tangent to the curve decreases, thereby defining the curvature as negative. Thus, if ρ is the radius of curvature for the curve shown, the expression for curvature is

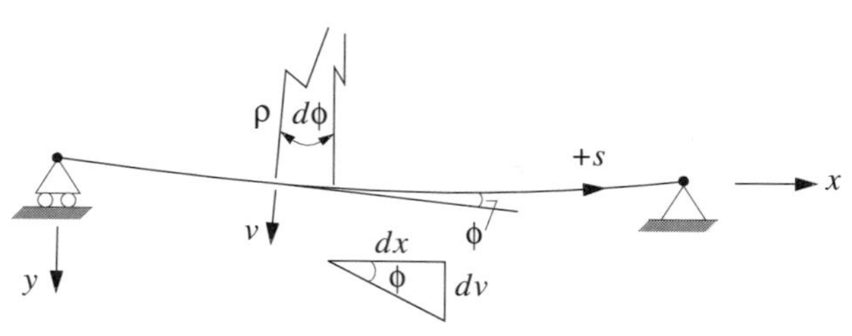

Figure 3.15 Elastic curve.

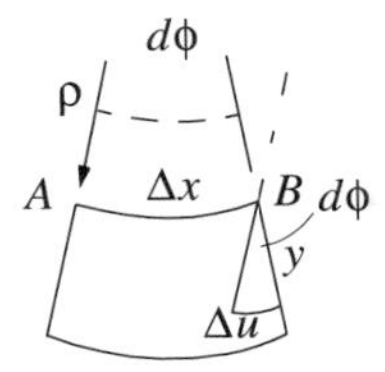

Figure 3.16 Segment Δx under assumption of plane sections.

$$\frac{1}{\rho} = -\frac{d\phi}{ds}$$

If only small deflections v are considered, $ds \simeq dx$ and $\phi \simeq dv/dx$; then, the expression for curvature becomes

$$\frac{1}{\rho} = -\frac{d^2v}{dx^2} = -v''$$

Figure 3.16 shows a segment Δx of the beam under the Bernoulli-Euler assumption of plane sections. The adoption of this assumption corresponds to the compatibility equations for simple beam theory. From this figure, the strain ϵ is defined as

$$\begin{aligned} \epsilon &= \frac{\Delta u}{\Delta x} \\ &= \frac{y\, d\phi}{\rho\, d\phi} \\ &= -yv'' \end{aligned} \tag{3.54}$$

If the beam is homogeneous, isotropic, and follows Hooke's law, the total moment M at any section is given as

$$\begin{aligned} M &= \int_A y\sigma\, dA \\ &= \int_A yE\epsilon\, dA \\ &= \int_A yE(-yv'')\, dA \\ &= -EIv'' \end{aligned}$$

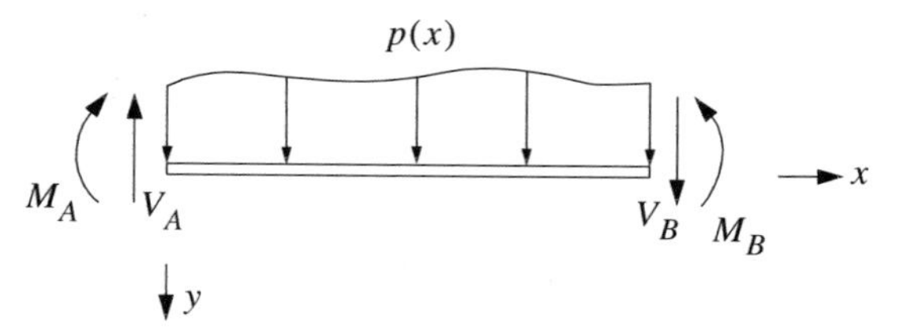

Figure 3.17 Free body of a beam.

where

I is the $\int_A y^2 \, dA$
E is the Young's modulus
σ is the longitudinal stress at any point in the beam's cross section

From these results, we obtain the following well-known relationship for stress:

$$\sigma = -Ev''y = \frac{My}{I}$$

It is instructive to apply virtual work to Eq. (3.53). Figure 3.17 represents a general free body of a beam (no displacements prescribed). For a differential element dx, the equation of equilibrium is given by Eq. (3.53). The virtual work is defined as

$$\delta W \stackrel{\text{def}}{=} \int_A^B \left(\frac{d^2M}{dx^2} + p \right) \overline{v} \, dx = 0$$

where $\overline{v}$ is the virtual displacement of the differential element. Integrating the above equation twice by parts, the following expression is obtained:

$$\delta W = \underbrace{\int_A^B M\overline{v}'' \, dx}_{-\delta U}$$

$$\underbrace{+ V_B\overline{v}_B - V_A\overline{v}_A - M_B\overline{v}'_B + M_A\overline{v}'_A + \int_A^B p\overline{v} \, dx}_{\delta W_e}$$

$$= 0$$

or

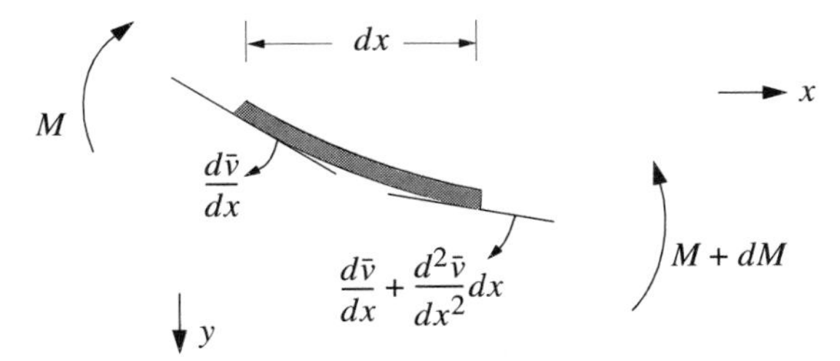

Figure 3.18 Work on a differential element.

$$\delta W = -\delta U + \delta W_e = 0$$

The relationship is identical to Eq. (3.6). Note, however, that we associated the minus sign with the integral itself. To expand on this, consider a small differential element shown in Fig. 3.18.

The figure describes the variation of bending moment M and slope $d\overline{v}/dx = \overline{v}'$ along a differential segment dx of a beam. Computing the work of the moments shown yields the following result:

$$M \frac{d\overline{v}}{dx} - (M + dM)\left(\frac{d\overline{v}}{dx} + \frac{d^2\overline{v}}{dx^2}\,dx\right)$$

which on expanding becomes

$$\underbrace{-M \frac{d^2\overline{v}}{dx^2}\,dx}_{\delta U} \qquad \underbrace{-dM \frac{d\overline{v}}{dx}}_{\text{force increment to produce } \Delta\overline{v}'} \qquad \underbrace{-\cdots}_{\text{higher-order terms}}$$

Thus we see that because of the sign conventions chosen for M and $\overline{v}'$ and the fact that $\overline{v}'$ is decreasing, a minus sign becomes associated with the work integral.

The above results will now be generalized for beams of nonsymmetrical cross section and loading in both the xy and xz plane. The loads and their reactions are assumed to act along the line of shear centers to avoid causing any torsion. This line will be located in the discussion of thin walled open beams for which the Bernoulli-Euler assumption of plane sections is a special case. In general, torsional loads acting on a beam cause warping of the cross section that, in turn, produces additional longitudinal stresses in the beam. This is why in most strength-of-material courses, only circular cross sections are considered when the beam is subject to torsion. The study of thin walled beams of both open and closed sections is essentially a study of the warping effect.

Figure 3.19 shows the undeformed and deformed axis of centroids for a

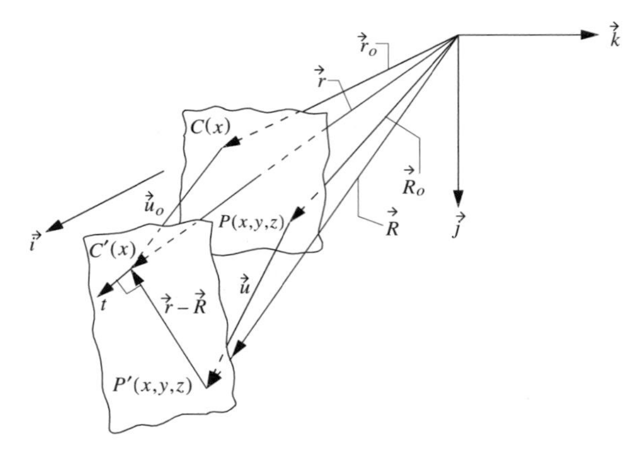

Figure 3.19 Deflection curve of a general straight beam.

beam. From this figure, the kinematics of the deflection of the centroid axis under the assumption of plane sections is determined as follows:

$\vec{r}_0 = x\vec{i}$ is the initial position of centroid $C(x)$

$\vec{u}_0 = u_0\vec{i} + v\vec{j} + w\vec{k}$ is the displacement of centroid $C(x)$

$\vec{r} = \vec{r}_0 + \vec{u}_0$ is the deformed position $C'(x)$ of centroid $C(x)$

$\vec{R}_0 = x\vec{i} + y\vec{j} + z\vec{k}$ is the initial position of a general point of the beam section $P(x, y, z)$

$\vec{u} = U\vec{i} + V\vec{j} + W\vec{k}$ is the displacement of $P(x, y, z)$

$\vec{R} = \vec{R}_0 + \vec{u}$ is the deformed position $P'(x, y, z)$ of $P(x, y, z)$

Using these relationships, $\vec{r}$ and $\vec{R}$ become

$$\vec{r} = (x + u_0)\vec{i} + v\vec{j} + w\vec{k}$$

and

$$\vec{R} = (x + U)\vec{i} + (y + V)\vec{j} + (z + W)\vec{k}$$

Also note that the tangent vector $\vec{t}$ to the deflection curve at $C'(x)$ is expressed as

$$\vec{t} = \frac{d\vec{r}/dx}{|d\vec{r}/dx|} = \frac{\vec{r}'}{|\vec{r}'|}$$

$$= \frac{(1 + u_0')\vec{i} + v'\vec{j} + w'\vec{k}}{\sqrt{(1 + u_0')^2 + (v')^2 + (w')^2}}$$

The assumption of plane sections requires that the points $C'(x)$ and $P'(x, y, z)$ lie in the same plane and that this plane be normal to $\vec{t}$. This requirement is expressed as

$$(\vec{r} - \vec{R}) \cdot \vec{t} = 0$$

or

$$(u_0 - U)(1 + u_0') + (v - V - y)v' + (w - W - z)w' = 0$$

Using the assumptions of small-strain theory, $u_0' \ll 1$, $(v - V)v'$, and $(w - W)w'$ are second-order (third-order if cross section is assumed not to deform in its plane) terms. Thus the longitudinal displacement is

$$U(x) = u_0 - yv' - zw' \tag{3.55}$$

and the longitudinal strain $\epsilon = dU/dx$ is

$$\epsilon(x) = u_0' - yv'' - zw'' \tag{3.56}$$

A virtual strain for the beam can be written as

$$\overline{\epsilon}(x) = \overline{u}_0' - y\overline{v}'' - zw'' \tag{3.57}$$

To continue the development, we will select Eq. (3.14) as the constitutive equation. Then we may write the expression for longitudinal stress σ as

$$\sigma(x) = E(u_0' - yv'' - zw'') - \alpha E \Delta T(x, y, z) \tag{3.58}$$

Before proceeding further, it is useful to define some integral expressions that will frequently occur in the text that follows. The total area and the centroid of a cross section are defined as

$$A = \int_A dA \tag{3.59a}$$

$$\bar{y}A = \int_A y \, dA = 0 \tag{3.59b}$$

$$\bar{z}A = \int_A z \, dA = 0 \tag{3.59c}$$

In our case, the Eqs. (3.59b, c) are always zero because we have already selected the centroid as the location of our reference axes.

The following three integrals define the second moments of area and the product of area of the y and z axes for arbitrarily shaped cross sections.

$$
\begin{aligned}
I_z &= \int_A y^2 \, dA \\
I_y &= \int_A z^2 \, dA \\
I_{yz} &= \int_A yz \, dA
\end{aligned}
\tag{3.60}
$$

For principal axes (which are easily determined by the use of Mohr's circle), the integral $I_{yz} = \int yz \, dA = 0$.

The following integrals represent stress resultants over a cross section:

$$
\begin{aligned}
N &= \int_A \sigma \, dA \\
M_z &= \int_A \sigma y \, dA \\
M_y &= \int_A \sigma z \, dA
\end{aligned}
\tag{3.61}
$$

A positive normal stress produces tension. The resultant $N = \int_A \sigma \, dA$ is positive for tension. The sign convention used for $M_y = \int_A \sigma z \, dA$ and $M_z = \int_A \sigma y \, dA$ is that on an element of the cross section in the positive yz quadrant, the moments are positive if they produce a positive normal stress.

The following integrals arise because of temperature variation and are the thermal loading equivalent of Eq. (3.61).

$$N_T = \int_A \alpha E \Delta T \, dA$$

$$M_{Tz} = \int_A \alpha E \Delta T y \, dA$$

$$M_{Ty} = \int_A \alpha E \Delta T z \, dA \tag{3.62}$$

The evaluation of these integrals depends on an explicit expression for the temperature variation $\Delta T(x, y, z)$.

We can now use virtual work to determine the stress σ in terms of the stress resultants and second moments of area. Consider the virtual work of the axial force $\int_A \sigma \, dA$ from the virtual displacement $\overline{U}$. This is expressed as follows:

$$\int_A \sigma \, dA(\overline{u}_0 - y\overline{v}' - z\overline{w}') = \int_A [E(u_0' - yv'' - zw'') - \alpha E \Delta T](\overline{u}_0 - y\overline{v}' - z\overline{w}') \, dA$$

Each of the virtual quantities $\overline{u}_0$, $\overline{v}'$, and $\overline{w}'$ represents a generalized coordinate. Collecting terms and placing the result into the form of Eq. (3.23) and using the expressions given by Eqs. (3.59)–Eqs. (3.62), we obtain three equations. These equations can be solved for u_0', v'', and w'' in terms of the stress resultants and area properties. The resulting expressions are

$$u_0' = \frac{N + N_T}{AE}$$

$$v'' = -\frac{(M_z + M_{Tz})I_y - (M_y + M_{Ty})I_{yz}}{E(I_y I_z - I_{yz}^2)}$$

$$w'' = -\frac{(M_y + M_{Ty})I_z - (M_z + M_{Tz})I_{yz}}{E(I_y I_z - I_{yz}^2)} \tag{3.63}$$

Substituting Eqs. (3.63) into Eq. (3.58), the general expression for the longitudinal stress σ is obtained as

$$\sigma = -\alpha E \Delta T + \frac{N + N_T}{A} + \frac{(M_z + M_{Tz})I_y - (M_y + M_{Ty})I_{yz}}{I_y I_z - I_{yz}^2} y + \frac{(M_y + M_{Ty})I_z - (M_z + M_{Tz})I_{yz}}{I_y I_z - I_{yz}^2} z \tag{3.64}$$

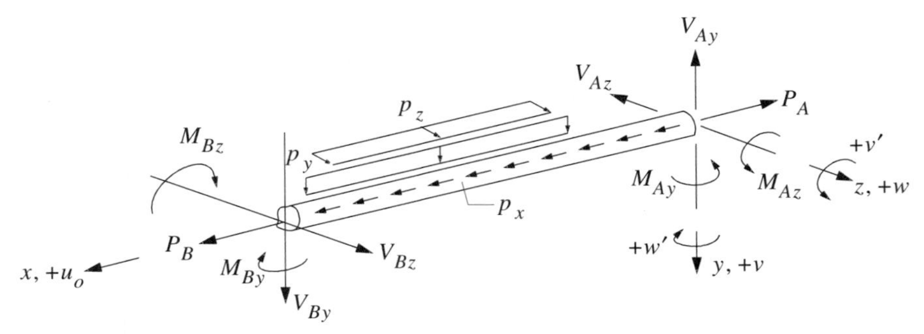

Figure 3.20 Beam loading.

If the principal axes are used, $I_{yz} = 0$, and Eq. (3.64) takes the following form:

$$\sigma = -\alpha E \Delta T + \frac{N + N_T}{A} + \frac{M_z + M_{Tz}}{I_z} y + \frac{M_y + M_{Ty}}{I_y} z \tag{3.65}$$

In general, we assume that the principal axes are used.

Equation (3.63) can be used to determine the explicit form of the deflection curve by expressing the stress resultants as functions of x. However, we would have to guess at the boundary conditions. By applying the priniple of virtual work, however, we explicitly obtain expressions for the boundary conditions. We now use Eq. (3.6) to determine the differential equations of equilibrium and their corresponding boundary conditions. For simplicity, consider the beam to be geometrically and materially linear.

Figure 3.20 shows a general applied loading condition for a beam with a cross section of arbitrary shape. The expression for virtual work is

$$\begin{aligned}
&\int [E(u_0' - yv'' - zw'') - \alpha E \Delta T](\overline{u}_0' - y\overline{v}'' - z\overline{w}'')\, dA\ dx \\
&\quad = \int (p_x \overline{U} + p_y \overline{v} + p_z \overline{w})\, dx - P_A \overline{U}(A) + P_B \overline{U}(B) \\
&\quad\quad - V_{Ay}\overline{v}(A) + V_{By}\overline{v}(B) - V_{Az}\overline{w}(A) + V_{Bz}\overline{w}(B) \\
&\quad\quad + M_{Az}\overline{v}'(A) - M_{Bz}\overline{v}'(B) - M_{Ay}\overline{w}'(A) + M_{By}\overline{w}'(B)
\end{aligned} \tag{3.66}$$

The preceding equation contains the generalized coordinates $\overline{u}_0$, $\overline{v}$, and $\overline{w}$ along with their derivatives. Thus, to get an independent set of generalized coordinates, terms containing $\overline{u}_0'$ in Eq. (3.66) must be integrated once by parts; terms containing $\overline{v}''$ and $\overline{w}''$, twice by parts.

Performing these integrations, we obtain the following equation:

$$
\begin{aligned}
&-\int [(AEu_0' - N_T)' + p_x]\overline{u}_0 \, dx \\
&+ \int [(EI_z v'' + EI_{yz} w'' + M_{Tz})'' - (p_y + yp_x')]\overline{v} \, dx \\
&+ \int [(EI_{yz} v'' + EI_y w'' + M_{Ty})'' - (p_z + zp_x')]\overline{w} \, dx \\
&+ [(AEu_0'(B) - N_T) - P_B]\overline{u}_0(B) \\
&+ [-(AEu_0'(A) - N_T) + P_A]\overline{u}_0(A) \\
&+ [(EI_z v''(B) + EI_{yz} w''(B) + M_{Tz}) + M_{Bz} + yP_B]\overline{v}'(B) \\
&+ [-(EI_z v''(A) + EI_{yz} w''(A) + M_{Tz}) - M_{Az} - yP_A]\overline{v}'(A) \\
&+ [-(EI_z v''(B) + EI_{yz} w''(B) + M_{Tz})' - V_{By} + yp_x]\overline{v}(B) \\
&+ [(EI_z v''(A) + EI_{yz} w''(A) + M_{Tz})' + V_{Ay} - yp_x]\overline{v}(A) \\
&+ [(EI_{yz} v''(B) + EI_y w''(B) + M_{Ty}) - M_{By} + zP_B]\overline{w}'(B) \\
&+ [-(EI_{yz} v''(A) + EI_y w''(A) + M_{Ty}) + M_{Ay} - zP_A]\overline{w}'(A) \\
&+ [-(EI_{yz} v''(B) + EI_y w''(B) + M_{Ty})' - V_{Bz} + zp_x]\overline{w}(B) \\
&+ [(EI_{yz} v''(A) + EI_y w''(A) + M_{Ty})' + V_{Az} - zp_x]\overline{w}(A) = 0
\end{aligned}
\tag{3.67}
$$

Following the discussion of Eq. (3.23), we can pick off the three differential equations and their boundary conditions from Eq. (3.67) as follows.

For longitudinal direction, we have

$$
-\frac{d}{dx}\left(AE\,\frac{du_0}{dx} - N_T\right) = p_x \tag{3.68}
$$

with the following boundary conditions:

$$
\begin{aligned}
u_0(B) = 0 \quad &\text{or} \quad AE\,\frac{du_0(B)}{dx} - N_T = P_B \\
u_0(A) = 0 \quad &\text{or} \quad AE\,\frac{du_0(A)}{dx} - N_T = P_A
\end{aligned}
\tag{3.69}
$$

For bending in the xy plane, we have

$$\frac{d^2}{dx^2}\left(EI_z \frac{d^2v}{dx^2} + EI_{yz} \frac{d^2w}{dx^2} + M_{Tz}\right) = p_y + y \frac{dp_x}{dx} \tag{3.70}$$

with the following boundary conditions:

$$\begin{aligned}
v'(B) = 0 \quad &\text{or} \quad \left(EI_z \frac{d^2v(B)}{dx^2} + EI_{yz} \frac{d^2w(B)}{dx^2} + M_{Tz}\right) = -M_{Bz} - yP_B \\
v'(A) = 0 \quad &\text{or} \quad -\left(EI_z \frac{d^2v(A)}{dx^2} + EI_{yz} \frac{d^2w(A)}{dx^2} + M_{Tz}\right) = M_{Az} + yP_A \\
v(B) = 0 \quad &\text{or} \quad -\frac{d}{dx}\left(EI_z \frac{d^2v(B)}{dx^2} + EI_{yz} \frac{d^2w(B)}{dx^2} + M_{Tz}\right) = V_{By} - yp_x \\
v(A) = 0 \quad &\text{or} \quad \frac{d}{dx}\left(EI_z \frac{d^2v(A)}{dx^2} + EI_{yz} \frac{d^2w(A)}{dx^2} + M_{Tz}\right) = -V_{Ay} + yp_x
\end{aligned} \tag{3.71}$$

For bending in the xz plane, we have

$$\frac{d^2}{dx^2}\left(EI_{yz} \frac{d^2v}{dx^2} + EI_y \frac{d^2w}{dx^2} + M_{Ty}\right) = p_z + z \frac{dp_x}{dx} \tag{3.72}$$

with the following boundary conditions:

$$\begin{aligned}
w'(B) = 0 \quad &\text{or} \quad \left(EI_{yz} \frac{d^2v(B)}{dx^2} + EI_y \frac{d^2w(B)}{dx^2} + M_{Ty}\right) = M_{By} - zP_B \\
w'(A) = 0 \quad &\text{or} \quad \left(EI_{yz} \frac{d^2v(A)}{dx^2} + EI_y \frac{d^2w(A)}{dx^2} + M_{Ty}\right) = -M_{Ay} + zP_A \\
w(B) = 0 \quad &\text{or} \quad -\frac{d}{dx}\left(EI_{yz} \frac{d^2v(B)}{dx^2} + EI_y \frac{d^2w(B)}{dx^2} + M_{Ty}\right) = V_{Bz} - zp_x \\
w(A) = 0 \quad &\text{or} \quad \frac{d}{dx}\left(EI_{yz} \frac{d^2v(A)}{dx^2} + EI_y \frac{d^2w(A)}{dx^2} + M_{Ty}\right) = -V_{Az} + zp_x
\end{aligned} \tag{3.73}$$

For beams of symmetric cross section, or if the principal axes are used, $I_{yz} = 0$, and all of the above equations reduce to a much simpler uncoupled form. However, rather than repeat the above equations in their simpler form, we will

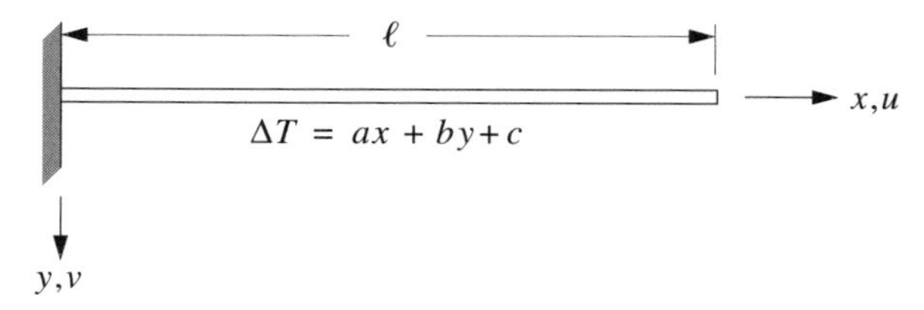

Figure 3.21 Cantilevered beam under a linear temperature.

just use the basic definition of virtual work to obtain the necessary equations for any explicit example.

Example 3.4 Figure 3.21 shows a symmetrical cantilevered beam under a linear temperature distribution $\Delta T = ax + by + c$, where a, b, and c are known specified constants. Assume the beam has an axial stiffness AE and a flexural stiffness $EI_z = EI$; then, using Eqs. (3.57) and (3.58) under the assumption of bending in the xy plane only, we can write Eq. (3.6) as

$$\begin{aligned}\delta U &= \int \sigma \bar{\epsilon}\, dV \\ &= \int E[u_0' - yv'' - \alpha(ax + by + c)](\bar{u}_0' - y\bar{v}'')\, dA\ dx \\ &= 0\end{aligned} \tag{3.74}$$

The external virtual work δW_e is zero because there are no external physically applied loads.

We can now expand Eq. (3.74) directly or take advantage of the fact that $\bar{u}_0$ and $\bar{v}$ are independent generalized coordinates. Taking advantage of the generalized coordinate nature of $\bar{u}_0$ and $\bar{v}$, we can say that

$$\bar{u}_0 \neq 0 \qquad \bar{v} \equiv 0$$

and using the fact that y is measured from the centroid $\int y\, dA = 0$. We then obtain the following expression:

$$\int [AE(u_0' - a\alpha x + c)]\bar{u}_0'\, dx = 0$$

Since there are no external loads, the expression contained in the [ ] represents the differential equation needed to determine u_0. However, by first integrating the above expression by parts, the required boundary conditions are also obtained. Thus, by performing the integration by parts and observing the subsets of $\overline{u}_0$, we obtain the two boundary conditions, from which the resulting differential equation is

$$u_0(0) = 0 \qquad u_0'(\ell) - a\alpha\ell + c = 0$$
$$u_0'' = a\alpha$$

with the resulting solution

$$u_0 = \frac{a\alpha x^2}{2} + cx$$

Next, let

$$\overline{u}_0 \equiv 0 \qquad \overline{v} \neq 0$$

and obtain the expression

$$\int [EI(v'' + \alpha b)]\overline{v}''\, dx = 0$$

Integrating twice by parts of this expression and selecting the appropriate subsets of $\overline{v}$ yields the following boundary conditions and differential equation:

$$v(0) = 0 \qquad v''(\ell) + \alpha b = 0$$
$$v'(0) = 0 \qquad v'''(\ell) = 0$$

and

$$v^{iv} = 0$$

where

$$v^{iv} = d^4v/dx^4$$

The resulting solution is

$$v = -\frac{\alpha b x^2}{2}$$

Substituting the appropriate derivatives of u_0 and v into Eq. (3.58) modified for bending only in the xy plane, we obtain the following expression for stress:

$$\sigma = \alpha E[(ax + c) + by - (ax + by + c)] = 0$$

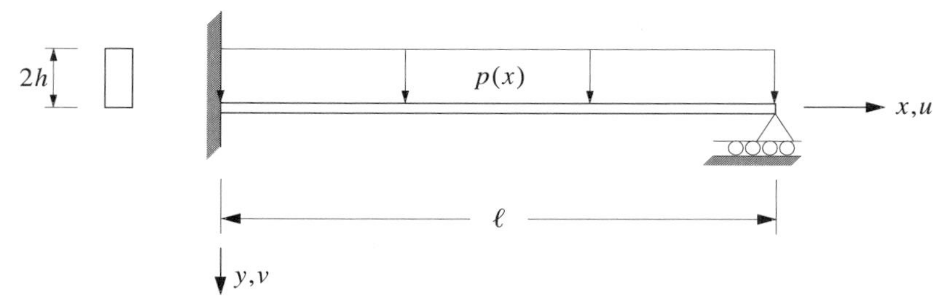

Figure 3.22 Cantilevered simply supported beam under a linear temperature.

This result is expected since the stress components are identically zero throughout any thermoelastic body, with stress-free boundaries under a linear temperature distribution with respect to rectangular Cartesian coordinates.

Example 3.5 As another example, consider the beam shown in Fig. 3.22 that depicts a symmetric beam clamped at $x = 0$, simply supported at end $x = \ell$, and loaded in the xy plane by a uniform load $p(x)$. The beam is also subject to a temperature distribution $\Delta T = T_0[(1 + y/(2h)]$, where T_0 is a known constant and $2h$ is the depth of the beam's cross section. The equation of virtual work is

$$\int E\left[u_0' - yv'' - \alpha T_0\left(1 + \frac{y}{2h}\right)\right](\overline{u}_0' - y\overline{v}'')\, dA\ dx = \int p(x)\overline{v}\, dx \qquad (3.75)$$

where the integral on the left is the internal virtual work and the integral on the right is the external virtual work of the applied load $p(x)$. If

$$\overline{u}_0 \equiv 0 \qquad \overline{v} \equiv 0$$

then Eq. (3.75) becomes

$$\int E\left[u_0' - yv'' - \alpha T_0\left(1 + \frac{y}{2h}\right)\right](\overline{u}_0')\, dA\ dx = 0$$

Integration once by parts yields the following boundary conditions:

$$u_0(0) = 0 \qquad u_0' - \alpha T_0 = 0$$

The resulting differential equation is

$$u_0'' = 0$$

with the following solution:

$$u_0 = \alpha T_0 x$$

If

$$\overline{u}_0 \equiv 0 \qquad \overline{v} \neq 0$$

then Eq. (3.75) becomes

$$\int E\left[u_0' - yv'' - \alpha T_0\left(1 + \frac{y}{2h}\right)\right](-y\overline{v}'')\, dA\ dx = \int p(x)\overline{v}\, dx$$

By integrating over the area, this becomes

$$\int \left[EIv'' - \frac{EI\alpha T_0}{2h}\right]\overline{v}''\, dA\, dx = \int p(x)\overline{v}\, dx$$

where we have used $\int y^2\, dA = I_z = I$ as well as the fact that y is measured from the centroid. Integrating this equation twice by parts yields the following boundary conditions:

$$v(0) = 0 \qquad v'(0) = 0$$

$$v(\ell) = 0 \qquad EIv''(\ell) + \frac{EI\alpha T_0}{2h} = 0$$

the resulting differential equation is

$$EIv^{iv} = p$$

with the following solution:

$$EIv = \frac{p}{48}\,(2x^4 - 5\ell x^3 + 3\ell^2 x^2) + \frac{EI\alpha T_0}{4h}\left(\frac{x^2}{2} - \frac{x^3}{2\ell}\right)$$

The slope at any location is

$$EIv' = \frac{p}{48}\,(8x^3 - 15\ell x^2 + 6\ell^2 x) + \frac{EI\alpha T_0}{4h}\left(x - \frac{3x^2}{2\ell}\right)$$

The curvature at any location is

$$EIv'' = \frac{p}{8}(4x^2 - 5\ell x + \ell^2) + \frac{EI\alpha T_0}{4h}\left(1 - \frac{3x}{\ell}\right)$$

The change in curvature at any location is

$$EIv''' = px - \frac{5p\ell}{8} - \frac{3EI\alpha T_0}{4h\ell}$$

From the relationship $M = \int \sigma y \, dA$, we have

$$EIv'' = -M - \frac{EI\alpha T_0}{2h}$$

and

$$EIv''' = -\frac{dM}{dx} = -V$$

Using these relationships, we can compute the end reaction as

$$V(0) = \frac{5p\ell}{8} + \frac{3EI\alpha T_0}{4h\ell}$$

$$V(\ell) = -\frac{3p\ell}{8} + \frac{3EI\alpha T_0}{4h\ell}$$

$$M(0) = -\frac{p\ell^2}{8} - \frac{3EI\alpha T_0}{4h}$$

$$M(\ell) = 0$$

Figure 3.23 shows a free-body diagram with these end reactions individually broken out.

Equation (3.58) determines the stress in the beam as

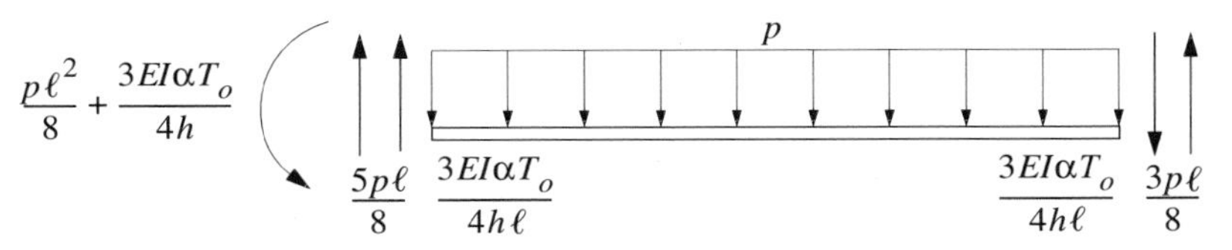

Figure 3.23 Free body of beam under a linear temperature.

$$\sigma = -\frac{py}{8I}(4x^2 - 5\ell x + \ell^2) + \frac{3\alpha E T_0 y}{4h}\left(\frac{x}{\ell} - 1\right)$$

Examination of the expression for σ shows that all of the stress is attributable to bending.

3.12 SHEAR STRESS IN SIMPLE BEAMS

The application of virtual work has for the beam yielded equations for the longitudinal stress in terms of stress resultants or curvature. However, it should be observed that its application has not yielded any results for the shear stress τ or its stress resultants V_y and V_z. In Section 3.6, we discussed the fact that unless all of the virtual work is included, the principle of virtual work does not provide the correct solution. In this section, we discuss a different problem that arises from the assumed strain functions.

The assumption of plane sections disallows the possibility of transverse shear strain in the beam; hence, the shear stress τ must be regarded as a reaction to internal constraints. Therefore, the only way to obtain expressions for shear without allowing for shear strain is to draw a free body of a portion of the beam and obtain the required equation of equilibrium directly from the figure. With the restriction of plane sections, the principle of virtual work cannot determine expressions for the shear stress.

Figure 3.24 represents a typical beam cross section and a slice of the cross section along the dimension b. The equation of equilibrium for the slice is

$$-\tau_{\text{ave}} b\, dx + \int_{A_b} d\sigma\, dA + \int q_x\, dx\, ds = 0$$

where

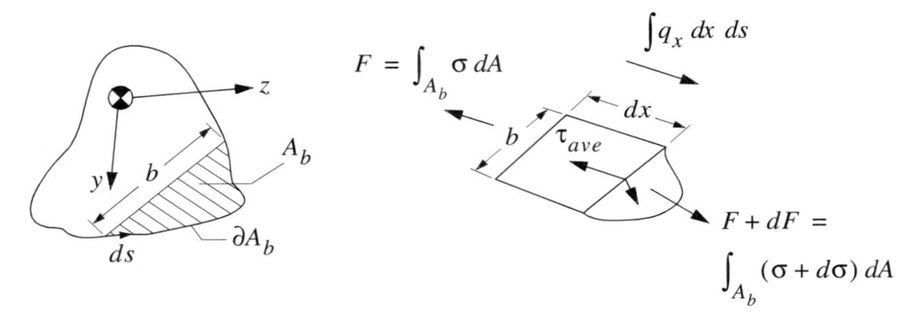

Figure 3.24 Free body to determine shear in a straight.

τ_{ave} is the average shear stress on the area $b\,dx$

$\int_{A_b} d\sigma\, dA$ is the increment in longitudinal force over the shaded area A_b

q_x is the portion of the longitudinal distributed load acting over the surface of the segment dx

In essence, $p_x = \int_{\text{around total boundary}} q_x\, ds$. Solving for τ_{ave}, we get

$$\tau_{ave} = \frac{1}{b}\int_{A_b} \sigma'\, dA + \frac{1}{b}\int_{\partial A_b} q_x\, ds \tag{3.76}$$

Next, substitute $\sigma' = d\sigma/dx$ using Eq. (3.58) into the expression for τ_{ave} to get

$$\tau_{ave} = \frac{E}{b}\int_{A_b} (u_0'' - yv''' - zw''' - \alpha\Delta T')\, dA + \frac{1}{b}\int_{\partial A_b} q_x\, ds \tag{3.77}$$

For what follows, we define the three integrals

$$A_b = \int_{A_b} dA$$

$$Q_z = \int_{A_b} y\, dA$$

$$Q_y = \int_{A_b} z\, dA \tag{3.78}$$

where the first expression is the area A_b and the last two expressions represent the first moments of area for the slice taken at b. The expression for τ_{ave} can now be expressed as

$$\tau_{ave} = \frac{E}{b}(A_b u_0'' - Q_z v''' - Q_y w''') - \frac{1}{b}\int_{A_b} E\alpha\Delta T'\, dA + \frac{1}{b}\int_{\partial A_b} q_x\, ds \tag{3.79}$$

Taking the derivatives of Eq. (3.63) with respect to x and substituting into Eq. (3.79), we get

$$
\begin{aligned}
\tau_{\text{ave}} = {} & \frac{A_b}{b}\,\frac{N' + N'_T}{A} \\
& + \frac{(M'_z + M'_{Tz})I_y - (M'_y + M'_{Ty})I_{yz}}{b(I_yI_z - I_{yz}^2)}\,Q_z \\
& + \frac{(M'_y + M'_{Ty})I_z - (M'_z + M'_{Tz})I_{yz}}{b(I_yI_z - I_{yz}^2)}\,Q_y \\
& - \frac{1}{b}\int_{A_b} E\alpha\Delta T'\,dA + \frac{1}{b}\int_{\partial A_b} q_x\,ds
\end{aligned}
\tag{3.80}
$$

Equation (3.80) may be rearranged to yield

$$
\begin{aligned}
\tau_{\text{ave}} = {} & \frac{A_b}{b}\,\frac{N' + N'_T}{A} \\
& + \frac{Q_zI_y - Q_yI_{yz}}{b(I_yI_z - I_{yz}^2)}\,(M'_z + M'_{Tz}) \\
& + \frac{Q_yI_z - Q_zI_{yz}}{b(I_yI_z - I_{yz}^2)}\,(M'_y + M'_{Ty}) \\
& - \frac{1}{b}\int_{A_b} E\alpha\Delta T'\,dA + \frac{1}{b}\int_{\partial A_b} q_x\,ds
\end{aligned}
\tag{3.81}
$$

To obtain a final expression for τ_{ave}, consider again Fig. 3.20, now regarded as a free body. For the length $AB = dx$, $P_A = N$, $P_B = N + dN$, $V_{Ay} = V_y$, $V_{By} = V_y + dV_y$, and so on; sum forces in the x, y, and z directions; and take moments at B around the y and z axes. Also, assume that the line of action of p_x is located at some (y, z) coordinate. The following five relationships are obtained after discarding second-order terms:

$$
\frac{dN}{dx} = -p_x
$$

$$
\frac{dV_y}{dx} = -p_y
$$

$$
\frac{dV_z}{dx} = -p_z
$$

$$\frac{dM_y}{dx} = V_z - zp_x$$

$$\frac{dM_z}{dx} = V_y - yp_x \tag{3.82}$$

If the first and last two of the terms in Eq. (3.82) are substituted into Eq. (3.81), the following general expression is obtained for the average shear τ_{ave}:

$$\begin{aligned}\tau_{\text{ave}} = {} & \frac{A_b}{b}\,\frac{-p_x + N_T'}{A} \\ & + \frac{Q_z I_y - Q_y I_{yz}}{b(I_y I_z - I_{yz}^2)}\,(V_y + M_{Tz}' - yp_x) \\ & + \frac{Q_y I_z - Q_z I_{yz}}{b(I_y I_z - I_{yz}^2)}\,(V_z + M_{Ty}' - zp_x) \\ & - \frac{1}{b}\int_{A_b} E\alpha\Delta T'\, dA + \frac{1}{b}\int_{\partial A_b} q_x\, ds\end{aligned} \tag{3.83}$$

If $p_x = 0$, no thermal loads are applied, and if the beam is symmetrical with loading only in the yx plane, the above equation reduces to the following familiar expression:

$$\tau = \frac{V_y Q_z}{bI_z}$$

Also, it should be noted that the direction of the slice determines the average shear stress found. For example, if b is parallel to the z axis, the stress found would be τ_{xy}. The shear found depends on the direction of the slice; however, implicit in the derivation is the fact that the stress resultant $V = \int_{A_b} \tau_{\text{ave}}\, dA$ corresponds to stresses that are everywhere in the section A_b parallel to V. This is not always the case; an obvious exception is a circular cross section where the shear must be tangent to the boundary. For such a cross section, Eq. (3.83) produces significant error.

3.13 SHEAR DEFLECTION IN STRAIGHT BEAMS

As mentioned in Section 3.12, the assumption that sections remain plane with the midplane axis normal to the plane of the cross section results in internal constraints. To see this more clearly, consider Fig. 3.25.

The figure shows a general free body of a beam in the xy plane. A section of the beam is shown to the right. The section is still assumed to remain plane,

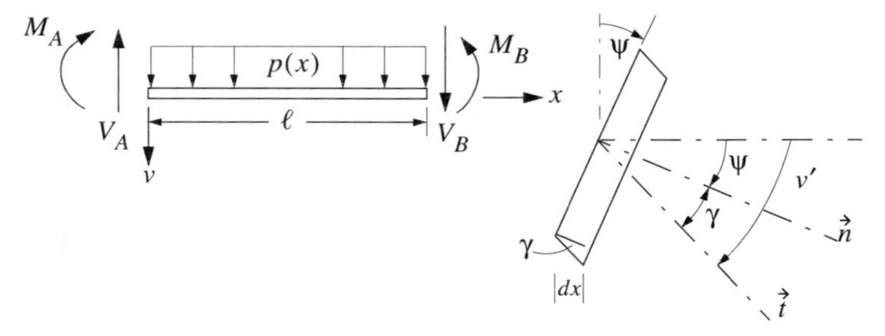

Figure 3.25 Beam deflection with transverse shear deformation.

but the tangent $\vec{t}$ to the deflection curve is no longer coincident with the normal $\vec{n}$ of the cross section. The angle ψ represents the rotation of the section from the vertical. This is the angle through which a moment M acting at the section would rotate. The slope of the deflection curve is v', and the angle γ represents the shear distortion of the section.

From the figure, we can deduce the following relationships:

$$u = -y\psi \tag{3.84}$$

$$\epsilon = u' = -y\psi' \tag{3.85}$$

and

$$\gamma = v' - \psi \tag{3.86}$$

Then, assuming a simple Hooke's law, the stress-strain relationships become

$$\sigma = E\epsilon = -yE\psi' \tag{3.87}$$

$$\tau = G\gamma = G(v' - \psi) \tag{3.88}$$

The principle of virtual work $\delta U = \delta W_e$ then takes the form

$$\int \sigma\bar{\epsilon}\, dV + \int \tau\bar{\gamma}\, dV = \int p\bar{v}\, dx - V_A\bar{v}_A + V_B\bar{v}_B + M_A\overline{\psi}_A - M_B\overline{\psi}_B \tag{3.89}$$

or

$$\int (-yE\psi')(-y\overline{\psi}')\, dA\ dx + \int G\gamma(\bar{v}' - \overline{\psi})\, dA\ dx$$
$$= \int p\bar{v}\, dx - V_A\bar{v}_A + V_B\bar{v}_B + M_A\overline{\psi}_A - M_B\overline{\psi}_B \tag{3.90}$$

Expanding, we get

$$\int EI\psi'\overline{\psi}'\,dx - \int G\gamma\overline{\psi}\,dx + \int G\gamma\overline{v}'\,dx$$

$$= \int p\overline{v}\,dx - V_A\overline{v}_A + V_B\overline{v}_B + M_A\overline{\psi}_A - M_B\overline{\psi}_B \tag{3.91}$$

Integrating the first and third integrals on the left by parts and by collecting terms, we get

$$EI\psi'\overline{\psi}\Big|_A^B - \int [(EI\psi')' + GA\gamma]\overline{\psi}\,dx + GA\gamma\overline{v}\Big|_A^B - \int (GA\gamma)'\overline{v}\,dx$$

$$= \int p\overline{v}\,dx - V_A\overline{v}_A + V_B\overline{v}_B + M_A\overline{\psi}_A - M_B\overline{\psi}_B \tag{3.92}$$

The above expression yields two differential equations

$$(EI\psi')' + GA\gamma = 0 \tag{3.93}$$

$$-(GA\gamma)' = p \tag{3.94}$$

and the four boundary conditions

$$\begin{aligned} \psi(B) = 0 \quad &\text{or} \quad EI\psi'(B) = -M_B \\ \psi(A) = 0 \quad &\text{or} \quad -EI\psi'(B) = -M_B \end{aligned} \tag{3.95}$$

$$\begin{aligned} v(B) = 0 \quad &\text{or} \quad GA\gamma(B) = V_B \\ v(A) = 0 \quad &\text{or} \quad -GA\gamma(A) = -V_A \end{aligned} \tag{3.96}$$

Take the derivative of Eq. (3.93) with respect to x and use Eq. (3.94) to arrive at the following third-order differential equation in ψ:

$$(EI\psi')'' - p = 0 \tag{3.97}$$

Next, express Eq. (3.93) as

$$GA(v' - \psi) = -(EI\psi')' \tag{3.98}$$

and solve for v' to get the following first-order differential equation:

$$v' = \psi - \frac{(EI\psi')'}{GA} \tag{3.99}$$

The relationship between the moment $M = \int_A \sigma y\,dA$ and the cross section rotation ψ can be obtained by considering the virtual work of the axial force $\int_A \sigma\,dA$ from a virtual displacement $\overline{u} = (-y\overline{\psi})$. This is expressed as

$$\int_A \sigma(-y\overline{\psi})\, dA = \int_A (-yE\psi')(-y\overline{\psi})\, dA \tag{3.100}$$

or

$$-M = EI\psi' \tag{3.101}$$

Similarly, the relationship between the shear force $V = \int_A \tau\, dA$ and the transverse displacement v can be obtained by considering the virtual work of the shear force $\int_A \tau\, dA$ from a virtual displacement $\overline{v}$. This is expressed as

$$\int_A \tau\overline{v}\, dA = \int_A G\gamma\overline{v}\, dA \tag{3.102}$$

or

$$\begin{aligned} V &= GA\gamma \\ &= GA(v' - \psi) \\ &= GA\left(\psi - \frac{(EI\psi')'}{GA} - \psi\right) \\ &= -(EI\psi')' \\ &= M' \end{aligned} \tag{3.103}$$

From the above, we also obtain various forms for the differential equation of equilibrium. For example:

$$V' = M'' = -(EI\psi')'' = -p$$

From Fig. 3.25, it is seen that at a given cross section, the shear strain γ is constant. If, however, we examine the local equation of equilibrium, we observe

$$\frac{\partial\sigma}{\partial x} + \frac{\partial\tau}{\partial y} = 0$$

or

$$\tau(y) = \int_y \frac{\partial\sigma}{\partial x}\, dy$$

Hence, τ must vary across the section. A compromise can be reached (as will be explained in Chapter 4) with the aid of complementary virtual work. The cross-sectional area A is then replaced with a reduced area A_s, defined as

$$A_s = \frac{A}{f_s}$$

where f_s is defined by Eq. (4.25). The first-order differential equation for v then becomes

$$v' = \psi - \frac{(EI\psi')'}{GA_s}$$

If $\gamma = 0$, then we have $v' = \psi$, which represents a kinematic constraint. From the above expression $V = GA\gamma$, we see that V is no longer a function of the deformation but must now be considered a constraint force needed to enforce the requirement $v' = \psi$. Hence τ is now obtainable only from the equilibrium of a free body.

3.14 BEAMS WITH INITIAL CURVATURE

In this section, we examine a beam with initial curvature. Such beams are used in the analysis of machine parts, crane hooks, aircraft-landing hooks for carrier-based aircraft, and for arch bridges. The basic assumption is that plane sections remain plane and no distortion in the plane of the cross section occurs.

Figure 3.26 shows the undeformed and deformed axis of centroids for a curved beam. The origin of the reference coordinate xYz is at the centroid of the curved beam. The initial curvature is assumed to lie in the xY plane; thus, we are considering a plane curve only. The distance along the curve is measured by the parameter s. At a distance s along the curve, we consider a coordinate system syz in which s is tangent to the curve and y is directed toward the center of curvature of the curve. The system syz forms a right-hand triad. At s, we also define the unit vectors $\vec{t}_0$, which is tangent to the curve; the principal normal

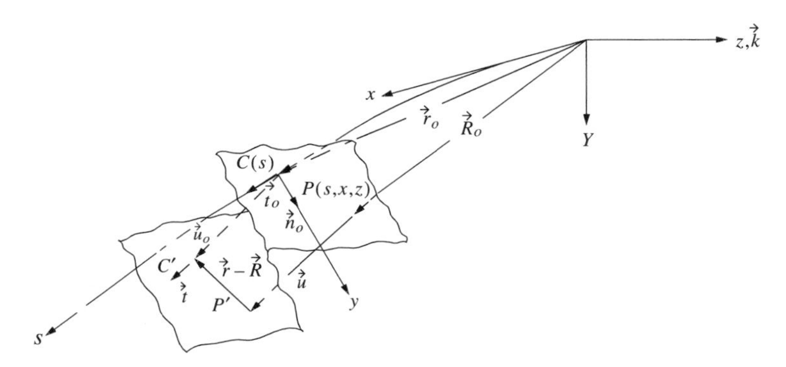

Figure 3.26 Deflection curve for a curved beam.

$\vec{n}_0$, which is directed from $C(s)$ toward the center of curvature; and $\vec{k}$, which for a general space curve is called the binormal and is related by the relationship $\vec{n}_0 = \vec{k} \times \vec{t}_0$. We define the scalar R as the radius of curvature at s.

From Fig. 3.26, the kinematics of the deflection of the centroid axis under the assumption of plane sections is determined as follows:

$\vec{r}_0 = \vec{r}_0(s)$ is the initial position of centroid $C(s)$

$\vec{u}_0 = u_0\vec{t}_0 + v\vec{n}_0 + w\vec{k}$ is the displacement of centroid $C(s)$

$\vec{r} = \vec{r}_0 + \vec{u}_0$ is the deformed position $C'(s)$ of centroid $C(s)$

$\vec{R}_0 = \vec{r}_0 + y\vec{n}_0 + z\vec{k}$ is the initial position of a general point of the beam section $P(s, y, z)$

$\vec{u} = U\vec{t}_0 + V\vec{n}_0 + W\vec{k}$ is the displacement of $P(s, y, z)$

$\vec{R} = \vec{R}_0 + \vec{u}$ is the deformed position $P'(s, y, z)$ of $P(s, y, z)$

Using these relationships, $\vec{r}$ and $\vec{R}$ become

$$\vec{r} = \vec{r}_0 + u_0\vec{t}_0 + v\vec{n}_0 + w\vec{k}$$

and

$$\vec{R} = \vec{r}_0 + U\vec{t}_0 + (y + V)\vec{n}_0 + (z + W)\vec{k}$$

The tangent vector $\vec{t}_0$ is, by definition, $d\vec{r}_0/ds = \vec{r}'_0$; on the other hand, the tangent vector $\vec{t}$ to the deflection curve at $C'(s)$ is, by definition, $\vec{r}'$ and may be expressed as

$$\vec{t} = \vec{r}'_0 + u'_0\vec{t}_0 + v'\vec{n}_0 + w'\vec{k} + u_0\vec{t}'_0 + v\vec{n}'_0$$

The derivative of the unit vectors $\vec{t}_0$ and $\vec{n}_0$ can be found by the Frenet formulas [29] for a plane curve as

$$\frac{d\vec{t}_0}{ds} = \frac{1}{R}\vec{n}_0$$

$$\frac{d\vec{n}_0}{ds} = -\frac{1}{R}\vec{t}_0$$

(For a plane curve, these relationships follow immediately by letting the parameter s be the time and by considering the time derivative of a unit vector in a plane.)

The final expression for $\vec{t}$ then becomes

$$\vec{t} = \left(1 - \frac{v}{R} + u_0'\right) \vec{t}_0 + \left(\frac{u_0}{R} + v'\right) \vec{n}_0 + w'\vec{k}$$

The assumption of plane sections requires that the points $C'(s)$ and $P'(s, y, z)$ lie in the same plane and that this plane be normal to $\vec{t}$. This requirement is expressed as

$$(\vec{r} - \vec{R}) \cdot \vec{t} = 0$$

or

$$(u_0 - U)\left(1 - \frac{v}{R} + u_0'\right) + (v - V - y)\left(\frac{u_0}{R} + v'\right) + (w - W - z)w' = 0$$

Using the assumptions of small-strain theory, $u_0' \ll 1$ and $v/R \ll 1$, as well as the fact that curved beams do not distort in the plane of the cross section ($V = v$ and $W = w$), the circumferential displacement is

$$U(s) = \left(1 - \frac{y}{R}\right) u_0 - yv' - zw' \tag{3.104}$$

The displacement $\vec{u}$ can now be expressed as

$$\vec{u} = \left[\left(1 - \frac{y}{R}\right) u_0 - yv' - zw'\right] \vec{t}_0 + v\vec{n}_0 + w\vec{k}$$

Again using the Frenet formulas, the derivatives of this expression yields

$$\frac{d\vec{u}}{ds} = \left[\left(1 - \frac{y}{R}\right) u_0' - \frac{v}{R} - yv'' - zw'' + \frac{y}{R^2}\,\frac{dR}{ds}\,u_0\right] \vec{t}_0$$
$$+ \left[\left(1 - \frac{y}{R}\right) \frac{u_0}{R} + \left(1 - \frac{y}{R}\right) v' - \frac{z}{R}\, w'\right] \vec{n}_0 + w'\vec{k}$$

We are now in a position to define the circumferential strain. Figure 3.27 shows a section ds along the beam and the fiber length dS of a fiber at a distance y from the centroid. From the figure, we can determine the relationship $dS = (1 - y/R)\, ds$. The circumferential strain is defined as the $\vec{t}_0$ component of $d\vec{u}/dS$, which can be written as

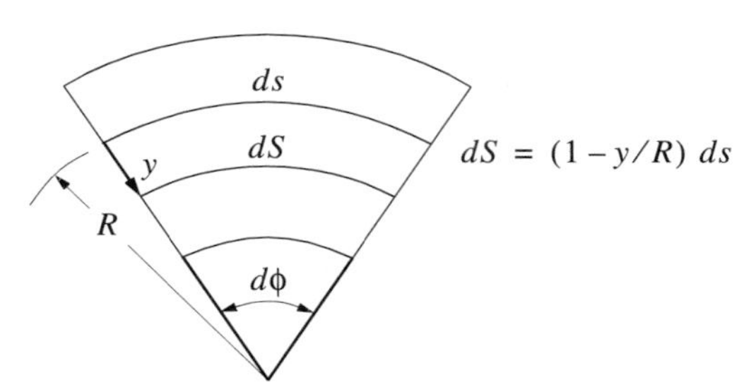

Figure 3.27 A curved beam section ds.

$$\frac{d\vec{u}}{dS} = \frac{d\vec{u}}{ds}\,\frac{ds}{dS}$$
$$= \frac{1}{(1 - y/R)}\,\frac{d\vec{u}}{ds}$$

Before we obtain an explicit expression for circumferential strain, however, we will make some further simplifying assumptions and observations. First, it is assumed that $(dR/ds)u_0/R \ll 1$, and second, while the formal differentiation yielded components for the $\vec{n}_0$ and $\vec{k}$, the assumption (which we have used in saying $V = v$) that the beam cannot deform in its own plane precludes the use of these terms as valid strain terms in the plane of the cross section. Thus the circumferential strain ϵ is

$$\epsilon(s) = \underbrace{u_0' - \frac{yv''}{(1 - y/R)} - \frac{zw''}{(1 - y/R)}}_{\text{From circumferential displacement}} \quad \underbrace{-\frac{v/R}{(1 - y/R)}}_{\text{From radial displacement}} \tag{3.105}$$

The relationship $1/(1 - y/R)$ appears in various forms; it is useful to define some expressions containing this relationship.

$$\frac{1}{1 - y/R} = 1 + \frac{y/R}{1 - y/R} \tag{3.106a}$$

$$\hat{I}_z = \int_A \frac{y^2}{1 - y/R}\,dA \tag{3.106b}$$

$$\hat{I}_y = \int_A \frac{z^2}{1 - y/R}\,dA \tag{3.106c}$$

$$\hat{I}_{yz} = \int_A \frac{yz}{1 - y/R}\,dA \tag{3.106d}$$

The three integrals look similar to the second moments of area and the product of area of the yz axes for arbitrarily shaped cross sections, and they indeed reduce to Eq. (3.60) for large R.

Substituting Eq. (3.106a) into the final term of Eq. (3.105), the expression for strain becomes

$$\epsilon(s) = \left(u_0' - \frac{v}{R}\right) - \left(\frac{v}{R^2} + v''\right) \frac{y}{1 - y/R} - \frac{zw''}{1 - y/R} \tag{3.107}$$

A virtual strain for the beam can be written as

$$\overline{\epsilon}(s) = \left(\overline{u}_0' - \frac{\overline{v}}{R}\right) - \left(\frac{\overline{v}}{R^2} + \overline{v}''\right) \frac{y}{1 - y/R} - \frac{z\overline{w}''}{1 - y/R} \tag{3.108}$$

As with the straight beam, we will select Eq. (3.14) as the constitutive equation. We may then write the expression for circumferential stress σ as

$$\sigma(s) = E\left[\left(u_0' - \frac{v}{R}\right) - \left(\frac{v}{R^2} + v''\right) \frac{y}{1 - y/R} - \frac{zw''}{1 - y/R}\right] - \alpha E \Delta T(s, y, z) \tag{3.109}$$

As with the straight beam, we can now use virtual work to determine the stress σ in terms of the stress resultants and second moments of area. Consider the virtual work of the axial force $\int_A \sigma\, dA$ from the virtual displacement $\overline{U}$. This is expressed as

$$\begin{aligned} &\int_A \sigma\, dA\left[\left(1 - \frac{y}{R}\right) \overline{u}_0 - y\overline{v}' - z\overline{w}'\right] \\ &\quad = \int_A \left[E\left\{\left(u_0' - \frac{v}{R}\right) - \left(\frac{v}{R^2} + v''\right) \frac{y}{1 - y/R} - \frac{zw''}{1 - y/R}\right\}\right. \\ &\qquad \left. -\alpha E \Delta T(s, y, z)\right] \left[\left(1 - \frac{y}{R}\right) \overline{u}_0 - y\overline{v}' - z\overline{w}'\right] dA \end{aligned}$$

By collecting the arguments of $\overline{u}_0$, $\overline{v}'$, and $\overline{w}'$, and by using Eqs. (3.61), (3.62), and (3.106b–d), Eq. (3.110) is obtained for $u_0' - v/R$, $v'' + v/R^2$, and w'' in terms of the stress resultants and area properties.

$$
\begin{aligned}
u_0' - \frac{v}{R} &= \frac{N + N_T}{AE} - \frac{M_z + M_{Tz}}{RAE} \\
v'' + \frac{v}{R^2} &= -\frac{(M_z + M_{Tz})\hat{I}_y - (M_y + M_{Ty})\hat{I}_{yz}}{E(\hat{I}_y \hat{I}_z - \hat{I}_{yz}^2)} \\
w'' &= -\frac{(M_y + M_{Ty})\hat{I}_z - (M_z + M_{Tz})\hat{I}_{yz}}{E(\hat{I}_y \hat{I}_z - \hat{I}_{yz}^2)}
\end{aligned} \tag{3.110}
$$

Substituting Eq. (3.110) into Eq. (3.109), the general expression for the circumferential stress σ is obtained as

$$
\begin{aligned}
\sigma = -\alpha E \Delta T &+ \frac{N + N_T}{A} - \frac{M_z + M_{Tz}}{RA} \\
&+ \frac{(M_z + M_{Tz})\hat{I}_y - (M_y + M_{Ty})\hat{I}_{yz}}{\hat{I}_y \hat{I}_z - \hat{I}_{yz}^2} \frac{y}{1 - y/R} \\
&+ \frac{(M_y + M_{Ty})\hat{I}_z - (M_z + M_{Tz})\hat{I}_{yz}}{\hat{I}_y \hat{I}_z - \hat{I}_{yz}^2} \frac{z}{1 - y/R}
\end{aligned} \tag{3.111}
$$

Equation (3.111) gives the stress at any point in the cross section of the beam and represents a hyperbolic distribution.

For curved beams having symmetrical cross sections with bending only in the xY plane but without thermal loads, Eq. (3.111) takes the following simpler form:

$$
\sigma = -\frac{M_z}{RA} + \frac{M_z}{\hat{I}_z} \frac{y}{1 - y/R} \tag{3.112}
$$

Rather than defining $\hat{I}_z$, many machine-design textbooks define the factor Z as

$$
\hat{I}_z = ZAR^2
$$

Then, Eq. (3.112) becomes

$$
\sigma = \frac{M_z}{AR} \left(-1 + \frac{1}{ZR} \frac{y}{1 - y/R} \right) \tag{3.113}
$$

For circular beams with solid, circular sections, Eq. (3.113) is exact. For noncircular curved beams with arbitrary solid cross sections (e.g., trapezoidal), Eq. (3.113) gives adequate results. As ref. [30] indicates, however, the value Z is difficult to compute. Hence many applications use the straight beam formula for stress with a correction factor K applied. The correction factor K is computed by considering the stress at the extreme fiber and computing the relationship as follows:

$$\sigma = \frac{M_z}{AR}\left(-1 - \frac{1}{ZR}\,\frac{c}{1 + c/R}\right) = -K\,\frac{M_z c}{I_z}$$

from which K is computed as

$$K = \frac{M_z}{AR}\left(1 + \frac{1}{ZR}\,\frac{c}{1 + c/R}\right) \bigg/ \frac{M_z c}{I_z}$$

In addition to giving tables for K for cross sections of various shapes, ref. [30] also indicates that for curved beams with **H**, **I**, or **T** cross sections, the curved beam theory presented here produces considerable nonconservative error because the flanges tend to undergo additional rotation in the plane of the cross section. Also, if the web of these sections is too thin, there is a tendency for it to fail by buckling.

Reference [31] arrives at Eq. (3.110) simplified for symmetrical arched bridges and notes that these equations must be iterated because the loading cannot be considered independent of the deflection and, hence, a nonlinear solution is required.

As with the straight beam, we now use Eq. (3.6) to determine the differential equations of equilibrium and the corresponding boundary conditions. Using a set of general applied loads shown in Fig. 3.28, we obtain the following expression for virtual work:

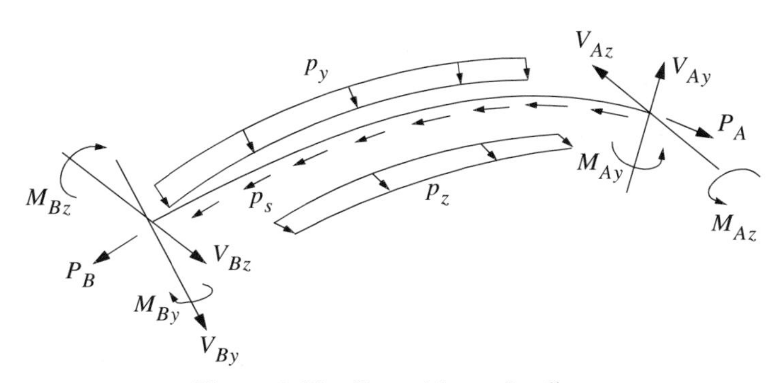

Figure 3.28 Curved beam loading.

$$
\begin{aligned}
&\int \left[E\left\{\left(u_0' - \frac{v}{R}\right) - \left(v'' + \frac{v}{R^2}\right) \frac{y}{1-y/R} - w'' \frac{z}{1-y/R}\right\} - \alpha E \Delta T\right] \\
&\qquad \left[\left(\overline{u}_0' - \frac{\overline{v}}{R}\right) - \left(\overline{v}'' + \frac{\overline{v}}{R^2}\right) \frac{y}{1-y/R} - \overline{w}'' \frac{z}{1-y/R}\right] dA(1-y/R)\, ds \\
&\quad = \int [p_s\overline{U} + p_y\overline{v} + p_z\overline{w}]\, ds - P_A\overline{U}(A) + P_B\overline{U}(B) \\
&\quad - V_{Ay}\overline{v}(A) + V_{By}\overline{v}(B) - V_{Az}\overline{w}(A) + V_{Bz}\overline{w}(B) \\
&\quad + M_{Az}\overline{v}'(A) - M_{Bz}\overline{v}'(B) - M_{Ay}\overline{w}'(A) + M_{By}\overline{w}'(B)
\end{aligned} \tag{3.114}
$$

where $\overline{U}$ is defined by Eq. (3.104). Also, we have defined the differential volume $dV = dy\, dz\, dS = dA\, dS = dA(1 - y/R)\, ds$.

Before expanding this expression, it is convenient to define the following three additional integral expressions:

$$
\begin{aligned}
\int_A \frac{dA}{1-y/R} &= A + \frac{\hat{I}_z}{R^2} \\
\int_A \frac{y\, dA}{1-y/R} &= \frac{\hat{I}_z}{R} \\
\int_A \frac{z\, dA}{1-y/R} &= \frac{\hat{I}_{yz}}{R}
\end{aligned} \tag{3.115}
$$

Expanding Eq. (3.114) by integration by parts and using the definitions given by Eq. (3.115), we get the following three differential equations and their related boundary conditions.

For circumferential direction, we have

$$
-\frac{d}{ds}\left[AE\left(\frac{du_0}{ds} - \frac{v}{R}\right) - N_T + \frac{M_{Tz}}{R}\right] = p_s(1 - y/R) \tag{3.116}
$$

with the following boundary conditions:

$$
\begin{aligned}
u_0(B) = 0 \quad &\text{or} \quad \left[AE\left(\frac{du_0}{ds} - \frac{v}{R}\right) - N_T + \frac{M_{Tz}}{R}\right]_B = P_B \\
u_0(A) = 0 \quad &\text{or} \quad \left[AE\left(\frac{du_0}{ds} - \frac{v}{R}\right) - N_T + \frac{M_{Tz}}{R}\right]_A = P_A
\end{aligned} \tag{3.117}
$$

For bending in the xY plane, we have

$$\frac{d^2}{ds^2}\left[E\hat{I}_z\left(\frac{d^2v}{ds^2}+\frac{v}{R^2}\right)+E\hat{I}_{yz}\frac{d^2w}{ds^2}+M_{Tz}\right]$$
$$+\left[-\frac{AE}{R}\left(\frac{du_0}{ds}-\frac{v}{R}\right)+\frac{E\hat{I}_z}{R^2}\left(\frac{d^2v}{ds^2}+\frac{v}{R^2}\right)\right.$$
$$\left.+\frac{E\hat{I}_{yz}}{R^2}\frac{d^2w}{ds^2}+\frac{N_T}{R}\right]=p_y+y\frac{dp_s}{ds} \tag{3.118}$$

with the following boundary conditions:

$$v'(B)=0 \quad \text{or} \quad \left[E\hat{I}_z\left(\frac{d^2v}{ds^2}+\frac{v}{R^2}\right)+E\hat{I}_{yz}\frac{d^2w}{ds^2}+M_{Tz}\right]_B=-M_{Bz}-yP_B$$
$$v'(A)=0 \quad \text{or} \quad -\left[E\hat{I}_z\left(\frac{d^2v}{ds^2}+\frac{v}{R^2}\right)+E\hat{I}_{yz}\frac{d^2w}{ds^2}+M_{Tz}\right]_A=M_{Az}+yP_A$$
$$v(B)=0 \quad \text{or} \quad -\frac{d}{ds}\left[E\hat{I}_z\left(\frac{d^2v}{ds^2}+\frac{v}{R^2}\right)+E\hat{I}_{yz}\frac{d^2w}{ds^2}+M_{Tz}\right]_B=V_{By}-yp_s$$
$$v(A)=0 \quad \text{or} \quad \frac{d}{ds}\left[E\hat{I}_z\left(\frac{d^2v}{ds^2}+\frac{v}{R^2}\right)+E\hat{I}_{yz}\frac{d^2w}{ds^2}+M_{Tz}\right]_A=-V_{Ay}+yp_s \tag{3.119}$$

For bending parallel to the xz plane, we have

$$\frac{d^2}{ds^2}\left[E\hat{I}_{yz}\left(\frac{d^2v}{ds^2}+\frac{v}{R^2}\right)+E\hat{I}_y\frac{d^2w}{ds^2}+M_{Ty}\right]=p_z+z\frac{dp_s}{ds} \tag{3.120}$$

with the following boundary conditions:

$$w'(B)=0 \quad \text{or} \quad \left[E\hat{I}_{yz}\left(\frac{d^2v}{ds^2}+\frac{v}{R^2}\right)+E\hat{I}_y\frac{d^2w}{ds^2}+M_{Ty}\right]_B=M_{By}-zP_B$$

$$w'(A) = 0 \quad \text{or} \quad -\left[E\hat{I}_{yz}\left(\frac{d^2v}{ds^2} + \frac{v}{R^2}\right) + E\hat{I}_y \frac{d^2w}{ds^2} + M_{Ty}\right]_A = -M_{Ay} + zP_A$$

$$w(B) = 0 \quad \text{or} \quad -\frac{d}{ds}\left[E\hat{I}_{yz}\left(\frac{d^2v}{ds^2} + \frac{v}{R^2}\right) + E\hat{I}_y \frac{d^2w}{ds^2} + M_{Ty}\right]_B = V_{Bz} - zp_s$$

$$w(A) = 0 \quad \text{or} \quad \frac{d}{ds}\left[E\hat{I}_{yz}\left(\frac{d^2v}{ds^2} + \frac{v}{R^2}\right) + E\hat{I}_y \frac{d^2w}{ds^2} + M_{Ty}\right]_A = -V_{Az} + zp_s$$

(3.121)

Equations (3.116)–(3.121) represent the general differential equations for determining the deflections of curved beams. For beams of solid cross section, the term v/R^2 can be neglected with reasonable errors of less than 5%. For beams with **I** and **T** sections, the aforementioned equations, though compelx, give considerable error because of the additional rotation of the flange. An alternative method for determining deflections using complementary virtual work is given in Chapter 4.

For $R/h \geq 10$, where h is the characteristic depth of a cross section, the straight beam formulas may be used. For a beam of rectangular cross section and an $R/h = 5$, the straight beam formula yields an error of only 7% in maximum bending stress.

Example 3.6 Figure 3.29 represents a curved beam of radius R bent through an arc of $\pi/2$ in the xY plane. End A is free; end B is clamped. The beam in the xY plane is subject to the following temperature distribution:

$$\Delta T = T_0\left[1 + a_1 \frac{y}{h} + a_2\left(\frac{y}{h}\right)^2\right]$$

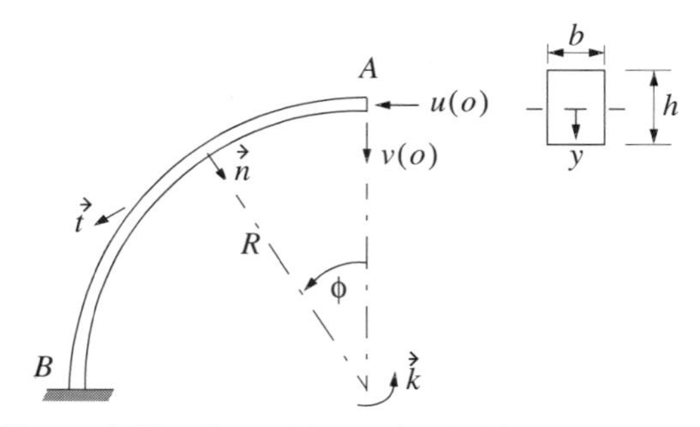

Figure 3.29 Curved beam loaded by temperature.

where

h is the depth of the rectangular cross section
a_1, a_2 are the known constants
T_0 is the given temperature

The cross-sectional area is $A = bh$, and the second moment of area of the cross section $I = bh^3/12$. The second moment associated with bending is $\hat{I}_z = \hat{I}$; it is computed as

$$\begin{aligned}\hat{I} &= \int_{-h/2}^{h/2} \frac{y^2\,dA}{1 - y/R}\\ &= Rb \int_{-h/2}^{h/2} \frac{y^2\,dy}{R - y}\\ &= -R^2 bh + R^3 b \ln\left(\frac{R + h/2}{R - h/2}\right)\end{aligned}$$

Using Young's modulus of E and the coefficient of thermal expansion α, the thermal loads are given as

$$\begin{aligned}N_T &= \alpha E T_0 \int \left[1 + a_1 \frac{y}{h} + a_2 \left(\frac{y}{h}\right)^2\right] dA\\ &= \alpha A E T_0 + \frac{\alpha a_2 E I T_0}{h^2}\\ M_T &= \alpha E T_0 \int \left[1 + a_1 \frac{y}{h} + a_2 \left(\frac{y}{h}\right)^2\right] y\,dA\\ &= \frac{\alpha a_1 E I T_0}{h}\end{aligned}$$

To perform the calculations, it is convenient to use the angle $d\phi$ in place of the arc length ds. Therefore, we have

$$\begin{aligned}ds &= R\,d\phi\\ \frac{d\phi}{ds} &= \frac{1}{R}\\ \frac{df}{ds} &= \frac{1}{R}\frac{df}{d\phi}\end{aligned}$$

For the circumferential direction, the differential equation takes the form

$$-\frac{1}{R}\frac{d}{d\phi}\left[AE\left(\frac{1}{R}\frac{du}{d\phi}-\frac{v}{R}\right)-N_T+\frac{M_T}{R}\right]=0$$

or

$$-\frac{AE}{R^2}(u''-v')=0$$

where $()' = d()/d\phi$, and so forth. Solving, we get the following differential equation:

$$u'-v=A_6$$

The boundary condition at A yields

$$\frac{AE}{R}(u'-v)-N_T+\frac{M_T}{R}=0; \qquad \phi=0$$

or

$$A_6=\frac{RN_T}{AE}-\frac{M_T}{AE}$$

The differential equation for circumferential-direction displacement becomes

$$u'-v=\frac{RN_T}{AE}-\frac{M_T}{AE}$$

The differential equation for xY bending becomes

$$\frac{1}{R^2}\frac{d^2}{d\phi^2}\left[\frac{E\hat{I}}{R^2}(v''+v)+M_T\right]-\frac{AE}{R^2}(u'-v)+\frac{E\hat{I}}{R^4}(v''+v)+\frac{N_T}{R}=0$$

By carrying out the derivatives, using the above expression for $(u'-v)$, and collecting the terms, we get

$$\frac{E\hat{I}}{R^4}(v^{iv}+2v''+v)=-\frac{M_T}{R^2}$$

for the final form of the differential equation for xY bending. The solution is

$$v=A_1\sin\phi+A_2\cos\phi+A_3\phi\sin\phi+A_4\phi\cos\phi-\frac{R^2M_T}{E\hat{I}}$$

The boundary conditions at A (where $\phi = 0$) yield

$$A_4 = A_3 = 0$$

The boundary conditions at B (where $\phi = \pi/2$) yield

$$A_1 = \frac{R^2 M_T}{E\hat{I}}; \qquad A_2 = 0$$

Therefore, the general solution for xY bending is

$$v = \frac{R^2 M_T}{E\hat{I}} (\sin \phi - 1)$$

Substituting the above solution for v back into the differential equation for circumferential-direction displacement yields

$$u = \frac{R^2 M_T}{E\hat{I}} (-\cos \phi - \phi) + \left(\frac{RN_T}{AE} - \frac{M_T}{AE} \right) \phi + A_5$$

The boundary condition at B yields

$$A_5 = \frac{\pi}{2} \frac{R^2 M_T}{E\hat{I}} - \frac{\pi}{2} \frac{RN_T}{AE} + \frac{\pi}{2} \frac{M_T}{AE}$$

Then, the solution for u is

$$u = \frac{R^2 M_T}{E\hat{I}} \left(\frac{\pi}{2} - \cos \phi - \phi \right) + \left(\frac{RN_T}{AE} - \frac{M_T}{AE} \right) \left(\phi - \frac{\pi}{2} \right)$$

The tip deflection is

$$u(0) = \frac{R^2 M_T}{E\hat{I}} \left(\frac{\pi}{2} - 1 \right) - \frac{\pi}{2} \frac{RN_T}{AE} + \frac{\pi}{2} \frac{M_T}{AE}$$

$$v(0) = -\frac{R^2 M_T}{E\hat{I}}$$

Consider the case where $a_1 = a_2 = 0$. The tip deflection then becomes

$$u(0) = -\frac{\pi}{2} R\alpha T_0$$

$$v(0) = 0$$

From elasticity theory, these results can be shown to be incorrect! In Section 3.15, we will correct these results.

3.15 THERMAL STRAIN CORRECTION IN CURVED BEAMS

Reference [32] (which is reprinted in ref. [33]) develops a "strength" approach to the calculation of stresses in curved beams that includes the effect of thermal strain on curvature. Reference [33] has excellent agreement with exact theory when $T(R, \theta) = T(R)$.

To obtain this effect, a simple modification to the restriction that curved beams do not distort in the plane of the cross section is required. Assume that instead of having $(V = v)$, we have $(V = v + \eta)$, where

$$\eta = \int_0^y \alpha \Delta T \, dy$$

Then, the position vector to the deformed position $P'(s, y, z)$ takes the following form:

$$\vec{R} = \vec{r_0} + U\vec{t_0} + (y + v + \eta)\vec{n_0} + (z + W)\vec{k}$$

Because $\eta \ll y$, Eq. (3.104) remains the same. However, the displacement $\vec{u}$ can now be expressed as

$$\vec{u} = \left[\left(1 - \frac{y}{R}\right) u_0 - yv' - zw'\right]\vec{t_0} + (v + \eta)\vec{n_0} + w\vec{k}$$

The expression for strain becomes

$$\epsilon(s) = \left(u_0' - \frac{v}{R}\right) - \left(\frac{v}{R^2} + v''\right) \frac{y}{1 - y/R} - \frac{zw''}{1 - y/R} - \frac{\eta}{(R - y)} \tag{3.122}$$

while the expression for virtual strain remains unchanged because η is a prescribed function of temperature.

The expression for stress becomes

$$\sigma(s) = E\left[\left(u_0' - \frac{v}{R}\right) - \left(\frac{v}{R^2} + v''\right) \frac{y}{1 - y/R} - \frac{zw''}{1 - y/R}\right] - \alpha E \Delta T(s, y, z) - \frac{n_t}{(R - y)} \tag{3.123}$$

where

$$n_t = E\eta = \int_0^y E\alpha\Delta T \, dy$$

and represents a thermal load per length.

Consider again the virtual work of the axial force $\int_A \sigma \, dA$ from the virtual displacement $\overline{U}$ and collecting the arguments of $\overline{u}_0$, $\overline{v}'$, and $\overline{w}'$, we obtain the Eq. (3.124) for $u_0' - v/R$, $v'' + v/R^2$, and w'' in terms of the stress resultants and area properties.

$$u_0' - \frac{v}{R} = \frac{N + \hat{N}_T}{AE} - \frac{M_z + \hat{M}_{Tz}}{RAE}$$

$$v'' + \frac{v}{R^2} = -\frac{(M_z + \hat{M}_{Tz})\hat{I}_y - (M_y + \hat{M}_{Ty})\hat{I}_{yz}}{E(\hat{I}_y\hat{I}_z - \hat{I}_{yz}^2)}$$

$$w'' = -\frac{(M_y + \hat{M}_{Ty})\hat{I}_z - (M_z + \hat{M}_{Tz})\hat{I}_{yz}}{E(\hat{I}_y\hat{I}_z - \hat{I}_{yz}^2)} \tag{3.124}$$

where

$$\hat{N}_T = N_T + \int_A \frac{n_t}{(R - y)} \, dA$$

$$\hat{M}_{Tz} = M_{Tz} + \int_A \frac{y n_t}{(R - y)} \, dA$$

$$\hat{M}_{Ty} = M_{Ty} + \int_A \frac{z n_t}{(R - y)} \, dA \tag{3.125}$$

The general expression for the circumferential stress σ becomes

$$\sigma = -\alpha E\Delta T - \frac{n_t}{(R - y)} + \frac{N + \hat{N}_T}{A} - \frac{M_z + \hat{M}_{Tz}}{RA}$$
$$+ \frac{(M_z + \hat{M}_{Tz})\hat{I}_y - (M_y + \hat{M}_{Ty})\hat{I}_{yz}}{\hat{I}_y\hat{I}_z - \hat{I}_{yz}^2} \frac{y}{1 - y/R}$$
$$+ \frac{(M_y + \hat{M}_{Ty})\hat{I}_z - (M_z + \hat{M}_{Tz})\hat{I}_{yz}}{\hat{I}_y\hat{I}_z - \hat{I}_{yz}^2} \frac{y}{1 - y/R} \tag{3.126}$$

These corrections can and *should* now be carried out over all of the differential equations of the previous section when applying thermal loading.

Example 3.7 Consider again the curved beam with the following temperature distribution:

$$\Delta T = T_0 \left[1 + a_1 \frac{y}{h} + a_2 \left(\frac{y}{h} \right)^2 \right]$$

The n_t of Eq. (3.123) becomes

$$n_t = \alpha E T_0 y + \frac{\alpha E a_1 T_0}{2h} y^2 + \frac{\alpha E a_2 T_0}{3h^2} y^3$$

Then, $\hat{N}_T$ and $\hat{M}_T$ become

$$\hat{N}_T = \alpha A E T_0 + \frac{\alpha E \hat{I} T_0}{R^2} + \frac{\alpha a_1 E \hat{I} T_0}{2hR} + \frac{\alpha a_2 E I T_0}{h^2} + \frac{\alpha a_2 E \hat{I}_4 T_0}{3h^2 R^2}$$

$$\hat{M}_{Tz} = \frac{\alpha E \hat{I} T_0}{R} + \frac{\alpha a_1 E I T_0}{h} + \frac{\alpha a_1 E \hat{I}_4 T_0}{2hR^2} + \frac{\alpha a_2 E \hat{I}_4 T_0}{3h^2 R}$$

where we have defined

$$\int_A \frac{y^3 \, dA}{1 - y/R} = \int_A y^3 \, dA + \frac{\hat{I}_4}{R}$$

and

$$\int_A y^3 \, dA = 0$$

from section symmetry.

Then, for the case where $a_1 = a_2 = 0$, the corrected tip deflection becomes

$$u(0) = -R\alpha T_0$$
$$v(0) = -R\alpha T_0$$

Note that for this case the stress σ is zero.

3.16 SHEAR AND RADIAL STRESS IN CURVED BEAMS

In addition to shear stress from variation in transverse loading, the curved beam also develops radial stress from the different fiber length caused by the initial curvature. Again, because of the assumption that plane sections remain plane, both the shear and radial stress must be considered as internal constraints. Hence

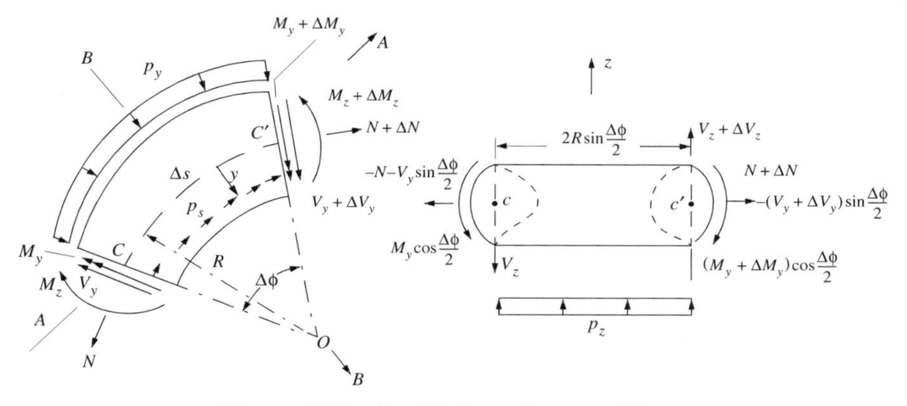

Figure 3.30 Equilibrium of a curved beam.

they have no conjugate variables, and the principle of virtual work will not account for them. Thus we must rely on free bodies and the direct formulation of the equations of equilibrium.

Figure 3.30 shows a partial free-body diagram of a section $\Delta\phi$ of a curved beam. If in Fig. 3.30(a) we sum forces in the *A–A* and *B–B* directions, take moments about *O*, and neglect a higher-order terms, the following three equilibrium equations will result:

$$\frac{dV_y}{ds} = -p_y - \frac{N}{R}$$

$$\frac{dN}{ds} = -p_s + \frac{V_y}{R}$$

$$\frac{dM_z}{ds} = V_y - yp_s \tag{3.127}$$

If in Fig. 3.30(b) we sum forces in the z direction and take moments about C', the following two equilibrium equations will result:

$$\frac{dV_z}{ds} = -p_z - \frac{N}{R}$$

$$\frac{dM_y}{ds} = V_z - zp_s + \frac{V_y}{R} \tag{3.128}$$

Figure 3.31 represents a typical beam cross section and a slice of the cross section along the dimension b. The equation of equilibrium for the slice is for the sum of forces in the vertical and horizontal directions, respectively.

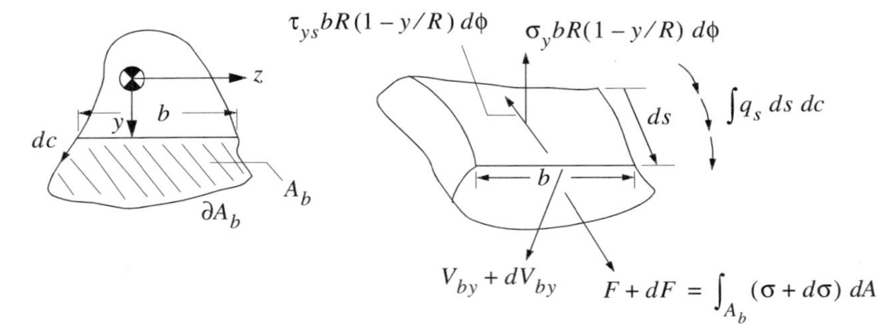

Figure 3.31 Equilibrium of a curved beam slice.

$$\sigma_y bR(1 - y/R)\, \frac{ds}{R} - \frac{F\, ds}{R} - d V_{by} = 0$$

$$-\tau_{ys} bR(1 - y/R)\, \frac{ds}{R} + dF - V_{by} + \int q_s \, ds \, dc = 0$$

where

σ_y is the average radial stress on the surface $b\ ds$

τ_{ys} is the average shear stress on the area $b\ ds$

$F = \int_{A_b} \sigma\, dA$ is the circumferential force over the shaded area A_b

$dF = \int_{A_b} d\sigma\, dA$ is the increment in circumferential force over the shaded area A_b

V_{by} is the stress resultant acting over the face of the cutout section

q_s is the portion of the circumferential distributed load acting over the surface of the segment ds

In essence, $p_s = \int_{\text{around total boundary } c} q_s \, dc$. Solving for σ_y and τ_{ys}, we get the following two expressions:

$$\sigma_y = \frac{1}{b(1 - y/R)} \left(\frac{1}{R} \int_{A_b} \sigma \, dA + \frac{d V_{by}}{ds} \right) \tag{3.129}$$

$$\tau_{ys} = \frac{1}{b(1 - y/R)} \left(\int_{A_b} \sigma' \, dA + \frac{V_{by}}{R} + \int_{\partial A_b} q_s \, dc \right) \tag{3.130}$$

Equations (3.129) and (3.130) each contain the additional unknown V_{by}. However, V_{by} is related to the shear stress by the following relationship:

$$V_{by} = \int_{A_b} \tau_{sy} \, dA \tag{3.131}$$

If Eq. (3.131) is substituted into Eq. (3.130), an integral equation results that then could by numerical techniques be solved for τ_{ys}. However, this effort is not warranted because, in general, shear is not often a controlling factor except in wood or metal beams that have thin webs or a small span-to-depth ratio. In the strength of material approach, it is customary to use an area factor to approximate Eq. (3.131). We use the following approximation:

$$V_{by} = \frac{A_b}{A} V_y \tag{3.132}$$

With the above approximation, we may now substitute σ and $\sigma' = d\sigma/ds$ using Eq. (3.126) into the expression for σ_y and τ_{ys} to get

$$\begin{aligned}\sigma_y = \frac{1}{b(1-y/R)} \Bigg[\frac{1}{R} \int_{A_b} E \Bigg\{ \left(u_0' - \frac{v}{R} \right) \\ - \left(\frac{v}{R^2} + v'' \right) \frac{y}{1-y/R} - \frac{zw''}{1-y/R} \Bigg\} dA \\ - \frac{1}{R} \int_{A_b} \left(\alpha E \Delta T + \frac{n_t}{(R-y)} \right) dA + \frac{A_b}{A} \frac{dV_y}{ds} \Bigg]\end{aligned} \tag{3.133}$$

$$\begin{aligned}\tau_{ys} = \frac{1}{b(1-y/R)} \Bigg[\int_{A_b} E \Bigg\{ \left(u_0' - \frac{v}{R} \right)' \\ - \left(\frac{v}{R^2} + v'' \right)' \frac{y}{1-y/R} - \frac{zw'''}{1-y/R} \Bigg\} dA \\ - \int_{A_b} \left(\alpha E \Delta T' + \frac{n_t'}{(R-y)} \right) dA - \frac{A_b}{AR} V_y + \int_{\partial A_b} q_s \, dc \Bigg]\end{aligned} \tag{3.134}$$

In taking the derivative of Eq. (3.126), we have not included $d(1/R)\, ds$ in the result. This is consistent with the derivative for strain; indeed, ref. [31] derives the expression for strain using the general expression for curvature in polar coordinates and arrives at the term $v'' + v/R^2$ by neglecting this derivative.

For what follows, we define the three integrals as follows:

$$A_b = \int_{A_b} dA$$

$$\hat{Q}_z = \int_{A_b} \frac{y}{1 - y/R} \, dA$$

$$\hat{Q}_y = \int_{A_b} \frac{z}{1 - y/R} \, dA \tag{3.135}$$

where the first expression is the area A_b and the last two expressions represent the "first" moments of area for the slice taken at b. The expressions for σ_y and τ_{ys} can now be expressed as

$$\sigma_y = \frac{1}{b(1 - y/R)} \left[\frac{E}{R} \left\{ A_b \left(u_0' - \frac{v}{R} \right) - \left(\frac{v}{R^2} + v'' \right) \hat{Q}_z - w'' \hat{Q}_y \right\} \right.$$
$$\left. - \frac{1}{R} \int_{A_b} \left(\alpha E \Delta T + \frac{n_t}{(R - y)} \right) dA + \frac{A_b}{A} \frac{dV_y}{ds} \right] \tag{3.136}$$

$$\tau_{ys} = \frac{1}{b(1 - y/R)} \left[E \left\{ A_b \left(u_0' - \frac{v}{R} \right)' - \left(\frac{v}{R^2} + v'' \right)' \hat{Q}_z - w''' \hat{Q}_y \right\} \right.$$
$$\left. - \int_{A_b} \left(\alpha E \Delta T' + \frac{n_t'}{(R - y)} \right) dA - \frac{A_b}{AR} V_y + \int_{\partial A_b} q_s \, dc \right] \tag{3.137}$$

Substituting Eq. (3.110) and its derivative with respect to s into Eqs. (3.136) and (3.137), we get

$$\sigma_y = \frac{1}{b(1 - y/R)} \left[\frac{A_b}{A} \left(\frac{N + \hat{N}_T}{R} - \frac{M_z + \hat{M}_{Tz}}{R^2} \right) \right.$$
$$- \frac{(M_z + \hat{M}_{Tz})\hat{I}_y - (M_y + \hat{M}_{Ty})\hat{I}_{yz}}{R(\hat{I}_y \hat{I}_z - \hat{I}_{yz}^2)} \hat{Q}_z$$
$$- \frac{(M_y + \hat{M}_{Ty})\hat{I}_z - (M_z + \hat{M}_{Tz})\hat{I}_{yz}}{R(\hat{I}_y \hat{I}_z - \hat{I}_{yz}^2)} \hat{Q}_y$$
$$\left. - \frac{1}{R} \int_{A_b} \left(\alpha E \Delta T + \frac{n_t}{(R - y)} \right) dA + \frac{A_b}{A} \frac{dV_y}{ds} \right] \tag{3.138}$$

$$\tau_{ys} = \frac{1}{b(1 - y/R)} \left[\frac{A_b}{A} \left((N' + \hat{N}'_T) - \frac{M'_z + \hat{M}'_{Tz}}{R} \right) \right.$$

$$- \frac{(M'_z + \hat{M}'_{Tz})\hat{I}_y - (M'_y + \hat{M}'_{Ty})\hat{I}_{yz}}{\hat{I}_y\hat{I}_z - \hat{I}^2_{yz}} \hat{Q}_z$$

$$- \frac{(M'_y + \hat{M}'_{Ty})\hat{I}_z - (M'_z + \hat{M}'_{Tz})\hat{I}_{yz}}{\hat{I}_y\hat{I}_z - \hat{I}^2_{yz}} \hat{Q}_y$$

$$\left. - \int_{A_b} \left(\alpha E \Delta T' + \frac{n'_t}{(R - y)} \right) dA - \frac{A_b}{AR} V_y + \int_{\partial A_b} q_s \, dc \right] \tag{3.139}$$

Eq. (3.139) may be rearranged to yield

$$\tau_{ys} = \frac{1}{b(1 - y/R)} \left[\frac{A_b}{A} \left((N' + \hat{N}'_T) - \frac{M'_z + \hat{M}'_{Tz}}{R} \right) \right.$$

$$+ \frac{\hat{Q}_z\hat{I}_y - \hat{Q}_y\hat{I}_{yz}}{\hat{I}_y\hat{I}_z - \hat{I}^2_{yz}} (M'_z + \hat{M}'_{Tz})$$

$$+ \frac{\hat{Q}_y\hat{I}_z - \hat{Q}_z\hat{I}_{yz}}{\hat{I}_y\hat{I}_z - \hat{I}^2_{yz}} (M'_y + \hat{M}'_{Ty})$$

$$\left. - \frac{A_b}{AR} V_y - \int_{A_b} \left(\alpha E \Delta T' + \frac{n'_t}{(R - y)} \right) dA + \int_{\partial A_b} q_s \, dc \right] \tag{3.140}$$

Using Eqs. (3.127) and (3.128), Eqs. (3.138) and (3.140) take the final form

$$\sigma_y = \frac{1}{b(1 - y/R)} \left[\frac{A_b}{A} \left(\frac{\hat{N}_T}{R} - \frac{M_z + \hat{M}_{Tz}}{R^2} \right) \right.$$

$$- \frac{(M_z + \hat{M}_{Tz})\hat{I}_y - (M_y + \hat{M}_{Ty})\hat{I}_{yz}}{R(\hat{I}_y\hat{I}_z - \hat{I}^2_{yz})} \hat{Q}_z$$

$$
- \frac{(M_y + \hat{M}_{Ty})\hat{I}_z - (M_z + \hat{M}_{Tz})\hat{I}_{yz}}{R(\hat{I}_y\hat{I}_z - \hat{I}_{yz}^2)} \hat{Q}_y
$$

$$
- \frac{1}{R} \int_{A_b} \left(\alpha E \Delta T + \frac{n_t}{(R - y)} \right) dA - \frac{A_b}{A} p_y \Bigg] \tag{3.141}
$$

$$
\tau_{ys} = \frac{1}{b(1 - y/R)} \left[\frac{A_b}{A} \left(\hat{N}'_T - \frac{\hat{M}'_{Tz}}{R} \right) \right.
$$

$$
+ \frac{\hat{Q}_z\hat{I}_y - \hat{Q}_y\hat{I}_{yz}}{\hat{I}_y\hat{I}_z - \hat{I}_{yz}^2} (V_y + \hat{M}'_{Tz} - yp_s)
$$

$$
+ \frac{\hat{Q}_y\hat{I}_z - \hat{Q}_z\hat{I}_{yz}}{\hat{I}_y\hat{I}_z - \hat{I}_{yz}^2} (V_z + \hat{M}'_{Ty} - zp_s)
$$

$$
- \frac{A_b}{AR} V_y - \int_{A_b} \left(\alpha E \Delta T' + \frac{n'_t}{(R - y)} \right) dA
$$

$$
\left. + \int_{\partial A_b} q_s \, dc \right] - \frac{A_b}{bA} p_s \tag{3.142}
$$

Equations (3.141) and (3.142) give the radial and shear stress in curved beams; however, some comments are in order. Reference [30] points out that while Eq. (3.111) is inaccurate for the case of **I** and **T** sections, Eq. (3.141) for pure bending gives good results. Thus for these types of sections, the radial stress is insensitive to the additional rotation of the flanges. Reference [34] indicates that it is not until a beam has sharp initial curvature $R/h = 2/3$ that the radial stress becomes equal to the smaller of the extreme fiber stress. For **I** and **T** sections, the radial stress may become critial when the web thickness is small but, in general, is not important for $R/h \geq 1$, provided the web thickness $t_w \geq 1/5$ flange breadth.

While Eq. (3.142) can be used to determine the shear stress in a curved beam, ref. [34] indicates that a good measure of shear is given by the straight beam approximation:

$$
\tau_{\max} = \text{form} \frac{V_y}{A}
$$

where form is a form factor (e.g., form $= 3/2$ for a rectangular section). For a curved beam with $R/h = 1$, the relationship $(3/2)V_y/A$ is only about 10% smaller then the actual maximum shear stress. The maximum shear occurs at the neutral axis, but because of the initial curvature of the beam, the neutral axis shifts inward toward the center of curvature and, hence, does not coincide with the centroid axis.

3.17 THIN WALLED BEAMS OF OPEN SECTION

In this section, we will briefly examine thin walled beams of open sections. It should be kept in mind, however, that entire textbooks have been written on the subject. Perhaps the most comprehensive basic text on the subject is *Thin-Walled Elastic Beams*, by V. Z. Vlasov [35]. Two other excellent texts are given in refs. [36] and [37].

A thin walled section is defined as open when the locus of points defining the centerline of the cross section is not a closed curve. The basic characteristic of thin walled sections is that the thickness of the component elements is small compared to that of the other dimensions. If t is the thickness of the section; d, any characteristic dimension of the cross section, such as height or width; and ℓ, the beam length, then when the ratios satisfy the relations

$$\frac{t}{d} \leq 0.1 \quad \text{and} \quad \frac{d}{\ell} \leq 0.1$$

the structure can be classified as a long, thin walled beam.

Figure 3.32 shows the centerline geometry of a thin walled section of general shape. Consider the kinematics of motion in the plane of the cross section under the assumption that the geometry of the cross section is unaltered during deformation.

This assumption means that the motion of any section in its own plane can be

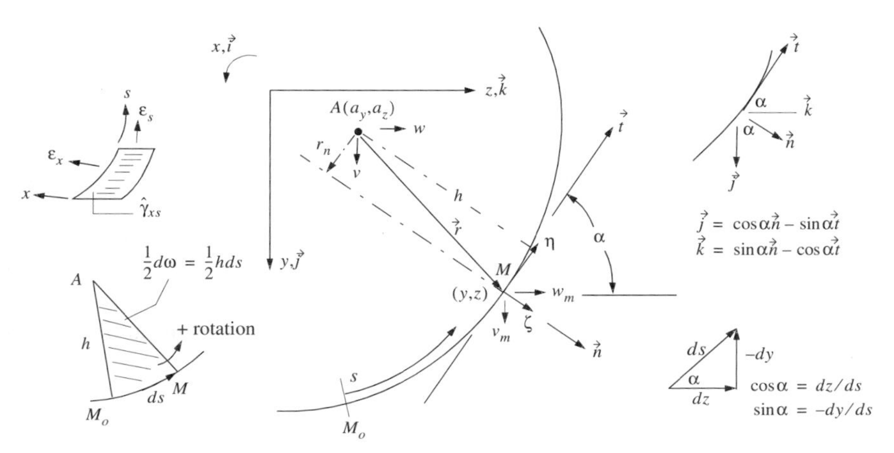

Figure 3.32 Centerline geometry of a thin walled section.

described by translations of points on the section plus rotations of these points about some point in the plane of the cross section.

Let A denote an arbitrary point on the cross section with coordinates (a_y, a_z) and describe the (y, z) displacement components by (v, w). If $\vec{\Omega} = \psi\vec{i}$ is the rotation of the cross section about some point C referred to as the center of twist or torsion center, then the displacement of the center of twist in the plane of the cross section is zero, and we have the following relationship:

$$\vec{U}_C = \vec{U}_A + \vec{\Omega} \times \vec{r}_C = 0$$

where

$$\vec{r}_C = (c_y - a_y)\vec{j} + (c_z - a_z)\vec{k}$$

and

$$\vec{U}_A = v\vec{j} + w\vec{k}$$

Hence we have for the location of the center of twist:

$$c_y = a_y - \frac{w(x)}{\phi(x)}$$

$$c_z = a_z + \frac{v(x)}{\phi(x)} \tag{3.143}$$

Notice that, in general, the location of C varies from point to point along the length of the beam and depends on the loading and boundary conditions. The center of twist will only under certain loading and boundary conditions coincide with the shear center (to be described later), which is a geometric property.

Consider next a point M on the median line or centerline of the cross section. Let M have the coordinates (y, z) with displacement components in the plane of the cross section of (v_M, w_M). We may then write

$$\vec{U}_M = \vec{U}_A + \vec{\Omega} \times \vec{r}$$

$$v_M\vec{i} + w_M\vec{k} = [v - (z - a_z)\phi]\vec{j} + [w + (y - a_y)\phi]\vec{k} \tag{3.144}$$

It will also be convenient to consider the components of displacement of M along the tangent to the median defined by the unit vector $\vec{t}$, with the normal to the tangent defined as the unit vector $\vec{n}$ positive outward as shown. Let the components of displacement in the tangent and normal direction be defined as (η, ζ), respectively. From Fig. 3.32, we see that

$$\vec{j} = \cos\,\alpha\vec{n} - \sin\,\alpha\vec{t}$$

$$\vec{k} = \sin\,\alpha\vec{n} + \cos\,\alpha\vec{t} \tag{3.145}$$

Notice also that $\vec{r}$ can be expressed as

$$\begin{aligned}\vec{r} &= h\vec{n} - r_n\vec{t} \\ &= (y - a_y)\vec{j} + (z - a_z)\vec{k} \\ &= [(y - a_y)\cos\alpha + (z - a_z)\sin\alpha]\vec{n} \\ &\quad + [-(y - a_n)\sin\alpha + (z - a_z)\cos\alpha]\vec{t}\end{aligned} \tag{3.146}$$

From the above equation, we can define h and r_n as

$$\begin{aligned}h &= (y - a_y)\cos\alpha + (z - a_z)\sin\alpha \\ r_n &= (y - a_y)\sin\alpha - (z - a_z)\cos\alpha\end{aligned} \tag{3.147}$$

Then, $\vec{U}_M$ can be expressed as

$$\eta\vec{t} + \zeta\vec{n} = [v - (z - a_z)\phi](\cos\alpha\vec{n} - \sin\alpha\vec{t}) + [w + (y - a_y)\phi](\sin\alpha\vec{n} + \cos\alpha\vec{t})$$

or, on collecting terms,

$$\begin{aligned}\eta &= -v\sin\alpha + w\cos\alpha + h\phi \\ \zeta &= v\cos\alpha + w\sin\alpha + r_n\phi\end{aligned} \tag{3.148}$$

From Fig. (3.32), we see that $\sin\alpha$ and $\cos\alpha$ may be replaced by $-dy/ds$ and dz/ds, respectively. Thus Eq. (3.148) becomes

$$\eta(x, s) = v\,\frac{dy}{ds} + w\,\frac{dz}{ds} + h\phi \tag{3.149a}$$

$$\zeta(x, s) = v\,\frac{dz}{ds} - w\,\frac{dy}{ds} + r_n\phi \tag{3.149b}$$

Also, the expression for h becomes

$$h(s) = (y - a_y)\,\frac{dz}{ds} - (z - a_z)\,\frac{dy}{ds} \tag{3.150}$$

Equation (3.149) represents the final form for the displacement of the point M at some location s along the median line of the cross section for a section located at x along the longitudinal axis of the beam.

If we denote the longitudinal displacement of a point on the centerline of the cross section as $u(x, s)$, we can express the shear strain $\hat{\gamma}_{xs}$ of the centerline as

$$\hat{\gamma}_{xs} = \frac{\partial u}{\partial s} + \frac{\partial \eta}{\partial x}$$

The fundamental assumption of thin walled beams of open sections is that the shear strain of the median surface is zero.

$$\hat{\gamma}_{xs} = \frac{\partial u}{\partial s} + \frac{\partial \eta}{\partial x} \stackrel{\text{def}}{=} 0 \tag{3.151}$$

It is important to realize, however, that while this is the fundamental assumption for thin walled beams of open sections, it is not true for thin walled beams of closed sections. Thus the results that follow are only for beams of open section.

Since the thickness dimensions of the cross section are small, we may assume a state of plane stress and use Eq. (3.10) as the constitutive equation. Further, the assumption of no distortion in the plane of the cross section requires that any strain in the transverse direction must be zero. Thus the extension ϵ_s along the median direction must be zero, and we have

$$\sigma_x = \frac{E}{1-\nu^2}\,\epsilon_x - \frac{\alpha}{1-\nu}\,E\Delta T$$

$$\sigma_s = \nu\,\sigma_x - \alpha E\Delta T$$

In the theory of thin walled beams of open sections, it is customary to neglect the effects of Poisson's ratio. Thus, in the following text, we will use the symbol E to represent either the reduced modulus or the Young's modulus, and we will use α to represent the reduced coefficient of thermal expansion or simply the coefficient of thermal expansion. Thus we will use the constitutive equation

$$\sigma_x = E\epsilon_x - \alpha E\Delta T \tag{3.152}$$

Using Eq. (3.151), we can solve $u(x, s)$. From Eq. (3.149a), we can write

$$\eta' = \frac{\partial \eta}{\partial x} = v'\,\frac{dy}{ds} + w'\,\frac{dz}{ds} + h\phi'$$

From Eq. (3.151), we can write

$$\begin{aligned} u(x, s) &= u_0(x) - \int_{M_0}^{M} \frac{\partial \eta}{\partial x}\,ds \\ &= u_0(x) - v'(x)y(s) - w'(x)z(s) - \phi'(x)\omega(s) \end{aligned} \tag{3.153}$$

where $u_0(x)$ describes the longitudinal displacement of M_0 and we have defined

$$\omega(s) = \int_{M_0}^{M} h(s)\, ds \tag{3.154}$$

as twice the area of the sector enclosed between the arc M_0M of the median line and the two lines AM_0 and AM. This area is referred to as the sectorial area.

Comparing Eq. (3.153) with Eq. (3.55), we see that the first three terms express the law of plane sections: u_0 determines the axial deformation of the point M_0 on the median line of the beam profile, while v' and w' describe the flexure of an arbitrary axis of the beam passing through point A at $y = a_y$, $z = a_z$.

The term ϕ' determines an additional displacement term that does not obey the law of plane sections, and it comes from the torsion ϕ of the section. The term $\omega\phi'$ is called the sectorial warping of the section.

Point A is called the pole of the sectorial area $\omega(s)$, whereas point M_0 on the median or profile line is called the sectorial origin.

From Eq. (3.153), we may define the strain ϵ_x as

$$\epsilon_x(x, s) = u_0'(x) - y(s)v''(x) - z(s)w''(x) - \omega(s)\phi''(x) \tag{3.155}$$

A virtual strain for the beam can be written as

$$\overline{\epsilon}_x(x, s) = \overline{u}_0'(x) - y(s)\overline{v}''(x) - z(s)\overline{w}''(x) - \omega(s)\overline{\phi}''(x) \tag{3.156}$$

Using Eq. (3.152), the longitudinal stress can be written as

$$\sigma_x(x, s) = E[u_0'(x) - y(s)v''(x) - z(s)w''(x) - \omega(s)\phi''(x)] - \alpha E\Delta T(x, s) \tag{3.157}$$

If we examine the deformation of the median surface under the stipulation $\epsilon_x = 0$, we obtain the following

$$1u_0'(x) - y(s)v''(x) - z(s)w''(x) - \omega(s)\phi''(x) = 0 \tag{3.158}$$

Equation (3.158) must be satisfied for any value of s. Also, the coefficients 1, y, z, and ω are linearly independent. Thus this equation represents the following four equations:

$$
\begin{aligned}
u_0' &= 0 \\
v'' &= 0 \\
w'' &= 0 \\
\phi'' &= 0
\end{aligned}
\tag{3.159}
$$

Integration yields the following:

$$
\begin{aligned}
u_0 &= u_0^0 \\
v &= v_0 + v_0' x \\
w &= w_0 + w_0' x \\
\phi &= \phi_0 + \phi_0' x
\end{aligned}
\tag{3.160}
$$

The constants of integration—u_0^0, v_0, and w_0—are the rigid-body displacements of a section $x = 0$. The constants v_0', w_0', and ϕ_0 are the angular rotations of the section at $x = 0$. Thus these six constants of integration represent the six rigid-body degrees of freedom of the section at $x = 0$ and contribute nothing to the deformation of the section.

The term ϕ_0' gives the warping of the section at $x = 0$ and constitutes a deformation of the section even under the stipulation that the relative longitudinal deformation ϵ_x is zero. The deformation is characterized by the constant relative-torsion angle.

$$
\phi' = \phi_0' \tag{3.161}
$$

Thus every beam with an inextensible longitudinal axis having a rigid cross section can still undergo a deformation normal to the plane of the cross section characterized by $\phi = \phi_0' x$. This is Saint Venant's torsion theory. Thus the theory of thin walled open beams under the assumption of Eq. (3.151) contains both the theory of plane sections and Saint Venant's theory as special cases.

Before proceeding further, it is useful to define some integral expressions that will frequently occur in the text that follows.

In addition to the integrals defined by Eq. (3.60), thin walled beams require the following four integrals to define sectorial properties of thin walled open beams:

$$
S_\omega = \int_A \omega \, dA
$$

$$
I_{\omega y} = \int_A \omega z \, dA
$$

$$I_{\omega z} = \int_A \omega y \, dA$$

$$I_\omega = \int_A \omega^2 \, dA \tag{3.162}$$

Just as for the case of the properties given by Eq. (3.59) and (3.60), principal sectorial properties can be defined. Setting the sectorial static moment S_ω to zero defines the sectorial centroid or sectorial zero point, and setting the two sectorial products of inertia $I_{\omega y}$ and $I_{\omega z}$ to zero defines the principal pole or shear center of the section. The sectorial moment of inertia is the sectorial counterpart to the second moments of area.

The following integral represents a bimoment and is similar to the stress resultants over a cross section defined by Eq. (3.61)

$$M_\omega = \int_A \sigma \, \omega \, dA \tag{3.163}$$

The bimoment (unit force multiplied by the length squared) corresponds to the sectorial warping of the section.

The following integral arises because of temperature variation:

$$M_{T\omega} = \int_A \alpha E \Delta T \omega \, dA \tag{3.164}$$

The evaluation of this integral depends on an explicit expression for the temperature variation $\Delta T(x, s)$.

We can now use virtual work to determine the stress σ_x in terms of the stress resultants, bimoments, second moments of area, and the sectorial properties. We again consider that our reference origin O is at the centroid of the section so that Eq. (3.59) again holds. Then, consider the virtual work of the axial force $\int_A \sigma_x \, dA$ from the virtual displacement $\overline{u}$, which is expressed as

$$\int_A \sigma_x \, dA(\overline{u}_0 - y\overline{v}' - z\overline{w}' - \omega\overline{\phi}')$$
$$= \int_A [E(u_0' - yv'' - zw'' - \omega\phi'') - \alpha E \Delta T](\overline{u}_0 - y\overline{v}' - z\overline{w}' - \omega\overline{\phi}') \, dA$$

Each of the virtual quantities $\overline{u}_0$, $\overline{v}'$, $\overline{w}'$, and $\overline{\phi}'$ represents a generalized coordinate. Collecting terms and placing the result into the form of Eq. (3.23) and using the expressions given by Eqs. (3.59)–(3.62) and Eqs. (3.162)–(3.164), we obtain the following four equations that contain the five unknowns u_0', v'', w'', ϕ'', and M_ω:

$$N + N_T = AEu_0' - ES_\omega \phi'' \tag{3.165a}$$
$$-M_z - M_{Tz} = EI_z v'' + EI_{yz} w'' + EI_{\omega z} \phi'' \tag{3.165b}$$
$$-M_y - M_{Ty} = EI_{yz} v'' + EI_y w'' + EI_{\omega y} \phi'' \tag{3.165c}$$
$$-M_\omega - M_{T\omega} = -ES_\omega u_0' + EI_{\omega z} v'' + EI_{\omega y} w'' + EI_\omega \phi'' \tag{3.165d}$$

Before we solve these equations for the four unknowns u_0', v'', w'', ϕ'', we need to make a brief digression to study some of the mathematical properties relating to the sectorial properties. For example, the sectorial area is defined as $\omega(s) = \int_0^s h\, ds$, where h is defined by Eq. (3.150) and $s = 0$ is measured from Fig. 3.32 at the point M_0. Then we can express ω as

$$\begin{aligned}\omega &= \int_0^s h\, ds \\ &= (y - a_y) \int_{z_0}^{z} dz - (z - a_z) \int_{y_0}^{y} dy \\ &= (y - a_y)(z - z_0) - (z - a_z)(y - y_0)\end{aligned} \tag{3.166}$$

where y_0, z_0 are the coordinates of the point M_0.

In a manner similar to the transfer formulas for the second moments of area, we can define a transfer formula for sectorial area. Figure 3.33 shows pole A and pole B used for the measurement of sectorial area. A common sectorial origin M_0 is used for the measurement of the sectorial areas. The increment of sectorial area $d\omega_A$ is given as

$$d\omega_A = h\, ds = (y - a_y)\, dz - (z - a_z)\, dz$$

and the increment of sectorial area $d\omega_B$ is given as

$$d\omega_B = h_B\, ds = (y - b_y)\, dz - (z - b_z)\, dy$$

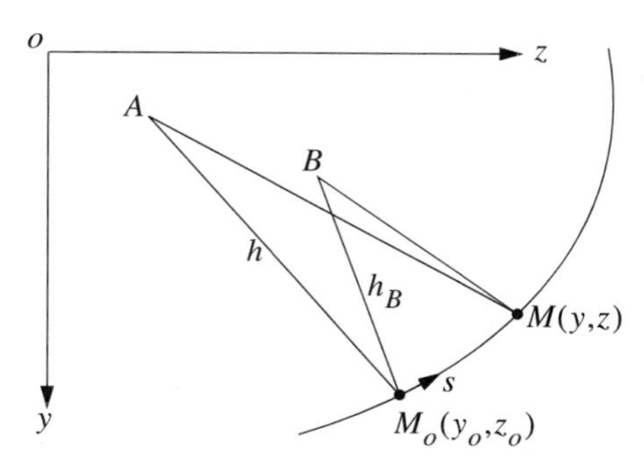

Figure 3.33 Sectorial area for two poles with a common origin.

The difference in increments is

$$d(\omega_A - \omega_B) = (a_z - b_z)\,dy - (a_y - b_y)\,dz$$

Integrating the left side from $s = 0$ at M_0 to s at M and the right side from M_0 at (y_0, z_0) to M at (y, z), we obtain

$$\omega_A = \omega_B + (a_z - b_z)(y - y_0) - (a_y - b_y)(z - z_0) \tag{3.167a}$$

$$= \omega_B + (a_z - b_z)y - (a_y - b_y)z \tag{3.167b}$$

$$+ \; [(a_y - b_y)z_0 - (a_z - b_z)y_0] \tag{3.167c}$$

Equation (3.167b) can be used to determine the location of the principal pole. Let A represent the principal pole; then, by definition,

$$I_{\omega A y} = \int_{M_0}^{M} \omega_A z\,dA = 0$$

$$I_{\omega A z} = \int_{M_0}^{M} \omega_A y\,dA = 0 \tag{3.168}$$

Substituting Eq. (3.167b) into Eq. (3.168) and using the fact that the origin of coordinates is at the centroid of the section, we obtain the following relationships:

$$a_y - b_y = \frac{I_z I_{\omega B y} - I_{yz} I_{\omega B z}}{I_y I_z - I_{yz}^2}$$

$$a_z - b_z = \frac{I_y I_{\omega B z} - I_{yz} I_{\omega B y}}{I_y I_z - I_{yz}^2} \tag{3.169}$$

If we move the pole B to the centroid, we have

$$a_y = \frac{I_z I_{\omega y} - I_{yz} I_{\omega z}}{I_y I_z - I_{yz}^2}$$

$$a_z = -\frac{I_y I_{\omega z} - I_{yz} I_{\omega y}}{I_y I_z - I_{yz}^2} \tag{3.170}$$

If principal axes are used, $I_{yz} = 0$, and we have

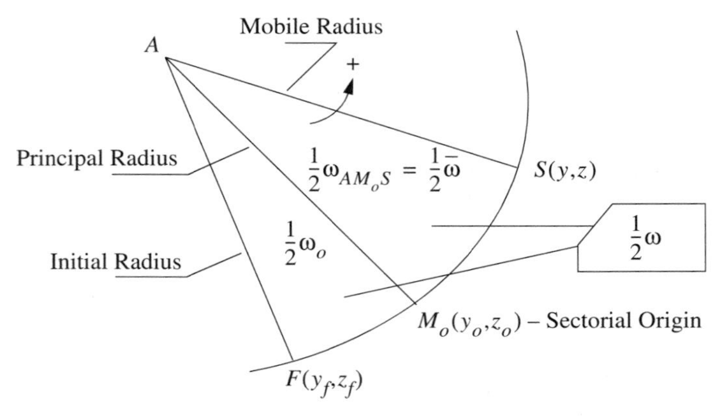

Figure 3.34 Sectorial centroid.

$$a_y = \frac{I_{\omega y}}{I_y}$$

$$a_z = -\frac{I_{\omega z}}{I_z} \tag{3.171}$$

For cross sections with one axis of symmetry, the principal pole always lies on the axis of symmetry; for cross sections with two axes of symmetry, the principal pole lies at the intersection of the symmetry axes. For cross sections consisting of a single bundle of thin rectangular plates with a common intersection, the principal pole lies at the intersection.

Using the expression $S_\omega = \int \omega \, dA$, we can define a sectorial centroid (or principal sectorial origin or principal sectorial zero point) for a sectorial area with pole A.

Figure 3.34 shows a sectorial area $\omega_{AFS} = \omega$ with pole A, origin $F(y_f, z_f)$, and initial radius AF. We wish to define a principal radius AM_0 with sectorial origin $M_0(y_0, z_0)$ so that $S_{\omega_{AM_0}S} = \int \omega_{AM_0S} \, dA = 0$. From Fig. 3.34, we define the constant sectorial area $\omega_{AFM_0} = \omega_0$ as

$$\omega_0 = \omega - \omega_{AM_0S} \tag{3.172}$$

and the sectorial static moment as

$$\int \omega_0 \, dA = \int \omega \, dA - \int \omega_{AM_0S} \, dA$$

Or, since $\int \omega_{AM_0S} \, dA = 0$ by definition,

$$\omega_0 \int dA = \int \omega \, dA = S_\omega$$

with the final result

$$\omega_0 = \frac{S_\omega}{A} \tag{3.173}$$

From Eq. (3.173), two useful identities to be used later are

$$\int (\omega_0 - \omega) \, dA = 0 \tag{3.174a}$$

$$\int (\omega_0^2 - \omega_0 \omega) \, dA = 0 \tag{3.174b}$$

For hipped cross sections (such as a channel section) with a single axis of symmetry, the sectorial centroid lies at the intersection of the symmetry axis and the cross section. For certain cross sections, such as an **I** section, there can be an infinite number of sectorial centroids (at every location along the web of an **I** section). In such a case, the sectorial centroid nearest to the principal pole is normally used.

For thin walled sections, the computation of section properties such as $\bar{z}$, I_z, $I_{\omega z}$, a_z, I_{ω_A}, and so forth are more easily computed if the following observations are made. First, since the wall thickness t is considered small when compared to other cross-sectional dimensions, the section properties can be characterized without significant loss of accuracy by their centerline dimensions. Second, consider two functions $y_1(s)$ and $y_2(s)$ in which y_1 is a general linear or nonlinear function of s and $y_2 = b + ms$ is a linear function of s. Next, consider the product

$$\begin{aligned} \int y_1 y_2 \, ds &= \int y_1(b + ms) \, ds \\ &= b \int y_1 \, ds + m \int s y_1 \, ds \\ &= bA_1 + mA_1 \bar{s}_1 \\ &= A_1(b + m\bar{s}_1) \\ &= A_1 \bar{y}_2 \end{aligned}$$

where

A_1 is the area under the curve $y_1(s)$

$\bar{s}_1$ locates the centroid of A_1

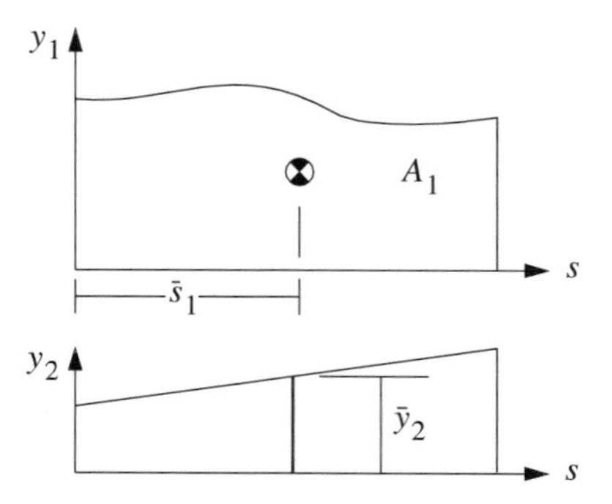

Figure 3.35 Evaluation of $\int y_1 y_2\, ds$.

$\bar{y}_2$ is the ordinate of y_2 measured at the location of the centroid $\bar{s}_1$ of the area A_1 (shown graphically in Fig. 3.35)

Example 3.8 As an example, consider the calculations of some of the section properties shown in Fig. 3.36. This figure shows a channel section and plots the variation of the zy coordinates, along with a plot of the sectorial area ω_c with a pole at the centroid C of the section and a plot of sectorial area ω_A using the principal pole. There is also a plot of a constant of value 1 over the section.

The area of the section is given as

$$A = 2t_f b + t_w h$$

Using the Yz axes shown, the centroid location $\bar{z}$ to the y axis uses the z and 1 plots computed as

$$\bar{z} = \frac{1}{A} \int z(t\, ds)$$

$$= -\frac{t_f b^2}{A}$$

Using the y plot twice (once as y_1, once as y_2), we can compute I_z as

$$I_z = \int y^2 t\, ds$$

$$= 2\left[t_w \frac{1}{2}\left(\frac{h}{2}\,\frac{h}{2}\right)\left(\frac{h}{3}\right) + t_f\left(b\,\frac{h}{2}\right)\left(\frac{h}{2}\right)\right]$$

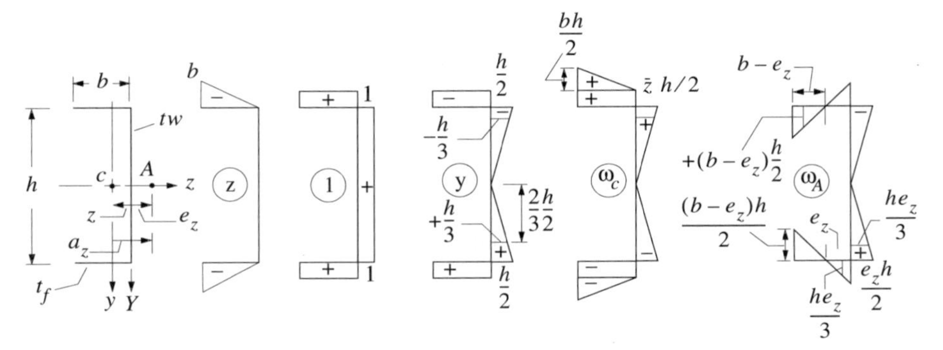

Figure 3.36 Some section properties for a channel section.

$$
\begin{aligned}
&= \frac{t_w h^3}{12} + \frac{t_f b h^2}{2} \\
&= I_{zw} + I_{zf}
\end{aligned}
$$

Using the y plot and the ω_c plot, we can compute $I_{\omega z}$ as

$$
\begin{aligned}
I_{\omega z} &= \int \omega_c y t \, ds \\
&= -2\Bigg[t_w \left(\frac{1}{2} \; \frac{\bar{z}h}{2} \; \frac{h}{2} \right) \left(\frac{h}{3} \right) \\
&\quad + t_f \left(b \; \frac{\bar{z}h}{2} \right) \left(\frac{h}{2} \right) + t_f \left(\frac{1}{2} \; b \; \frac{bh}{2} \right) \left(\frac{h}{2} \right) \Bigg] \\
&= -\frac{\bar{z} t_w h^3}{12} - \frac{\bar{z} t_f b h^2}{2} - \frac{t_f b^2 b h^2}{4} \\
&= -\bar{z} I_z - \frac{b}{2} \, I_{zf}
\end{aligned}
$$

The location of the principal pole is then given as

$$
\begin{aligned}
a_z &= -\frac{I_{\omega z}}{I_z} \\
&= \bar{z} + \frac{b}{2} \; \frac{I_{zf}}{I_z} \\
&= \bar{z} + e_z
\end{aligned}
$$

where

$$
\begin{aligned}
e_z &= \frac{b}{2}\,\frac{I_{zf}}{I_z} \\
&= \frac{t_f b^2 h^2}{4I_z}
\end{aligned}
$$

The sectorial centroid relative to the principal pole A lies at the intersection of the z axis and the web of the channel section. The sectorial moment of inertia $I_\omega = I_{\omega_A}$ is computed as

$$
\begin{aligned}
I_\omega &= \int \omega_A^2 t\, ds \\
&= 2\left[t_w\left(\frac{1}{2}\,\frac{e_z h}{2}\,\frac{h}{2}\right)\left(\frac{e_z h}{3}\right)\right. \\
&\quad \left. + t_f\left(\frac{1}{2}\,\frac{e_z h}{2}\,e_z\right)\left(\frac{e_z h}{3}\right) + t_f\left(\frac{1}{2}\,\frac{(b-e_z)h}{2}\,(b-e_z)\right)\left(\frac{(b-e_z)h}{3}\right)\right] \\
&= \frac{e_z^2 t_w h^3}{12} + \frac{1}{6}\,(b-3e_z)t_f b^2 h^2 + e_z^2\,\frac{t_f b h^2}{2} \\
&= \frac{1}{3}\,b^2 I_{zf} - \frac{b^2 I_{zf}}{4I_z} \\
&= \frac{t_f b^3 h^2}{12}\,\frac{2t_w h + 3t_f b}{t_w h + 6t_f b}
\end{aligned}
$$

Equation (3.173) locates the sectorial centroid in terms of the sectorial area ω_0; thus for unsymmetric sections the actual y_0, z_0 coordinates are difficult to compute. There are then two choices of action. The first is that after ω_0 is computed using an arbitrary origin, ω for each part of a cross section can be computed using the same arbitrary origin and can then be recomputed using Eq. (3.172) rewritten in the following form:

$$
\overline{\omega} = \omega - \omega_0 \tag{3.175}
$$

where we have written $\overline{\omega} = \omega_{AM_0S}$. The advantage of this method is that by using all principal sectorial properties, Eq. (3.165) becomes completely uncoupled. The disadvantage is that it involves additional recalculations.

The second course of action is to use the principal pole—always of an arbitrary sectorial origin, which in this case is usually measured from a free edge of the cross section. The disadvantage of this method is that not all of the terms of Eq. (3.165) become uncoupled.

In the following text, we will derive expressions for the longitudinal stress using both courses of action.

If we use both the principal sectorial centroid and the principal pole, then Eq. (3.165) takes the form

$$\begin{aligned} N + N_T &= AEu_0' \\ -M_z - M_{Tz} &= EI_z v'' + EI_{yz} w'' \\ -M_y - M_{Ty} &= EI_{yz} v'' + EI_y w'' \\ -M_\omega - M_{T\omega} &= EI_\omega \phi'' \end{aligned} \tag{3.176}$$

Solving Eq. (3.176), we get

$$u_0' = \frac{N + N_T}{AE} \tag{3.177a}$$

$$v'' = -\frac{(M_z + M_{Tz})I_y - (M_y + M_{Ty})I_{yz}}{E(I_y I_z - I_{yz}^2)} \tag{3.177b}$$

$$w'' = -\frac{(M_y + M_{Ty})I_z - (M_z + M_{Tz})I_{yz}}{E(I_y I_z - I_{yz}^2)} \tag{3.177c}$$

$$\phi'' = -\frac{M_\omega + M_{T\omega}}{EI_\omega} \tag{3.177d}$$

Substituting Eq. (3.177) into Eq. (3.157), the general expression for the longitudinal stress σ_x is obtained as

$$\begin{aligned} \sigma_x = -\alpha E \Delta T &+ \frac{N + N_T}{A} \\ &+ \frac{(M_z + M_{Tz})I_y - (M_y + M_{Ty})I_{yz}}{I_y I_z - I_{yz}^2} y \\ &+ \frac{(M_y + M_{Ty})I_z - (M_z + M_{Tz})I_{yz}}{I_y I_z - I_{yz}^2} z \\ &+ \frac{M_\omega + M_{T\omega}}{I_\omega} \omega \end{aligned} \tag{3.178}$$

If the principal axes are used, $I_{yz} = 0$, and Eq. (3.178) takes the form

$$\sigma_x = -\alpha E \Delta T + \frac{N + N_T}{A} + \frac{M_z + M_{Tz}}{I_z} y + \frac{M_y + M_{Ty}}{I_y} z + \frac{M_\omega + M_{T\omega}}{I_\omega} \omega \tag{3.179}$$

The new term

$$\sigma_W = \frac{M_\omega + M_{T\omega}}{I_\omega}$$

represents the warping stress in the beam cross section. It is important to remember that Eqs. (3.177)–(3.178) contain the still-unsolved bimoment for M_ω. The evaluation of this unknown requires the solution of a differential equation to be derived later.

If we do not locate the sectorial centroid, Eq. (3.165) takes the form

$$N + N_T = AEu_0' - ES_\omega \phi'' \tag{3.180a}$$

$$-M_z - M_{Tz} = EI_z v'' + EI_{yz} w'' \tag{3.180b}$$

$$-M_y - M_{Ty} = EI_{yz} v'' + EI_y w'' \tag{3.180c}$$

$$-M_\omega - M_{T\omega} = -EI_\omega u_0' + EI_\omega \phi'' \tag{3.180d}$$

Let us take the Eqs. (3.180a) and (3.180d) and rewrite them in their basic integral form as

$$\int \sigma \, dA = Eu_0' \int dA - E\phi'' \int \omega \, dA - \int \alpha E \Delta T \, dA \tag{3.181a}$$

$$\int \sigma \omega \, dA = -Eu_0' \int \omega \, dA + E\phi'' \int \omega^2 \, dA + \int \alpha E \omega \Delta T \, dA \tag{3.181b}$$

If we multiply Eq. (3.181a) by ω_0 and add Eq. (3.181b), we obtain the single expression

$$\int \sigma(\omega_0 - \omega) \, dA = E\phi'' \int (\omega^2 - \omega_0 \omega) \, dA - \int \alpha(\omega_0 - \omega) E \Delta T \, dA \tag{3.182}$$

where we have used the identity given by Eq. (3.174a) to eliminate the integral term containing u_0'.

Using Eq. (3.174b), we can write

$$\int \sigma(\omega_0 - \omega) \, dA = E\phi'' \int (\omega^2 - \omega_0 \omega) \, dA + E\phi'' \int (\omega_0^2 - \omega_0 \omega) \, dA - \int \alpha(\omega_0 - \omega) E \Delta T \, dA \tag{3.183}$$

The second and third terms in the above expression complete the square; hence, we may write

$$\int \sigma(\omega_0 - \omega)\, dA = E\phi'' \int (\omega_0 - \omega)^2\, dA - \int \alpha(\omega_0 - \omega)E\Delta T\, dA \tag{3.184}$$

From Eq. (3.184), we define the following three expressions:

$$\begin{aligned}
M_{\omega 0} &= \int \sigma(\omega_0 - \omega)\, dA \\
I_{\omega 0} &= \int (\omega_0 - \omega)^2\, dA \\
&= \int \omega^2\, dA - \omega_0^2 A \\
M_{T\omega 0} &= \int \alpha(\omega_0 - \omega)E\Delta T\, dA
\end{aligned} \tag{3.185}$$

where

$M_{\omega 0}$ is the bimoment using an arbitrary point for the sectorial origin

$I_{\omega 0}$ is the corresponding sectorial principal moment of inertia

$M_{T\omega 0}$ is the corresponding thermal moment

Equation (3.180) now becomes:

$$\begin{aligned}
N + N_T &= AEu_0' - AE\omega_0\phi'' \\
-M_z - M_{Tz} &= EI_z v'' + EI_{yz} w'' \\
-M_y - M_{Ty} &= EI_{yz} v'' + EI_y w'' \\
-M_{\omega 0} - M_{T\omega 0} &= EI_{\omega 0}\phi''
\end{aligned} \tag{3.186}$$

where we have used Eq. (3.173) to eliminate S_ω in the first equation.
Solving these equations, we obtain the result

$$u_0' = \frac{N + N_T}{AE} + \omega_0 \phi''$$

$$v'' = -\frac{(M_z + M_{Tz})I_y - (M_y + M_{Ty})I_{yz}}{E(I_y I_z - I_{yz}^2)}$$

$$w'' = -\frac{(M_y + M_{Ty})I_z - (M_z + M_{Tz})I_{yz}}{E(I_y I_z - I_{yz}^2)}$$

$$\phi'' = -\frac{M_{\omega 0} + M_{T\omega 0}}{EI_{\omega 0}} \tag{3.187}$$

Substituting Eq. (3.187) into Eq. (3.157), the general expression for the longitudinal stress σ_x is obtained as

$$\begin{aligned}\sigma_x = -\alpha E \Delta T &+ \frac{N + N_T}{A} \\ &+ \frac{(M_z + M_{Tz})I_y - (M_y + M_{Ty})I_{yz}}{I_y I_z - I_{yz}^2} y \\ &+ \frac{(M_y + M_{Ty})I_z - (M_z + M_{Tz})I_{yz}}{I_y I_z - I_{yz}^2} z \\ &+ \frac{M_{\omega 0} + M_{T\omega 0}}{I_{\omega 0}} (\omega_0 - \omega)\end{aligned} \tag{3.188}$$

If the principal axes are used, $I_{yz} = 0$, and Eq. (3.188) takes the form

$$\begin{aligned}\sigma_x = -\alpha E \Delta T + \frac{N + N_T}{A} + \frac{M_z + M_{Tz}}{I_z} y + \frac{M_y + M_{Ty}}{I_y} z \\ + \frac{M_{\omega 0} + M_{T\omega 0}}{I_{\omega 0}} (\omega_0 - \omega)\end{aligned} \tag{3.189}$$

From Eqs. (3.185) and (3.175), we observe

$$M_{\omega 0} = -M_\omega$$
$$I_{\omega 0} = I_\omega$$
$$M_{T\omega 0} = -M_{T\omega}$$

Thus, Eq. (3.188) is identical to Eq. (3.178). The nomenclature I_ω, $I_{\omega 0}$, M_ω, $M_{\omega 0}$, and so on is used only to call attention to where the sectorial origin is located.

We now use the principle of virtual work to obtain the necessary differential

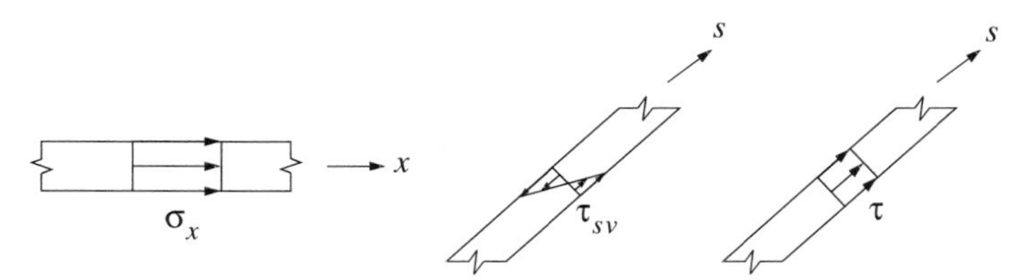

Figure 3.37 Stresses on a section.

equations to solve for ϕ'', but before doing so, we need to examine the complete state of stress on a typical section of the thin walled beam.

Figure 3.37 shows the distribution of stress on a section of a thin walled beam. The figure shows a direct stress σ_x, which we have already discussed; a shear stress $\tau_{xs} = \tau$ from transverse loading (and warping), which, because of the assumption of plane sections inherent in the thin walled beam theory, does not possess a conjugate variable; and a shear stress $\tau_{xs} = \tau_{SV}$ from torsion. Because of the thinness of the beam wall, τ_{xs} and its two components τ and τ_{SV} are assumed to be everywhere parallel to the wall, with the distributions shown in Fig. 3.37.

Figure 3.38 shows point P located on the wall of a thin walled beam in the neighborhood of point M located on the centerline. Using Eq. (3.148), we can determine the normal and tangential displacements of P and their derivatives as

$$
\begin{aligned}
\zeta_P &= (r_n - s)\phi \\
\eta_P &= (h + n)\phi \\
\zeta'_P &= (r_n - s)\phi' \\
\eta'_P &= (h + n)\phi'
\end{aligned}
\tag{3.190}
$$

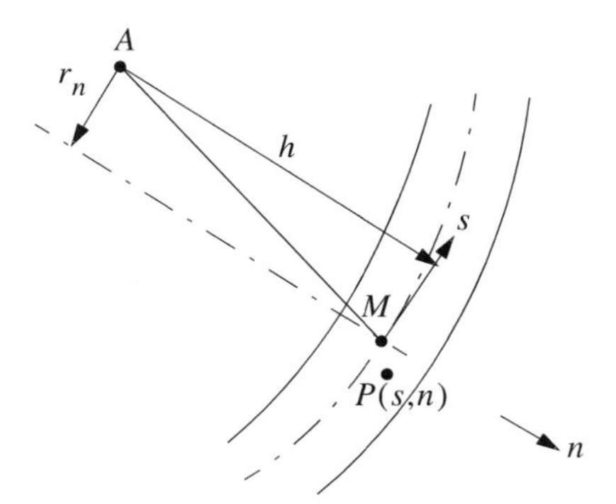

Figure 3.38 Geometry of a non-centerline point P.

By the assumption that the torsional shears are everywhere parallel to the side walls of the section, $\tau_{xn} = 0$, and we may write

$$\begin{aligned}\gamma_{xn} &= \frac{\partial u}{\partial n} + \frac{\partial \zeta_P}{\partial x} \\ &= \frac{\partial u}{\partial n} + r_n\phi' - s\phi' \\ &= 0\end{aligned} \tag{3.191}$$

Holding x and s fixed, we may integrate the above equation to obtain

$$u = -nr_n\phi' + ns\phi' + f(x, s) \tag{3.192}$$

At point M, n is equal to zero, and in the absence of all external loading (except torsion), $u = u_0 - \omega\phi'$ by virtue of Eq. (3.153). Then, Eq. (3.192) becomes

$$u = -nr_n\phi' + ns\phi' + u_0 - \omega\phi' \tag{3.193}$$

(The term $-nr_n\phi'$ represents a secondary bending of the wall of the beam section with a bending axis along the wall's centerline. This term will be used later to establish a correction term to I_ω.)

Noting that

$$\frac{\partial u}{\partial s} = -n\frac{\partial r_n}{\partial s}\phi' + n\phi' - h\phi'$$

where we have used $\partial\omega/\partial s = h$ from Eq. (3.154); thus, the shear strain γ_{xs} can be written as

$$\begin{aligned}\gamma_{xs} &= \frac{\partial u}{\partial s} + \frac{\partial \eta_P}{\partial x} \\ &= 2n\phi' - n\frac{\partial r_n}{\partial s}\phi' \\ &= 2n\phi'\end{aligned} \tag{3.194}$$

where we have used the assumption $(\partial r_n/\partial s)\phi' \ll 1$.

Since γ_{xs} is zero at the section median line, this definition of shear strain does not violate the fundamental assumption that $\hat{\gamma}_{xs} = 0$. With this definition for γ_{xs}, we can define the torsion shear stress as

$$\tau_{SV} = G\gamma_{xs} \tag{3.195}$$

where G is Young's modulus in torsion.

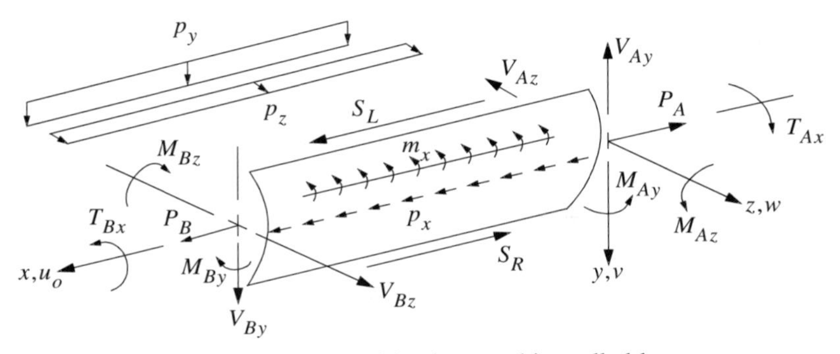

Figure 3.39 External loads on a thin walled beam.

Using Eq. (3.6) and the set of general applied loads shown in Fig. 3.39, we obtain the following expression for virtual work

$$
\begin{aligned}
&\int (\tau_{SV}\overline{\gamma}_{xs} + \sigma_x \overline{\epsilon}_x)\, dA \;\, dx \\
&\quad = \int (p_x \overline{u} + p_y \overline{v} + p_z \overline{w} + m_x \overline{\phi})\, dx - P_A \overline{u}(A) + P_B \overline{u}(B) \\
&\quad - V_{Ay}\overline{v}(A) + V_{By}\overline{v}(B) - V_{Az}\overline{w}(A) + V_{Bz}\overline{w}(B) \\
&\quad + M_{Az}\overline{v}'(A) - M_{Bz}\overline{v}'(B) - M_{Ay}\overline{w}'(A) + M_{By}\overline{w}'(B) \\
&\quad - T_{Ax}\overline{\phi}(A) + T_{Bx}\overline{\phi}(B)
\end{aligned}
\tag{3.196}
$$

where $\overline{u}$ is defined by Eq. (3.153), and p_x has the surface tractions S_L and S_R lumped so that

$$p_x = p_x + S_L - S_R \tag{3.197}$$

and m_x consists of a distributed moment m_x per unit length plus the moment from the transverse applied distributed loads p_y and p_z. Hence, m_x is the lumped moment in which

$$m_x = m_x + y p_z - z p_y \tag{3.198}$$

Substituting Eq. (3.195) for τ_{SV} and Eq. (3.157) for σ_x, the above equation becomes

$$
\begin{aligned}
&\int [E(u_0' - yv'' - zw'' - \omega\phi'') - \alpha E \Delta T] \\
&\qquad \cdot [\overline{u}_0' - y\overline{v}'' - z\overline{w}'' - \omega\overline{\phi}''] \, dA \; dx \\
&\qquad + \int G(2n\phi')(2n\overline{\phi}') \, dn \; ds \; dx \\
&\qquad = \int [p_x(\overline{u}_0 - y\overline{v}' - z\overline{w}' - \omega\overline{\phi}') + p_y\overline{v} + p_z\overline{w} + m_x\overline{\phi}] \, dx \\
&\qquad - P_A(\overline{u}_0 - y\overline{v}' - z\overline{w}' - \omega\overline{\phi}')(A) \\
&\qquad + P_B(\overline{u}_0 - y\overline{v}' - z\overline{w}' - \omega\overline{\phi}')(B) \\
&\qquad - V_{Ay}\overline{v}(A) + V_{By}\overline{v}(B) - V_{Az}\overline{w}(A) + V_{Bz}\overline{w}(B) \\
&\qquad + M_{Az}\overline{v}'(A) - M_{Bz}\overline{v}'(B) - M_{Ay}\overline{w}'(A) + M_{By}\overline{w}'(B) \\
&\qquad - T_{Ax}\overline{\phi}(A) + T_{Bx}\overline{\phi}(B)
\end{aligned}
\tag{3.199}
$$

Before evaluating this equation, we define the following integral,

$$
\begin{aligned}
J &= \int_0^s \left(\int_{-t/2}^{+t/2} 4n^2 \, dn \right) ds \\
&= \int_0^s \frac{t^3}{3} \, ds
\end{aligned}
\tag{3.200}
$$

as the torsion constant for a thin walled beam of open section. For sections made up of straight constant-thickness plate elements, J can be computed as

$$
J = \frac{1}{3} \sum_{i=1}^{n} b_i t_i^3 \tag{3.201}
$$

Equation (3.200) neglects the effects of the ends of the cross section as well as the effect of plate intersection points. Beam manufacturers provide correction terms in tabulated form for such shapes as rolled, wide flange beams and extruded sections.

We may now integrate Eq. (3.199) by parts and collect the coefficients of the virtual quantities to define four differential equations and their boundary conditions. We assume, however, that the *principal pole is always used*. Therefore, the differential equations for bending in the xy plane and bending in the xz plane, as well as their respective boundary conditions, reduce to Eqs. (3.70)–(3.73).

We point out that the computation of the principal centroid axes ($I_{yz} = 0$) is a relatively simple task; hence, it would make little sense to solve these equations in such a general form. Additionally, we should realize that in the derivation of the thin walled beam of open section theory, the section properties such as $\omega(s)$ or $t(s)$ were taken as functions of s only. Thus terms such as $d^2/dx^2(EI_z\, d^2v/dx^2)$ should be written as $EI_z\, d^4v/dx^4 = EI_z v^{iv}$.

If the sectorial centroid ($S_\omega = 0$) is used, the differential equation for the longitudinal displacement and its boundary conditions reduce to Eqs. (3.68) and (3.69). Also, for the case $S_\omega = 0$ we obtain a differential equation for the angle of rotation ϕ, which is

$$EI_\omega \phi^{iv} - JG\phi'' + M''_{T\omega} = m_x + \omega p'_x \tag{3.202}$$

with the following boundary conditions:

$$\begin{aligned}
\phi'(B) = 0 \quad &\text{or} \quad EI_\omega \phi''(B) + M_{T\omega}(B) = -\omega P_B && (3.203a)\\
\phi'(A) = 0 \quad &\text{or} \quad -EI_\omega \phi''(A) - M_{T\omega}(A) = \omega P_A && (3.203b)\\
\phi(B) = 0 \quad &\text{or} \quad -EI_\omega \phi'''(B) + JG\phi'(B) - M_{T\omega}(B) = T_{Bx} - \omega p_x(B) && (3.203c)\\
\phi(A) = 0 \quad &\text{or} \quad EI_\omega \phi'''(A) - JG\phi'(A) + M_{T\omega}(A) = -T_{Ax} + \omega p_x(A) && (3.203d)
\end{aligned}$$

When evaluating these equations, terms such as $\omega p'_x$ should take the following typical form when expanded:

$$\omega p' + \omega_L S'_L - \omega_R S'_R$$

This also holds true for similar terms, such as yp'_x, in Eqs. (3.72), (3.73), and so on.

If the sectorial centroid is not used, $S_\omega = \omega_0 A$, and we get the differential equation

$$EI_{\omega 0} \phi^{iv} - JG\phi'' - M''_{T\omega 0} = m_x - (\omega_0 - \omega) p'_x \tag{3.204}$$

with the following boundary conditions:

$$\begin{aligned}
\phi'(B) = 0 \quad &\text{or} \quad EI_{\omega 0} \phi''(B) - M_{T\omega 0}(B) = (\omega_0 - \omega) P_B\\
\phi'(A) = 0 \quad &\text{or} \quad -EI_{\omega 0} \phi''(A) + M_{T\omega 0}(A) = -(\omega_0 - \omega) P_A\\
\phi(B) = 0 \quad &\text{or} \quad -EI_{\omega 0} \phi'''(B) + JG\phi'(B) + M_{T\omega 0}(B) = T_{Bx} + (\omega_0 - \omega) p_x(B)\\
\phi(A) = 0 \quad &\text{or} \quad EI_{\omega 0} \phi'''(A) - JG\phi'(A) - M_{T\omega 0}(A) = -T_{Ax} - (\omega_0 - \omega) p_x(A)
\end{aligned} \tag{3.205}$$

The differential equation for u_0 is

$$AEu_0'' = AE\omega_0\phi''' + N_T' - p_x \tag{3.206}$$

with the following boundary conditions:

$$\begin{aligned} u_0(B) = 0 \quad &\text{or} \quad AEu_0'(B) = AE\omega_0\phi''(B) + N_T(B) + P_B \\ u_0(A) = 0 \quad &\text{or} \quad AEu_0'(A) = AE\omega_0\phi''(A) + N_T(A) + P_A \end{aligned} \tag{3.207}$$

If we divide either Eq. (3.202) or (3.204) by the coefficient of the ϕ^{iv}, it can be put into the following form:

$$\phi^{iv} - \frac{k^2}{\ell^2}\phi'' = f(x) \tag{3.208}$$

where

$$k^2 = \frac{GJ}{EI_\omega}\ell^2 \tag{3.209}$$

in which ℓ is the length of the beam. We can assume a homogeneous solution of the form

$$\phi = e^{\lambda x}$$

yielding the general solution

$$\phi(x) = C_1 + C_2 x + C_3 \sinh\frac{k}{\ell}x + C_4 \cosh\frac{k}{\ell}x + \phi_0(x) \tag{3.210}$$

where $\phi_0(x)$ is the particular solution.

The torsional-boundary constraints are determined in general from Eq. (3.203); however, the common ones are the following:

1. For a clamped boundary, $\phi = 0$ and $\phi' = 0$.
2. For a simply supported or hinged boundary, no twist is allowed; hence, the boundary condition is $\phi = 0$ and $\phi'' = 0$ (the latter is equivalent to $M_\omega = 0$).
3. For a free end, no loads are allowed to act; hence, $\phi'' = 0$ and $\phi''' = (k^2/\ell^2)\phi'$ (both are equivalent to $M_\omega = 0$ and $T_x = 0$).

The torsional rigidity GJ is proportional to the cube of the thickness; hence, when

$$\frac{t}{d} \le 0.02$$

where d is a characteristic dimension of the cross section, then the differential equation for torsion can without appreciable error be written as

$$EI_{\omega}\phi^{iv} = m_x \tag{3.211}$$

with the appropriate boundary-condition equations adjusted accordingly.

The bimoment is a statically zero (or self-equilibrating) system. To see such a system, consider, for example, Eq. (3.188), from which the stress from the bimoment is

$$\sigma_{W_0} = \frac{M_{\omega 0}}{I_{\omega 0}} (\omega_0 - \omega)$$

In the absence of thermal loading, Eq. (3.187) becomes

$$\phi'' = -\frac{M_{\omega 0}}{EI_{\omega}}$$

or

$$\sigma_{W_0} = -E\phi''(\omega_0 - \omega)$$

We can compute the total contribution of force over a cross section produced by σ_{W_0} as

$$\begin{aligned} \int \sigma_{W_0}\, dA &= -E\phi'' \int (\omega_0 - \omega)\, dA \\ &= E\phi'' \int \overline{\omega}\, dA \\ &= 0 \end{aligned}$$

There are cases when the bimoment can be determined by inspection. Consider, for example, Fig. 3.40(a), which shows an idealized channel section of length ℓ where all the longitudinal stress–carrying capability is in the four flange stringers of equal area, and the connecting web pannels of thickness t carry only the shear stress. An axial tensile load is applied to the flange area A_3 as shown. Figures 3.40(b)–(e) show the applied load distributed to all of the flange areas. The load is distributed as a uniform axial load N, a bending moment M_z, a bending moment M_y, and a bimoment M_{ω}. From Fig. 3.40(e), we see that the bimoment can be considered the product of two equal but opposite moments multiplied by the distance that separates their planes of action. We obtain from Fig. 3.40 the following equation:

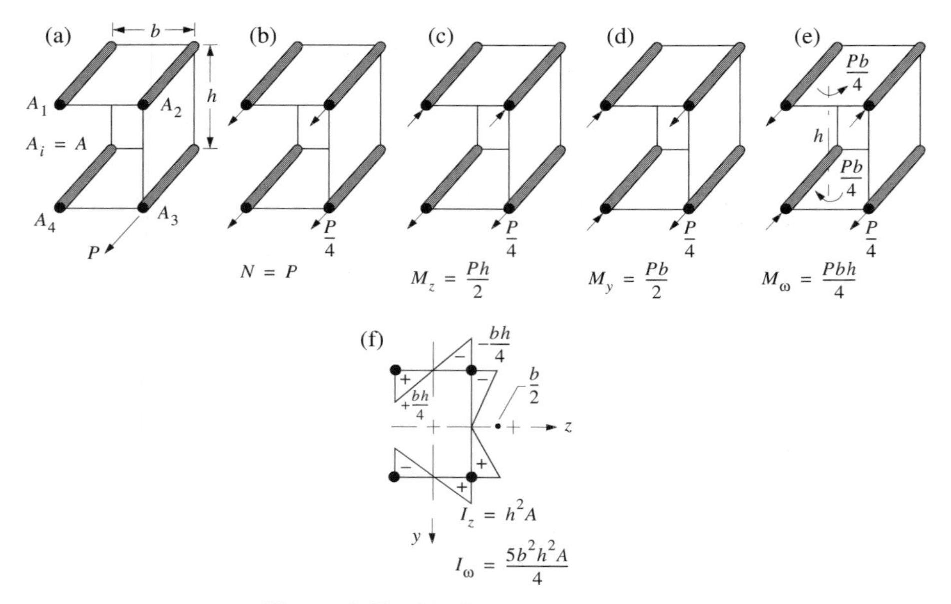

Figure 3.40 Idealized channel section.

$$M_\omega = \frac{Pb}{4}\, h$$

Figure 3.40(f) represents the principal sectorial area of the section using the principal pole located $b/2$ to the right of the vertical web and the sectorial centroid located at the intersection of the z axis and the vertical web. Direct use of Eq. (3.163) gives the bioment as

$$M_\omega = P\omega_{PM_0S}$$

$$= \frac{Pbh}{4}$$

The rotation can be obtained from Eq. (3.210). Looking first at Eq. (3.202), we observe that the right side is zero. Hence Eq. (3.202) becomes

$$EI_\omega\phi''' - JG\phi' = 0$$

where we have used the forced boundary condition Eq. (3.203c). The C_2 term and the particular solution term in Eq. (3.210) are therefore zero.

Using the following valid boundary conditions:

$$EI_\omega \phi''(\ell) = -\omega P$$
$$\phi'(0) = 0$$
$$\phi(0) = 0$$

we can determine the remaining three constants and obtain the solution for the angle of twist as

$$\phi = \frac{k^2}{\ell^2} \frac{\omega P}{EI_\omega} \frac{1}{\cosh k} \left(1 - \cosh \frac{k}{\ell} x \right)$$

We should observe that the boundary condition for $\phi''(\ell)$ is identical to Eq. (3.177d) in the absence of thermal effects, and it is again equal to $Pbh/4$.

Reference [35] shows that the bimoment for any grouping of discrete axial loads acting over the contour (idealized or nonidealized) of a thin walled open section is given by

$$M_\omega = \sum P_k \omega_k \tag{3.212}$$

where ω_k is the sectorial area computed at the location of the load P_k. Reference [35] also shows that the bimoment of a bending moment acting in a plane parallel to the longitudinal axis of the beam (and normal to either the z or y axis) is given by

$$M_\omega = eM \tag{3.213}$$

where e is the distance from the shear center to the plane of the applied bending moment M.

The implications of Eq. (3.212) are important. While virtually all textbooks on the strength of materials discuss the twisting of beams caused by the application of transverse loading not through the shear center, virtually all fail to mention that a direct load applied parallel to the longitudinal axis of the beam can also cause twisting. This twisting can have serious implications in the resulting stress distribution and twisting behavior of thin walled beams if end-fasteners are improperly placed.

3.18 SHEAR IN OPEN SECTION BEAMS

We will consider the shear stress τ on a thin walled open section to be composed of three components: the shear τ_{SV} from the unrestrained twisting of the section [see Fig. 3.37(b)], the section shear τ_{ave} from the loading under the assumption of plane sections, and the shear τ_W from the restraint of warping. The last two are depicted in Fig. 3.37(c) as a uniform shear parallel to the walls of the section.

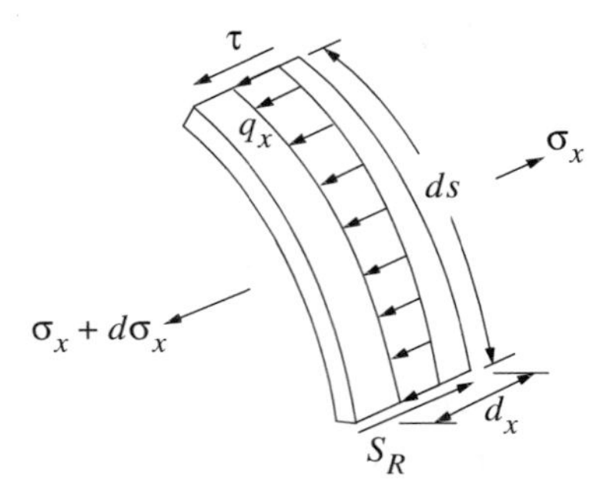

Figure 3.41 Equilibrium of a slice *dx ds*.

The moment T_{SV} of the shear τ_{SV} about the principal pole located at A is given as

$$T_{SV} = \int_0^s [\tau_{SV}(h+n) - \tau_{SV}(h-n)]\, dn\ ds \tag{3.214a}$$

$$= \int_0^s \tau_{SV}(2n)\, dn\ ds \tag{3.214b}$$

$$= G\phi' \int_0^s 4n^2\, dn\ ds \tag{3.214c}$$

$$= GJ\phi' \tag{3.214d}$$

where we have used Eqs. (3.194), (3.195), and (3.200). Equation (3.161) showed that in the case of Saint Venant's torsion, ϕ' is constant.

Figure 3.41 represents a slice of the cross section along the dimension $dx \times ds$. The equation of equilibrium for the slice is

$$\tau t\ dx + \int_{A(s)} d\sigma_x t\ ds + \int_{A(s)} q_x\ dx\ ds - S_R\ dx = 0$$

where

τ is the shear stress on the area $t\ dx$

$\int_{A(s)} d\sigma_x t\ ds$ is the increment in longitudinal force over the area $t\ ds$

$S_R\ dx$ is the force from an applied surface traction at the right end of the section wall

q_x is the portion of the longitudinal distributed load acting over the surface of the segment $dx\ ds$

In essence, $p_x = \int_0^s q_x(x, s)\, ds$. Solving for τ we get

$$\tau = \frac{S_R}{t} - \frac{1}{t} \int_{A(s)} \sigma_x' \, dA - \frac{1}{t} \int_{A(s)} q_x(x, s)\, ds \tag{3.215}$$

Next, substitute $\sigma_x' = d\sigma_x/dx$ using Eq. (3.157) into the expression for τ to get

$$\tau = \frac{S_R}{t} - \frac{E}{t} \int_{A(s)} (u_0'' - yv''' - zw''' - \omega\phi''' - \alpha\Delta T')\, dA - \frac{1}{t} \int_{A(s)} q_x(x, s)\, ds \tag{3.216}$$

Since the bimoment represents a zero-force system, it cannot contribute to the equilibrium Eq. (3.82). Thus these equations remain valid for a thin walled beam of open section. We can, therefore, define τ as

$$\tau = \tau_{\text{ave}} + \tau_W$$

where by using Eq. (3.83), τ_{ave} is given as

$$\begin{aligned}
\tau_{\text{ave}} = {} & \frac{S_R}{t} - \frac{A(s)}{t} \frac{-p_x + N_T'}{A} - \frac{Q_z I_y - Q_y I_{yz}}{t(I_y I_z - I_{yz}^2)} (V_y + M_{Tz}' - yp_x) \\
& - \frac{Q_y I_z - Q_z I_{yz}}{t(I_y I_z - I_{yz}^2)} (V_z + M_{Ty}' - zp_x) \\
& + \frac{1}{t} \int_{A(s)} E\alpha\Delta T' \, dA - \frac{1}{t} \int_{A(s)} q_x(x, s)\, ds
\end{aligned} \tag{3.217}$$

where V_y and V_z are defined for transverse loading as

$$V_h = \int \tau t \, dy$$

$$V_z = \int \tau t \, dz$$

and τ_W is given by

$$\tau_W = \frac{E}{t} \int_A \omega\phi''' \, dA \tag{3.218a}$$

$$= \frac{E}{t} S_\omega(s)\phi''' \tag{3.218b}$$

$$= -\frac{S_\omega(s)}{tI_\omega} (M'_\omega + M'_{T\omega}) \tag{3.218c}$$

where we have used the derivative of Eq. (3.177d) to get the last form of the above equation.

Consider the moment T_ω about pole A of the shear τ when the section is subject to torsion only without any temperature effects; then consider $\tau = \tau_W$. We can then write

$$T_\omega = \int_A \tau_W \, ht \, ds$$

$$= E\phi''' \int_A S_\omega(s) \, d\overline{\omega} \tag{3.219}$$

where we have used the first term of Eq. (3.218) and the fact that, by definition, $h\,ds = d\overline{\omega}$. The bar over the ω is to emphasize that we are using the principal sectorial centroid as well as the principal pole in the current derivation. Taking note of the fact that $dS_\omega(s) = \overline{\omega}\,dA$, we can integrate by parts as follows:

$$\int_A S_\omega \, d\overline{\omega} = S_{\overline{\omega}}\overline{\omega} - \int_A dS_\omega(s)\overline{\omega} = -I_\omega$$

Then, T_ω becomes the following equation after using the derivative of Eq. (3.177d) in the absence of temperature.

$$T_\omega = -EI_\omega\phi'''$$

$$= \frac{dM_\omega}{dx} \tag{3.220}$$

If we now use the alternate notation,

$$Q_\omega = S_\omega(s)$$
$$V_\omega = T_\omega \tag{3.221}$$

then Eq. (3.218c) takes a form similar to Eq. (3.217)—namely,

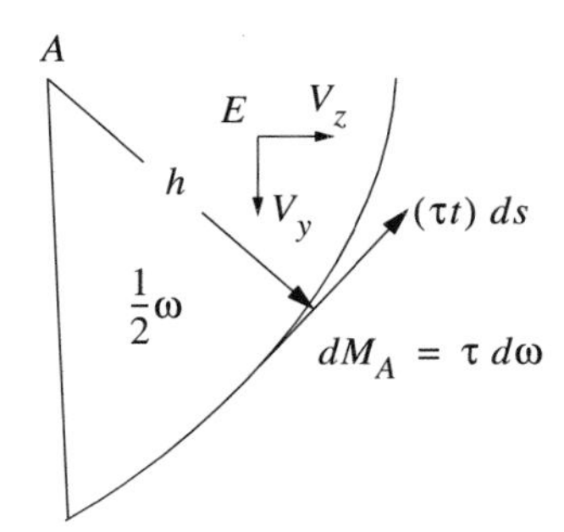

Figure 3.42 Determination of shear center.

$$\tau_W = -\frac{Q_\omega}{tI_\omega}\,(V_\omega + M'_{T\omega}) \tag{3.222}$$

The total longitudinal moment T_x can now be expressed as

$$\begin{aligned} T_x &= T_{SV} + T_\omega \\ &= GJ\phi' - EI_\omega\phi''' \end{aligned} \tag{3.223}$$

The derivative of the above equation is identical to the left side of Eq. (3.202).

In the case in which a sectorial origin other than the sectorial centroid is used, the development is parallel to the development of Eqs. (3.219)–(3.223), with the exception that

$$\tau_W = \frac{E}{t}\int_A [-\omega_0 A(s) + S_\omega(s)]\,dA$$

Next, consider the case in which only transverse loads act on a section of the beam. Let the stress resultants V_y and V_z be located at point $E(e_y, e_z)$, as shown in Fig. 3.42. The moments of the forces shown about pole A are

$$-V_y(e_z - a_z) + V_z(e_y - a_y) + \int_A (\tau t)\,d\omega = 0$$

From Eq. (3.83) the above takes the following form:

$$-V_y(e_z - a_z) + V_z(e_y - a_y) + \int_0^s \left[\frac{Q_z I_y - Q_y I_{yz}}{(I_y I_z - I_{yz}^2)}\,V_y + \frac{Q_y I_z - Q_z I_{yz}}{(I_y I_z - I_{yz}^2)}\,V_z\right] d\omega = 0$$

Also, note that integration by parts gives

$$\int Q_y \, d\omega = -I_{\omega y}$$

$$\int Q_z \, d\omega = -I_{\omega z}$$

Then, by collecting the coefficients of V_y and V_z, we get the following two relationships:

$$e_y - a_y = \frac{I_z I_{\omega Ay} - I_{yz} I_{\omega Az}}{I_y I_z - I_{yz}^2}$$

$$e_z - a_z = -\frac{I_z I_{\omega Az} - I_{yz} I_{\omega Ay}}{I_y I_z - I_{yz}^2}$$

which from Eq. (3.169) defines point E as the principal pole. To determine the location of the shear center from equilibrium considerations, we consider only transverse loading. Hence the shear center or principal sectorial pole is often called the center of flexure.

Considering next Eqs. (3.165b, c) evaluated at pole A and under the conditions of no transverse loading or thermal effects, we get

$$0 = EI_z v'' + EI_{yz} w'' + EI_{\omega Az} \phi''$$
$$0 = EI_y v'' + EI_y w'' + EI_{\omega Ay} \phi''$$

that when solved for v'' and w'' yield

$$w'' = -\frac{I_z I_{\omega Ay} - I_{yz} I_{\omega Az}}{I_y I_z - I_{yz}^2} \phi''$$

$$v'' = -\frac{I_y I_{\omega Az} - I_{yz} I_{\omega Ay}}{I_y I_z - I_{yz}^2} \phi''$$

The right side of the above equation locates the shear center. Thus, if a beam undergoes pure twisting about the shear center in the absence of transverse loads, the beam will also undergo flexure. We can write this equation as

$$v'' = (e_z - a_z)\phi''$$
$$w'' = -(e_y - a_y)\phi''$$

or

$$(e_z - a_z) = \frac{v''}{\phi''}$$

$$(e_y - a_y) = -\frac{w''}{\phi''} \tag{3.224}$$

Equation (3.224) was obtained under the assumption of torsion with no transverse loads, and for this case, the left side locates the shear center. Notice that Eq. (3.224) for the case of transverse load but without twist would locate the points $(e_y - a_y)$ and $(e_z - a_z)$ at infinity. Thus only for the special case considered do the coordinates located by Eq. (3.224) fall at the shear center. Equation (3.224) locates a point called the center of warping. Vlasov shows that under general loading conditions of simultaneously applied axial load, transverse load, and torsion load, Eq. (3.155) takes the following form:

$$\epsilon_x = \omega_W \phi''$$

The pole for ω_W is located by Eq. (3.224).

For our special loading condition, however, Eq. (3.224) does locate the shear center. Next, consider Eq. (3.143) for the center of twist. If the center of twist lies at the shear center, then for our present loading assumption,

$$\frac{v}{\phi} = \frac{v''}{\phi''}$$

$$\frac{w}{\phi} = \frac{w''}{\phi''}$$

This can happen only for the special case of a fixed end condition or a sinusoidal deflection. Thus, in general the center of twist does not lie at the shear center.

As mentioned in the derivation of Eq. (3.193), the term $-nr_n\phi'$ represents a secondary bending displacement. If we define it as

$$u^s = -nr_n\phi'$$

a secondary stress

$$\sigma_x^s = E\,\frac{\partial u^s}{\partial x}$$

$$= -Enr_n\phi''$$

can be determined. Considering only the direct stress σ_x^s and the shear stress τ_{xn} to be acting, the equation of equilibrium Eq. (2.21) becomes

$$\frac{\partial \sigma_x^s}{\partial x} + \frac{\partial \tau_{xn}}{\partial n} = 0$$

Integrating this equation and taking into consideration that the shear τ_{xn} must vanish at $n = \pm t/2$, we get the following expression:

$$\tau_{xn} = \frac{E r_n}{2} \left(n^2 - \frac{t^2}{4} \right) \phi'''$$

We can now define a stress resultant V^s per unit length s as

$$\begin{aligned} V^s &= \int_{-t/2}^{t/2} \tau_{xn} \, dn \\ &= -\frac{E t^3 r_n}{12} \phi''' \end{aligned}$$

We can now define a secondary twisting moment as

$$\begin{aligned} T_\omega^s &= \int_A V^s r_n \, ds \\ &= -E\phi''' \int_A \frac{t^3 r_n}{12} \, ds \\ &= -E I_\omega^s \phi''' \end{aligned} \tag{3.225}$$

where we have defined

$$I_\omega^s = \int_A \frac{t^3 r_n}{12} \, ds \tag{3.226}$$

The total longitudinal moment T_x can now be expressed as

$$\begin{aligned} T_x &= T_{SV} + T_\omega + T_\omega^s \\ &= GJ\phi' - E(I_\omega + I_\omega^s)\phi''' \end{aligned} \tag{3.227}$$

The effect of I_ω^s is usually important only in very thin cross sections consisting of a single bundle of thin rectangular plates with a common intersection, for these types of cross section do not in general exhibit primary warping.

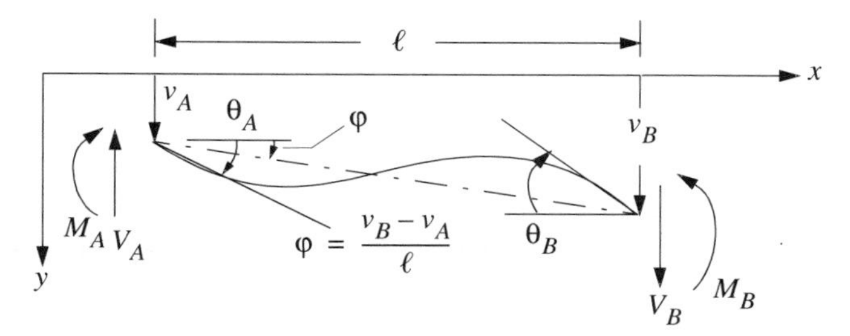

Figure 3.43 Beam displacement and joint rotations.

3.19 SLOPE-DEFLECTION EQUATIONS

We now briefly discuss the slope-deflection equations or the method of initial parameters and how these relationships along with the principle of virtual work will allow us to determine the stiffness matrix for a structural frame.

Figure 3.43 shows the displacement and joint rotations of a beam of length ℓ lying in the xy plane. Also, in the following we will assume principal centroid axes ($I_{yz} = 0$). The beam is loaded only by the stress resultants applied at the ends as shown. For the stated conditions, Eq. (3.70) becomes

$$EI_z v^{iv} = 0$$

with the following general solution:

$$v = c_0 + c_1 x + c_2 \frac{x^2}{2} + c_3 \frac{x^3}{6} \tag{3.228}$$

We can now determine the shape v as a function of the end displacements $v(0) = v_A$ and $v(\ell) = v_B$ and the end rotations $v'_A = \theta_A$ and $v'_B = \theta_B$. The result is

$$\begin{aligned} c_0 &= v_A \\ c_1 &= \theta_A \\ c_3 &= \frac{2}{\ell}\,(3\varphi - 2\theta_A - \theta_B) \\ c_4 &= \frac{6}{\ell^2}\,(\theta_A + \theta_B - 2\varphi) \end{aligned} \tag{3.229}$$

where, for convenience, we have defined

$$\varphi = \frac{v_B - v_A}{\ell} \tag{3.230}$$

The resulting solution for v is

$$v = v_A + \theta_A x - (2\theta_A + \theta_B - 3\varphi)\,\frac{x^2}{\ell} + (\theta_A + \theta_B - 2\varphi)\,\frac{x^3}{\ell^2} \tag{3.231}$$

which is similar to what we did in Eq. (3.38) when we determined the displacement $\tilde{u}$ of the rod as a function of the two unknown parameters u_1^e and u_2^e, which represent the axial displacements of node 1 and node 2, respectively. Thus we see the similarity between the finite element method, used in the example of the rod, and the current method. Indeed, what evolves out of the current method is the classic stiffness method, from which structural engineers in the mid-1950s created the current finite element method.

Determining the second and third derivatives of Eq. (3.231) at points A and B, we get

$$v''(A) = -\frac{2}{\ell}\,(2\theta_A + \theta_B - 3\varphi)$$

$$v''(B) = \frac{2}{\ell}\,(\theta_A + 2\theta_B - 3\varphi)$$

$$v'''(A) = \frac{6}{\ell^2}\,(\theta_A + \theta_B - 2\varphi)$$

$$v'''(B) = \frac{6}{\ell^2}\,(\theta_A + \theta_B - 2\varphi)$$

and substituting the results into Eq. (3.71) and adopting the new sign convention shown in Fig. 3.44, we get the following equations, which represent the stress resultants in terms of the unknown displacement and rotation parameters. [Again, note the similarity with Eq. (3.40).]

$$M_A = \frac{2EI}{\ell}\,(2\theta_A + \theta_B - 3\varphi)$$

$$M_B = \frac{2EI}{\ell}\,(\theta_A + 2\theta_B - 3\varphi)$$

$$V_A = \frac{6EI}{\ell^2}\,(\theta_A + \theta_B - 2\varphi)$$

$$V_B = \frac{6EI}{\ell^2}\,(\theta_A + \theta_B - 2\varphi) \tag{3.232}$$

Comparing Fig. 3.20 and Fig. 3.44, we see that M_B and V_A have changed signs. Thus, with this new-sign convention, all positive directions correspond to positive coordinate directions, which helps to automate calculations on a computer. Also, for convenience, we have dropped the z subscript.

If we now apply the unit-displacement method described in Section 3.8, we

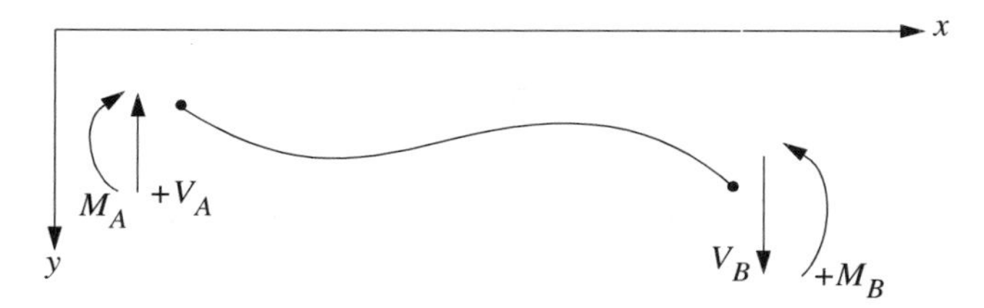

Figure 3.44 Sign convention for the slope-deflection method.

can determine the stiffness matrix of a beam in bending. For example, let $v_A = 1$ and $v_B = \theta_A = \theta_B = 0$. Then, $\varphi = -1/\ell$, and Eq. (3.232) becomes

$$M_A = \frac{6EI}{\ell^2}$$

$$M_B = \frac{6EI}{\ell^2}$$

$$V_A = \frac{12EI}{\ell^3}$$

$$V_B = -\frac{12EI}{\ell^3}$$

This result is shown in Fig. 3.45 with the correct signs applied. That these are the components of a stiffness matrix comes directly from Eq. (3.34), which, in our present case, takes the following form:

$$\begin{aligned}(\overline{u} = 1) \cdot k_{uv} &= \int_V \lfloor \sigma \rfloor_v \{\overline{\epsilon}\}_u \, dV \\ &= \int_A^B EIv''\overline{v}'' \, dx\end{aligned}$$

This last form is Eq. (3.67), and for our present case, the only loadings are the end stress resultants. Thus, for our present case Eq. (3.67) becomes

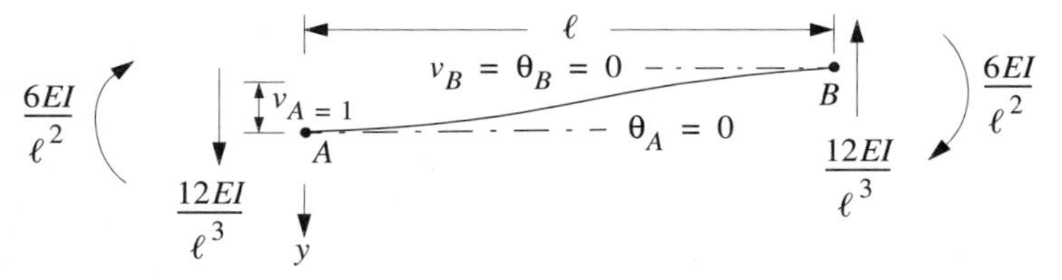

Figure 3.45 Stiffness components for $v_A = 1$ and $v_B = \theta_A = \theta_B = 0$.

$$[(EI_z v''(B) + M_{Bz}]\overline{v}'(B) + [-(EI_z v''(A) - M_{Az}]\overline{v}'(A) + [-(EI_z v'''(B) - V_{By}]\overline{v}(B) + [(EI_z v'''(A) + V_{Ay}]\overline{v}(A) = 0$$

or on substituting the second and third derivatives of v and switching to our new sign convention,

$$\begin{aligned} &\frac{2EI}{\ell}\left[\theta_A + 2\theta_B - \frac{3}{\ell}(v_B - v_A) - M_B\right]\overline{\theta}_B \\ &+ \frac{2EI}{\ell}\left[2\theta_A + \theta_B - \frac{3}{\ell}(v_B - v_A) - M_A\right]\overline{\theta}_A \\ &- \frac{6EI}{\ell^2}\left[\theta_A + \theta_B - \frac{2}{\ell}(v_B - v_A) - V_B\right]\overline{v}_B \\ &+ \frac{6EI}{\ell^2}\left[\theta_A + \theta_B - \frac{2}{\ell}(v_B - v_A) - V_A\right]\overline{v}_A = 0 \end{aligned} \tag{3.233}$$

Each square bracket represents a stress (resultant) as a function of the end-point generalized coordinate. Each generalized coordinate can be given a unit value while the others are set to zero. Then, by simply collectng the terms we obtain the stiffness term k_{uv} of the assembled element stiffness matrix.

$$\frac{EI}{\ell^3}\begin{bmatrix} 12 & 6\ell & -12 & 6\ell \\ 6\ell & 4\ell^2 & -6\ell & 2\ell^2 \\ -12 & -6\ell & 12 & -6\ell \\ 6\ell & 2\ell^2 & -6\ell^2 & 4\ell^2 \end{bmatrix}\begin{Bmatrix} v_A \\ \theta_A \\ v_B \\ \theta_B \end{Bmatrix} = \begin{Bmatrix} V_A \\ M_A \\ V_B \\ M_B \end{Bmatrix} \tag{3.234}$$

Through the interpretation of k_{uv} as a work term, we can assemble a stiffness matrix for a complete structure simply by adding the individual work terms into the appropriate slot of the stiffness matrix.

Example 3.9 Figure 3.46 shows a frame structure built into the foundation at A and D. We will adopt the common assumption for frames (when buckling is of no concern) that displacement along the axis of a frame member may be neglected. Then, the generalized coordinates that characterize the motion of the frame are rotations u_1 and u_2 and the displacement u_3, as shown in Fig. 3.46. The following table correlates the frame coordinates to the coordinates given in Eq. (3.233):

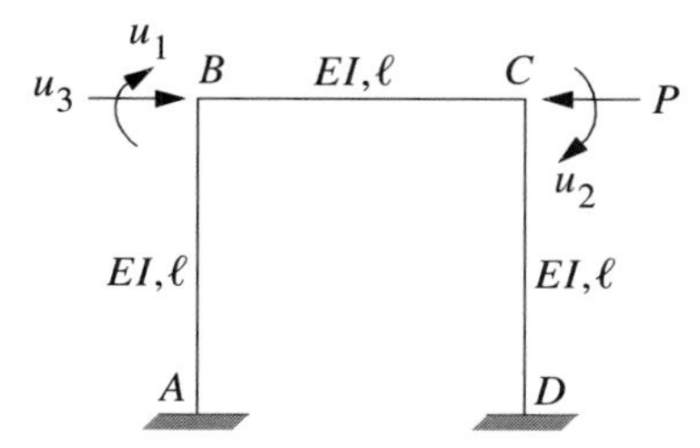

Figure 3.46 Simple plane frame.

	v_A	θ_A	v_B	θ_B
AB	0	0	u_3	u_1
BC	0	u_1	0	u_2
DC	0	0	u_3	u_2

and Eq. (3.233) takes the form for the complete structure:

$$\frac{2EI}{\ell}\left(4u_1 + u_2 - \frac{3}{\ell}\,u_3\right)\overline{u}_1 + \frac{2EI}{\ell}\left(u_1 + 4u_2 - \frac{3}{\ell}\,u_3\right)\overline{u}_2$$
$$-\frac{6EI}{\ell^2}\left(u_1 + u_2 - \frac{4}{\ell}\,\overline{u}_3\right)\overline{u}_3 = -P\overline{u}_3$$

where the last term comes from the virtual work of the applied load P. This expression results in

$$\frac{2EI}{\ell^3}\begin{bmatrix} 4\ell^2 & \ell^2 & -3\ell \\ \ell^2 & 4\ell^2 & -3\ell \\ -3\ell & -3\ell & 12 \end{bmatrix}\begin{Bmatrix} u_1 \\ u_2 \\ u_3 \end{Bmatrix} = \begin{Bmatrix} 0 \\ 0 \\ -P \end{Bmatrix}$$

with the following solutions:

$$u_1 = u_2 = -\frac{P\ell^2}{28EI}$$

and

$$u_3 = -\frac{5P\ell^3}{84EI}$$

Figure 3.47 shows the beam of Fig. 3.43 with a concentrated load P applied at the midpoint. For this case, we can write Eq. (3.70) as

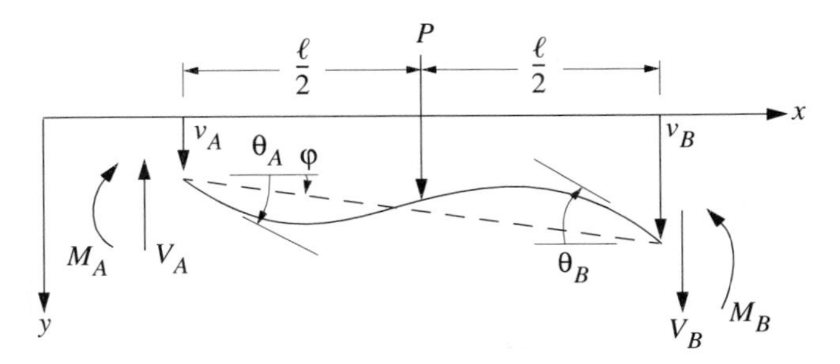

Figure 3.47 Beam with a point load.

$$EI\check{v}^{iv} = P\langle x - \ell/2\rangle^{-1}$$

where $\langle x - \ell/2\rangle^{-1}$ is a generalized distribution function[4] and represents the point load P as a large distributed force P/ϵ acting over a small interval ϵ. A concentrated moment in this notation would be written as $M\langle x - \ell/2\rangle^{-2}$. In the equation, we have dropped the subscript z from the moment of inertia. The reason for the ($\check{}$) over the displacement will become apparent in the text that follows.

Integration of this expression proceeds as follows:

$$EI\check{v}''' = P\langle x - \ell/2\rangle^{0} + \check{c}_3$$

$$EI\check{v}'' = P\langle x - \ell/2\rangle^{-1} + \check{c}_3 x + \check{c}_2$$

$$EI\check{v}' = \frac{P\langle x - \ell/2\rangle^{2}}{2} + \frac{\check{c}_3 x^2}{2} + \check{c}_2 x + \check{c}_1$$

$$EI\check{v} = \frac{P\langle x - \ell/2\rangle^{3}}{6} + \frac{\check{c}_3 x^3}{6} + \frac{\check{c}_2 x^2}{2} + \check{c}_1 x + \check{c}_0$$

We can now determine the shape $\check{v}$ as a function of the end displacements and the end rotations. The somewhat interesting result is

[4]Reference [38], pp. 47–53 gives a discussion of their application to beams, whereas ref. [6], vol. I, pp. 1–90 gives a general discussion and ref. [39], pp. 422–439 discusses general applications. The expression $\langle x - \ell/2\rangle^n$ from ref. [38] is defined as

$$\langle x - \ell/2\rangle^n = \begin{cases} 0 & 0 < x \leq \ell/2 \\ (x - \ell/2)^n & \ell/2 \leq x < \infty \end{cases}$$

and $n \geq 0$ is an integer. For $n \geq 0$, the function is bounded and the rules of normal integration apply.

$$\check{c}_0 = c_0$$
$$\check{c}_1 = c_1$$
$$\check{c}_3 = c_2 + \frac{P\ell}{8}$$
$$\check{c}_4 = c_3 - \frac{P\ell}{8} \tag{3.235}$$

The resulting solution for $\check{v}$ is

$$\check{v} = v + \frac{P\ell^2 x}{16EI} - \frac{Px^3}{12EI} + \frac{P\langle x - \ell/2\rangle^3}{6EI}$$
$$\check{v}' = v' + \frac{P\ell x}{8EI} - \frac{Px^2}{4EI} + \frac{P\langle x - \ell/2\rangle^2}{2EI}$$
$$\check{v}'' = v'' + \frac{P\ell}{8EI} - \frac{Px}{2EI} + \frac{P\langle x - \ell/2\rangle^1}{EI}$$
$$\check{v}''' = v''' - \frac{P}{2EI} + \frac{P\langle x - \ell/2\rangle^0}{EI} \tag{3.236}$$

where the c's are defined in Eq. (3.229) and v is defined by Eq. (3.231). From these relationships, we see that

$$\check{v}(A) = v(A)$$
$$\check{v}(B) = v(B)$$
$$\check{v}'(A) = v'(A)$$
$$\check{v}'(B) = v'(B)$$
$$\check{v}''(A) = v''(A) + \frac{P\ell}{8EI}$$
$$\check{v}''(B) = v''(B) + \frac{P\ell}{8EI}$$
$$\check{v}'''(A) = v'''(A) - \frac{P}{2EI}$$
$$\check{v}'''(B) = v'''(B) + \frac{P}{2EI} \tag{3.237}$$

From this, we reach the important result that the joint displacements and rotations are identical in Figs. 3.43 and 3.47. Also, because the terms in Eq. (3.236) containing the concentration load are prescribed, their variation is zero and we have

$$\bar{\check{v}} = \bar{v} \tag{3.238}$$

Then, for the case of a concentrated load at midspan, Eq. (3.232) becomes

$$\begin{aligned}
M_A &= \frac{2EI}{\ell}(2\theta_A + \theta_B - 3\varphi) - \frac{P\ell}{8EI} \\
M_B &= \frac{2EI}{\ell}(\theta_A + 2\theta_B - 3\varphi) + \frac{P\ell}{8EI} \\
V_A &= \frac{6EI}{\ell^2}(\theta_A + \theta_B - 2\varphi) - \frac{P}{2EI} \\
V_B &= \frac{6EI}{\ell^2}(\theta_A + \theta_B - 2\varphi) - \frac{P}{2EI}
\end{aligned} \tag{3.239}$$

The only difference between the moments in Eqs. (3.232) and (3.239) are the terms containing $P\ell/8$. These terms are called fixed end moments. Likewise, the additive terms in the deflected shape are called the fixed end deflections, because they represent the deflections that would exist if the beam were clamped at its ends with the lateral loads applied. These ideas are shown in Fig. 3.48.

From this, we obtain the following general method of solution:

1. Using the beam stiffness obtained from Eq. (3.233), solve the following system of equations:

$$[K]\{u\} = \{P\}$$

This will give the correct joint displacements.

2. To obtain the correct internal moments, simply add the fixed end moments to those obtained from the joint-displacement solution.

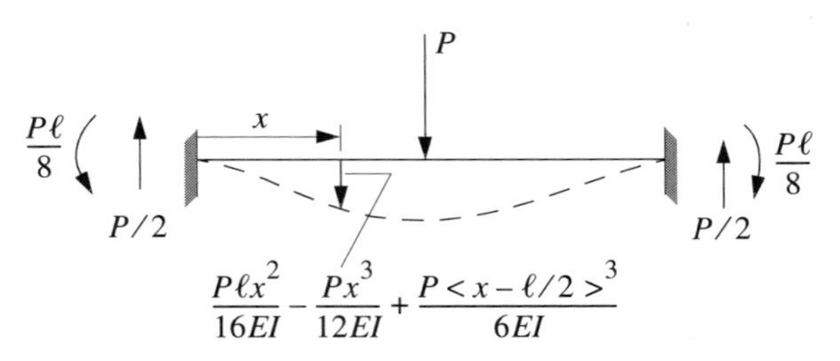

Figure 3.48 Fixed end conditions.

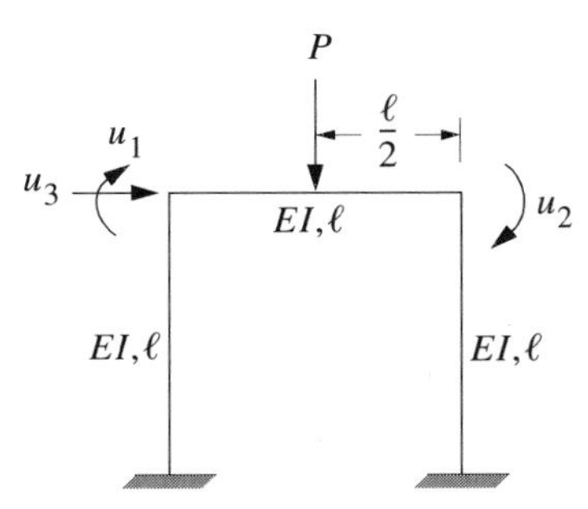

Figure 3.49 Simple plane frame with a vertical load.

Example 3.10 Consider Fig. 3.49, which represents a frame with a point load at $\ell/2$ of the horizontal member. The frame-stiffness matrix is the same as that given in the previous example. Because of Eq. (3.238), we can write

$$\int_0^\ell P\langle x - \ell/2\rangle^{-1}\bar{v}\,dx = P\bar{v}(\ell/2)$$

$$= P\left(\frac{1}{2}\,\bar{v}_A + \frac{\ell}{8}\,\bar{\theta}_A + \frac{1}{2}\,\bar{v}_B - \frac{\ell}{8}\,\bar{\theta}_B\right)$$

where the first expression comes from both the fundamental definition of a distribution function and common sense. The second expression comes from using Eq. (3.231) for the virtual displacement and its direct evaluation at $x = \ell/2$. We now obtain the result

$$\frac{2EI}{\ell^3}\begin{bmatrix} 4\ell^2 & \ell^2 & -3\ell \\ \ell^2 & 4\ell^2 & -3\ell \\ -3\ell & -3\ell & 12 \end{bmatrix}\begin{Bmatrix} u_1 \\ u_2 \\ u_3 \end{Bmatrix} = \begin{Bmatrix} \dfrac{P\ell}{8} \\ -\dfrac{P\ell}{8} \\ 0 \end{Bmatrix}$$

with the following solutions:

$$u_1 = \frac{P\ell^2}{48EI}$$

and

$$u_2 = \frac{P\ell^2}{48EI}$$

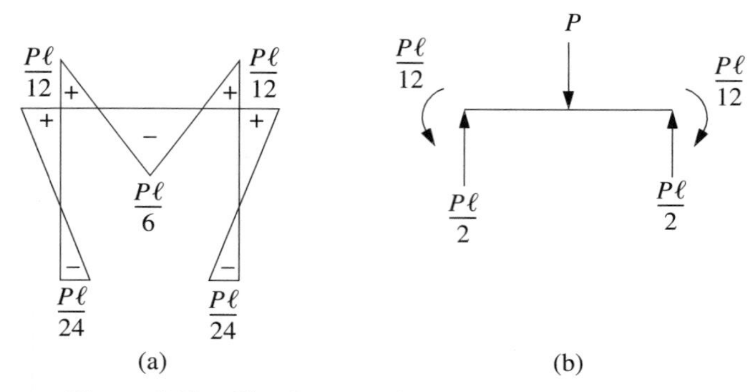

Figure 3.50 Simple plane frame with a vertical load.

and

$$u_3 = 0$$

These are the correct joint displacements. Using Eq. (3.232), we obtain the end moments for the joint rotations; to these, we add the fixed end moments that we have shown to be $\pm P\ell/8$. The following table gives the results:

	M_A	M_B	M_A^{fixed}	M_B^{fixed}
AB	$P\ell/24$	$P\ell/12$	—	—
BC	$P\ell/24$	$-P\ell/24$	$-P\ell/8$	$P\ell/8$
DC	$-P\ell/24$	$-P\ell/12$	—	—

Figure 3.50(a) shows the final moment distribution where for the purpose of plotting we have assumed that a positive moment causes tension on the outside of the frame. This is a completely arbitrary convention. Figure 3.50(b) shows a free body of member BC, from which the moment of value $P\ell/6$ was obtained. Reference [40] gives many examples using this method. The method developed here is the classical stiffness method. As noted above, it is from this method that the direct stiffness or displacement finite element method was developed.

We may now use Eq. (3.49) to transform the element-stiffness matrix given by Eq. (3.234) into its corresponding degrees of freedom in the global stiffness. Figure 3.51 shows a beam element at an arbitrary orientation in a two-dimensional structural system. The degrees of freedom of the beam in its element coordinate system are represented by (v_1, θ_1) at end 1 and (v_2, θ_2) at end 2. In the structural system, let the degrees of freedom be represented by (u_1^1, u_2^1, u_6^1) at end 1 and (u_1^2, u_2^2, u_6^2) at end 2. Then, from the figure, we can determine the transformation matrix $[T_{eg}]$ as

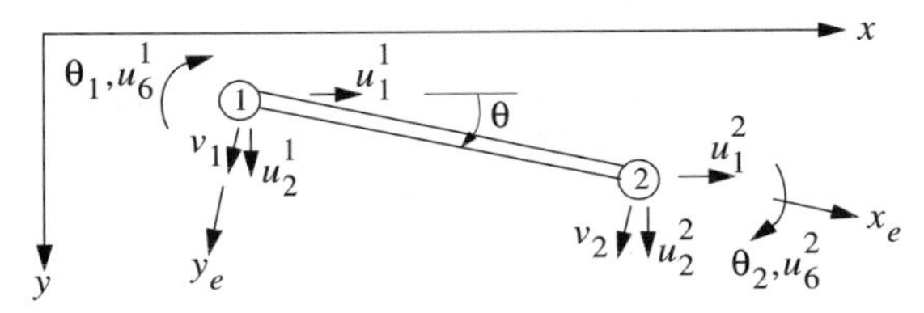

Figure 3.51 Element system to global system—two-dimensional beam.

$$\begin{Bmatrix} v_1 \\ \theta_1 \\ v_2 \\ \theta_2 \end{Bmatrix} = \begin{bmatrix} -\sin\theta & \cos\theta & 0 & 0 & 0 & 0 \\ 0 & 0 & 1 & 0 & 0 & 0 \\ 0 & 0 & 0 & -\sin\theta & \cos\theta & 0 \\ 0 & 0 & 0 & 0 & 0 & 1 \end{bmatrix} \begin{Bmatrix} \begin{bmatrix} u_1^1 \\ u_2^1 \\ u_6^1 \\ u_1^2 \\ u_2^2 \\ u_6^2 \end{bmatrix} \end{Bmatrix}$$

where

$$[T_{eg}] = \begin{bmatrix} -\sin\theta & \cos\theta & 0 & 0 & 0 & 0 \\ 0 & 0 & 1 & 0 & 0 & 0 \\ 0 & 0 & 0 & -\sin\theta & \cos\theta & 0 \\ 0 & 0 & 0 & 0 & 0 & 1 \end{bmatrix}$$

For three dimensions, the transformation would be similar with eight additional rows and the appropriate directions. Also, Eq. (3.234) would be a 12 × 12 if we accounted for axial displacement and torsion.

3.20 APPROXIMATE METHODS

The principle of virtual work provides us with an approximate method of solution. In general, we assume expressions for $u(x, y, z)$, $v(x, y, z)$ and $w(x, y, z)$, which satisfy the kinematic or prescribed boundary conditions. These expressions are in terms of some generalized coordinates a_i, b_i, c_i $i = 1$ to n and are substituted into the expression for δW. In addition, they obtain integrals relating stress and displacement gradient and use some stress-strain relationship to establish the stresses as functions of the displacement gradients. We then have $3n$ simultaneous equations with respect to the $3n$ unknowns: a_i, b_i, and c_i.

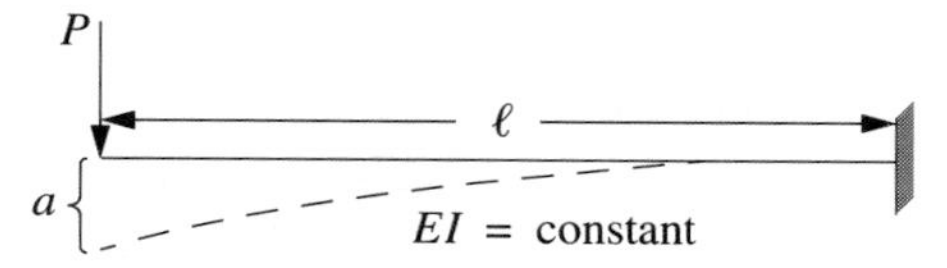

Figure 3.52 Cantilever beam under a concatenated end load.

Example 3.11 As a simple example, consider the beam shown in Fig. 3.52 with the following approximate solution:

$$v = a\left(1 - \sin \frac{\pi x}{2\ell}\right)$$

This function is compatible with the displacement boundaries, since at $x = \ell$, $v(\ell) = 0$ and $v'(\ell) = 0$. The parameter a takes on the properties of displacement, and the magnitude of v depends upon the parameter a. In this case, the relation for v represents an infinity of equations of constraint relating the deflections at each point to the single parameter a. The parameter a represents the least number of variables necessary to specify a configuration and hence is the generalized coordinate. We have reduced the beam from an infinite number of degrees of freedom to a system of one degree of freedom represented by the tip deflection a.

The virtual displacement is then represented as

$$\overline{v} = \overline{a}\left(1 - \sin \frac{\pi x}{2\ell}\right)$$

The virtual work becomes

$$\delta U = \delta W_e$$

$$EI \int_0^\ell v''\overline{v}''\, dx = P\overline{a}$$

$$\overline{a}\, \frac{aEI\pi^4}{16\ell^4} \int_0^\ell \sin^2 \frac{\pi x}{2\ell}\, dx = P\overline{a}$$

Using the relationship

$$\xi = \frac{\pi x}{2\ell}$$

the integral becomes

$$\int_0^\ell \sin^2 \frac{\pi x}{2\ell}\,dx = \frac{2\ell}{\pi}\int_0^{\pi/2} \sin^2 \xi\,d\xi = \frac{\ell}{2}$$

and collecting the coefficients of $\overline{a}$ that must equal zero, we obtain the solution for a as follows:

$$a = \frac{32P\ell^3}{\pi^4 EI}$$

The deflection curve is represented by the expression

$$v = \frac{32P\ell^3}{\pi^4 EI}\left(1 - \sin\frac{\pi x}{2\ell}\right)$$

At $x = 0$, the solution is

$$0.3285\,\frac{P\ell^3}{EI}$$

which is approximately 1% of the strength solution of

$$0.3333\,\frac{P\ell^3}{EI}$$

The moment is given by

$$M = -EIv''$$

which at $x = \ell$ has the value $0.8106P\ell$, or approximately 19% of the exact moment of value $P\ell$.

Looking at the foregoing results, it is natural to raise some questions about the convergence of approximate solutions. The above solution represents the first term in a sequence of approximate solutions:

$$v_1, v_2, v_3, \ldots$$

where, for example,

$$v_2 = v_1 + \sum_{j=2}^{m} a_j \phi_j(x)$$

where $\phi_j(x)$ are shape functions and the a_j are additional generalized coordinates. If $\tilde{v}$ represents the exact solution, point convergence is defined as

$$|\tilde{v} - v_n| \longrightarrow 0 \quad \text{as} \quad n \longrightarrow \infty$$

Point convergence depends on the concept of the Cauchy sequence [41], stated as

Let $\{P_n\}$ be a sequence of points. Let $d\{P_m, P_n\}$ denote the distance between P_m and P_n. A necessary and sufficient condition that the sequence be convergent is that $d(P_m, P_n) \to 0$ as $m, n \longrightarrow \infty$.

The Cauchy condition implies that the sequence is converging to some limit, but it does not guarantee that it is the correct limit. The definition of point convergence implies that the sequence is converging to the correct limit. Another important consideration is uniform convergence, which requires that

$$\max |\tilde{v} - v_n| \longrightarrow 0 \quad \text{as} \quad n \longrightarrow \infty$$

Uniform convergence also has important implications on the term-by-term integration and differentiation of sequences.

In theoretical work, convergence in energy is often considered. This measures the closeness of solution over the entire domain and represents a global average. The measure used is an energy error defined as

$$\text{Energy error} = \left[\int_{\text{domain}} [E(x)]^{\mathrm{T}} [A] [E(x)] \, dx \right]^2$$

where

$$E(x) = \tilde{v}(x) - v(x)$$

and A is the differential operator (e.g., $d^2/dx^2(EI)d^2/dx^2$). An infinite sequence of functions with the above property is said to be complete in energy, meaning that the energy error can be made as close to zero as desired. An example of functions complete in energy are the following functions:

$$\phi_n(x) = \sin \frac{n\pi x}{\ell}$$

One of the most useful features of the sine function is its orthogonality properties represented by the following relationship:

$$\int_0^{\pi} \sin mx \sin nx \, dx = \begin{cases} 0 & m \neq n \\ \pi/2 & m = n \end{cases}$$

Example 3.12 Next, consider the simply supported beam shown in Fig. 3.53 with the assumed displacement

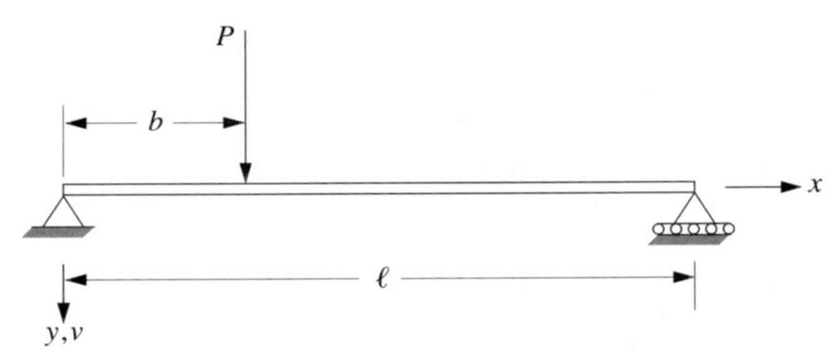

Figure 3.53 Simple beam with a load at b.

$$v = \sum_{n=1}^{k} a_n \sin \frac{n\pi x}{\ell}$$

Let the virtual displacement be

$$\overline{v} = \sum_{n=1}^{k} \overline{a}_n \sin \frac{n\pi x}{\ell}$$

and the virtual displacement under the applied load P is

$$\overline{v}(b) = \sum_{n=1}^{k} \overline{a}_n \sin \frac{n\pi b}{\ell}$$

The virtual work becomes

$$\delta U = \delta W_e$$

$$EI \int_0^{\ell} v'' \overline{v}'' \, dx = P\overline{v}(b)$$

$$\sum_{n=1}^{k} a_n \overline{a}_n \frac{EI n^4 \pi^4}{\ell^4} \int_0^{\ell} \sin^2 \frac{n\pi x}{\ell} \, dx = P \sum_{n=1}^{k} \overline{a}_n \sin \frac{n\pi b}{\ell}$$

Collecting the coefficients of $\overline{a}_n$ and solving the resulting equation for a_n, we get

$$a_n = \frac{2P\ell^3}{n^4 EI \pi^4} \sin \frac{n\pi b}{\ell}$$

which yields the solution for v of

$$v = \frac{2P\ell^3}{EI\pi^4} \sum_{n=1}^{k} \frac{1}{n^4} \sin \frac{n\pi b}{\ell} \sin \frac{n\pi x}{\ell}$$

Because the sine function is complete in energy, we can make this solution as close to the exact as we please; however, a few terms usually suffice.

The method outlined above using virtual work is often referred to as the Ritz or Rayleigh-Ritz method of approximation. Strictly speaking, the assumed trial functions ϕ_n must be defined over the whole domain and satisfy all the domains' boundary conditions—an often impossible task for complex domains. The finite element method removes the restriction of the requirement that the assumed trial or shape functions be defined over the whole domain; instead, it requires that the shape function be defined over the element and that they need not satisfy any boundary condition, only certain continuity conditions.

Finite elements are grouped into three categories.

- h-elements (currently the predominant type), where the number of shape functions is fixed for each finite element. The error of approximation is controlled by mesh refinement, and such elements should pass certain convergence tests (called "Patch tests"), which indicate that by refining the mesh the error will get smaller. The h comes from the fact that such elements are noted by the characteristic dimension h.
- p-elements, where the polynomial degree of the element can be varied over a wide range. Analysis using these methods attempts to get a global measure in energy and recommends that the analysis be repeated several times with the sequence of trial polynomials increased sequentially as a test for convergence. These elements can at the node point and boundary intersection also begin to get a measure on point convergence. The p comes from polynomial, and in most cases, the mesh stays fixed.
- Boundary elements, which, in general, are defined by integral equations representing the model characteristics as elements distributed on the surface of the domain. These elements are most useful in infinite domain problems.

Perhaps the most efficient (and certainly the most sophisticated) type is a combination of h-convergent and p-convergent elements.

Thus far in seeking approximate solutions, we have used variational techniques. If no physical principle giving a variational statement for the problem is apparent, but the differential equations and their boundary conditions are known, we can try to manipulate the differential equations and their boundary conditions to seek a variational statement (this is done in Chapter 5). If manipulation does not work, we can of course attempt to construct an appropriate Hilbert space; however, if we are more interested in obtaining an immediate solution, we are often better off obtaining approximate solutions via one of the several techniques of method-of-weighted residuals. Of these methods, Galerkin's method is the one in most common usage. This method forms an error function called the residual of the equation, usually denoted by $R(x; a)$. We then try to find numerical values for each a that make $R(x; a)$ as close to zero as possible for all values of x (for a one-dimensional problem) throughout the entire domain (in Chapter 5, we pursue this technique in more detail). If we use the Galerkin method for a problem for which a variational formulation is possible, then for the same trial solutions, the Ritz and Galerkin methods will yield identical results.

Example 3.13 We have stated that the method of virtual work is valid for nonlinear problems. As an example that demonstrates how the principle of virtual work can be used with a material nonlinear problem, consider again as a simple example the beam shown in Fig. 3.52. Assume that the stress-strain law takes the following form:

$$\sigma = \pm B\epsilon^{1/2}$$

where the plus sign is associated with tensile strain and the minus sign is associated with compressive strain,[5] and only positive values of ϵ are used in the expression evaluation. The value B is a known constant. For variety, let the assumed displacement be given as

$$v = \frac{a}{2\ell^3}(x^3 - 3\ell^2 x + 2\ell^3)$$

At the boundary $x = \ell$, both $v(\ell)$ and $v'(\ell)$ satisfy the required conditions (as expected, since the shape function assumed is that of a cantilevered beam with a concentrated end load). Using this displacement function, the curvature is

$$v'' = \frac{3ax}{\ell^3}$$

and a convenient virtual curvature is

$$\bar{v}'' = \frac{3\bar{a}x}{\ell^3}$$

The virtual work is

[5] The signum function $\mathrm{sgn}(\epsilon) = \epsilon/|\epsilon|$ could be used in place of the $\pm$.

$$\delta U = \delta W_e$$

$$\int_0^{\ell} \sigma \overline{\epsilon}\, dA \;\, dx = P\overline{v}(0)$$

In evaluating the expression for virtual work, it is important to remember that the only restriction on $\overline{\epsilon}$ is that it be consistent with the constraints. Therefore, for convenience, we take the virtual strain as

$$\overline{\epsilon} = -y\overline{v}''$$

The magnitude of the true strain is yv''. Hence for $\epsilon^{1/2}$ we use

$$\epsilon^{1/2} = -y^{1/2}(v'')^{1/2}$$

where the (−) is to ensure positive strain for negative curvature in accordance with our sign convention. Before carrying out the integration, define the moment of inertia $\breve{I}$ as

$$\breve{I} = \int_{-h/2}^{h/2} y^{3/2}\, dA = b\int_{-h/2}^{h/2} y^{3/2}\, dy = \frac{bh^{5/2}}{5\sqrt{2}}$$

where we have specified the cross section of the beam to be rectangular with height h and width b.

The expression for virtual work now becomes (written for positive strain)

$$B\int \epsilon^{1/2}\overline{\epsilon}\, dA \;\, dx = P\overline{v}(0)$$

$$\int (-B)y^{1/2}(v'')^{1/2}(-y)\overline{v}''\, dA\, dx = P\overline{v}(0)$$

$$\frac{3\sqrt{3}B\breve{I}}{\ell^{9/2}}\, a\overline{a}\int_0^{\ell} x^{3/2}\, dx = P\overline{a}$$

$$\left(\frac{6\sqrt{3}B\breve{I}}{5\ell^2}\, a^{1/2} - P\right)\overline{a} = 0$$

The expression for a is now obtained as

$$a = \frac{25P^2\ell^4}{108B^2\breve{I}^2} = \frac{625P^2\ell^4}{54B^2b^2h^5}$$

and the deflection curve for the beam is

$$v = \frac{625P^2 \ell}{108B^2b^2h^5} (x^3 - 3\ell^2 x + 2\ell^3)$$

PROBLEMS

3.1 The total virtual work of an elastic body is given by the expression

$$\delta U = \int_S (\sigma_x \, \delta\epsilon_x + \sigma_y \, \delta\epsilon_y + \tau_{xy} \, \delta\gamma_{xy}) \, dS$$

If $\delta U = \delta W_e$, then show that the two-dimensional differential equations of equilibrium follow.

3.2 For the compound beam shown in Fig. 3.54, determine by the principle of virtual work the moment reaction at the built-in end.

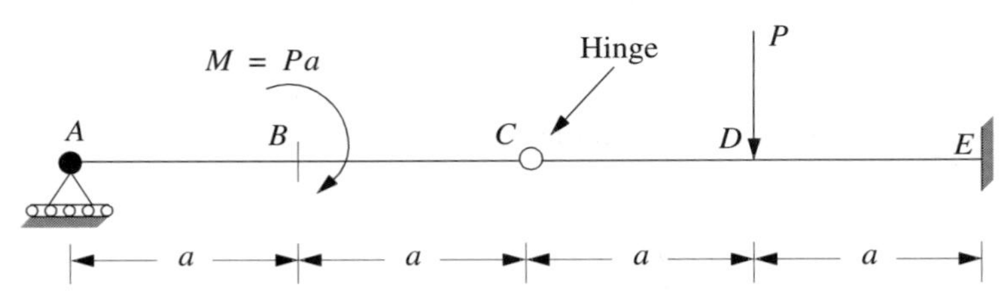

Figure 3.54 Problem 3.2.

3.3 A rigid bar of length a and weight W leans against a frictionless wall at B and rests on a frictionless floor at A. Force P acts on the base at A as shown in Fig. 3.55. By the method of virtual displacements, compute the angle α for which the bar will be in equilibrium.

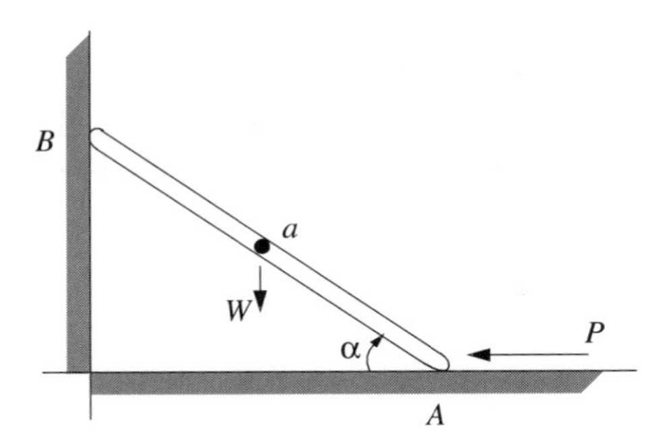

Figure 3.55 Problem 3.3.

3.4 By the principle of virtual work and by neglecting friction, determine the mechanical advantage (W/P) of the differential pully shown in Fig. 3.56. Do not disassemble; instead, try virtual velocity and power concepts.

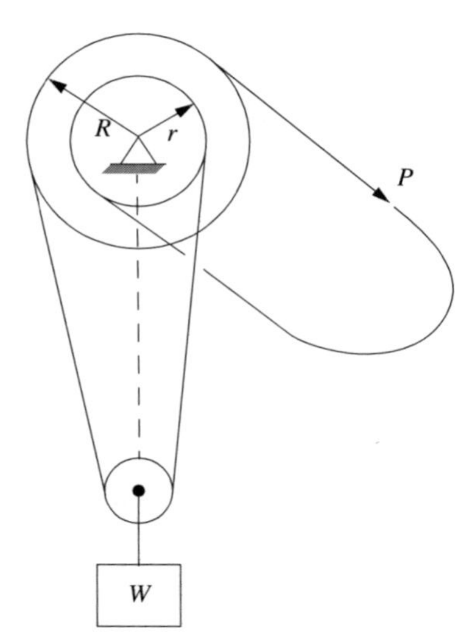

Figure 3.56 Problem 3.4.

3.5 The square pin–jointed truss is loaded as shown in Fig. 3.57. A load-measuring device indicates that member AD has a zero load. By the principle of virtual work, determine the force in member BC.

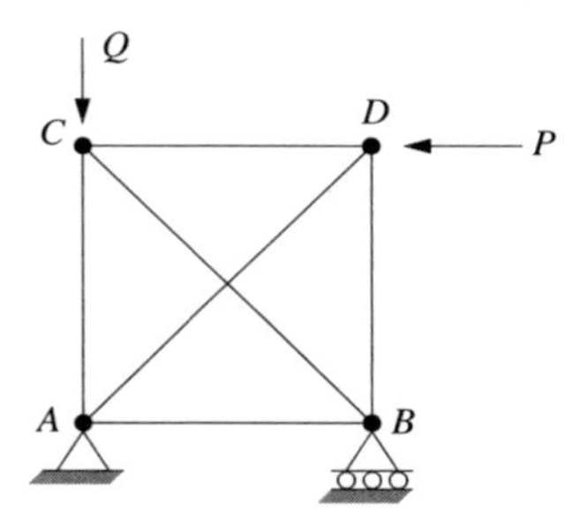

Figure 3.57 Problem 3.5.

3.6 For the pin-jointed truss shown in Fig. 3.58, by the principle of virtual displacements, find the reaction at R from the load P.

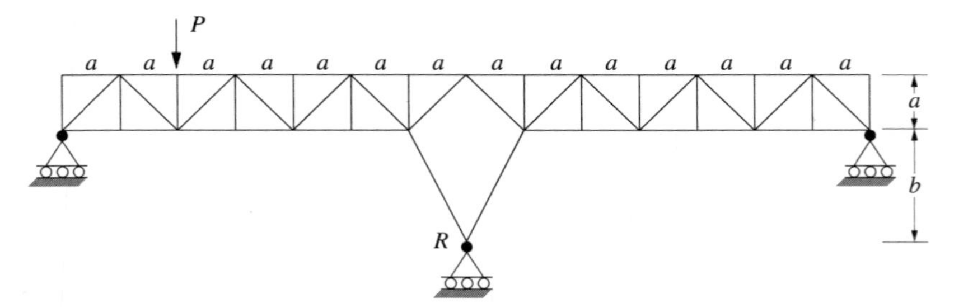

Figure 3.58 Problem 3.6.

3.7 A pin-jointed truss is loaded with a load W as shown in Fig. 3.59. Measurements show that the member AB carries a compressive force P. By virtual work, determine the reaction at the center support.

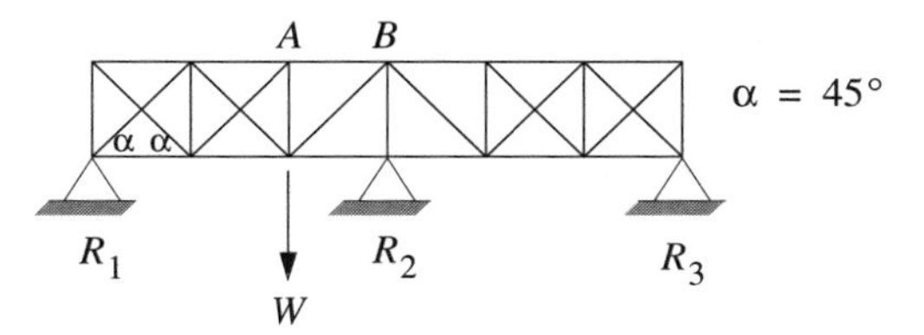

Figure 3.59 Problem 3.7.

3.8 The pin-jointed truss is in equilibrium under the load P as shown in Fig. 3.60. By virtual work, determine the load carried by the strut BD.

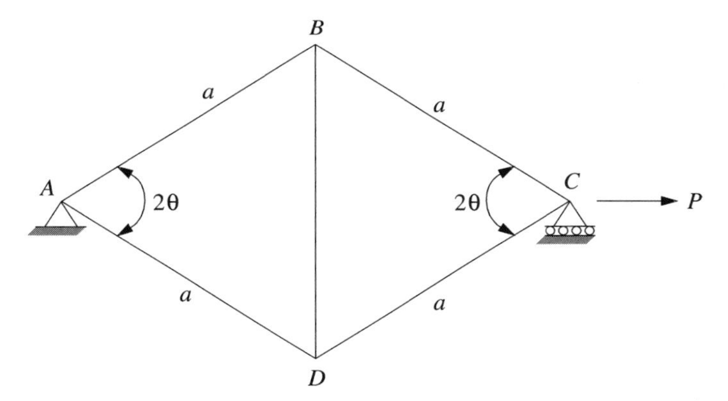

Figure 3.60 Problem 3.8.

3.9 A rigid bar AB of length ℓ is suspended in a horizontal position by two vertical wires attached at each end as shown in Fig. 3.61. Both wires have the same length and the same cross-sectional area. The wire at A has a modulus E_1; the wire at B, a modulus E_2. By neglecting bar weight and using virtual work, determine the two equilibrium equations and the distance x if the bar is to remain horizontal under the load P located as shown in Fig. 3.61.

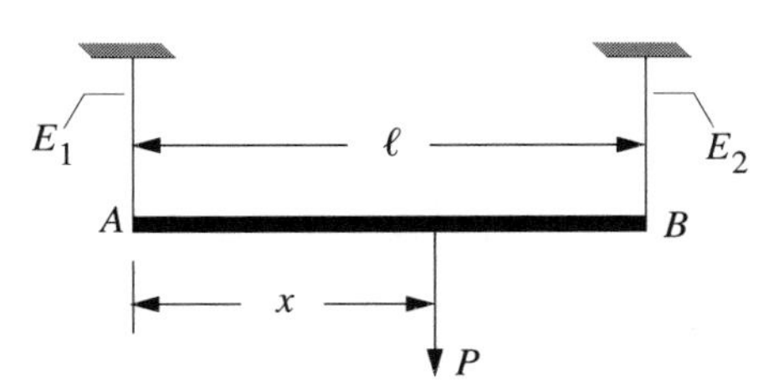

Figure 3.61 Problem 3.9.

3.10 A bar of square cross section is constructed of two bars of different materials having moduli of elasticity E_1 and E_2 ($E_1 > E_2$) as shown in Fig. 3.62. The end plates are rigid, and both bars have the same cross-sectional areas. By virtual work, determine the eccentricity e of the load P so that the bars are in uniform tension.

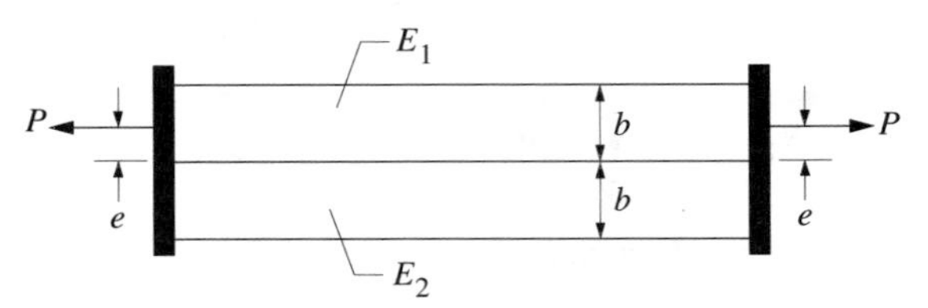

Figure 3.62 Problem 3.10.

3.11 A weight W is attached to an arm of length ℓ as shown in Fig. 3.63. The arm has a weight per unit volume of γ and rotates in a horizontal plane about a vertical piviot as shown in the figure. The arm and weight rotate at a constant angular velocity ω. Using virtual work, determine

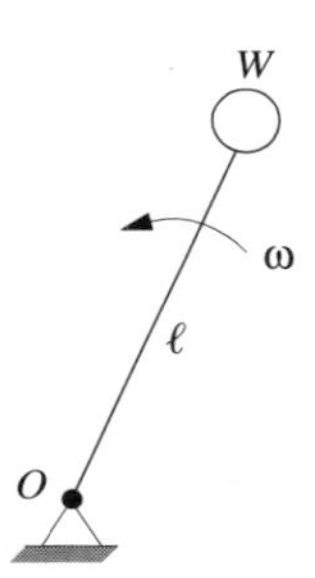

Figure 3.63 Problem 3.11.

(a) A formula for the required cross-sectional area A of the arm if the allowable stress in the arm at the pivot (neglecting stress concentration) is σ_W.

(b) For a given cross-sectional area A, the total elongation $u(\ell)$ of the bar.

3.12 The bar AB of constant cross section A and Young's modulus E is

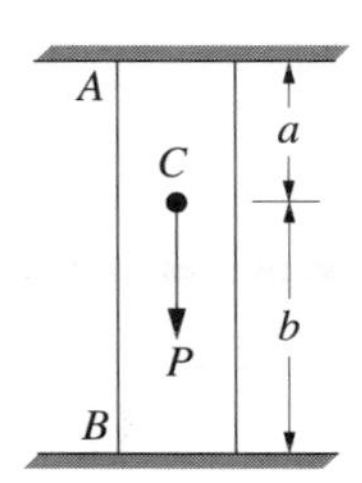

Figure 3.64 Problem 3.12.

attached at both ends to rigid supports and loaded axially by force P at the intermediate point C as shown in Fig. 3.64. By virtual work, determine the equation of equilibrium and the deflection of point C.

3.13 A rigid bar AB is hinged at A and supported in a horizontal position by two identical vertical wires at C and D as shown in Fig. 3.65. The stress-strain law for the wire is $\sigma = E\epsilon^n$. By virtual work, determine the tensile force in the wire attached at D induced by a vertical load P applied as shown at B.

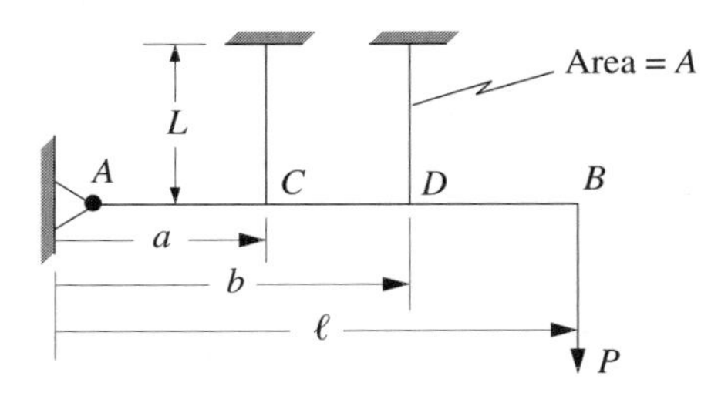

Figure 3.65 Problem 3.13.

3.14 A rectangular block ℓ units high consists of a steel section of modulus E_s, an area A_s, and a stress-strain relationship $\sigma_s = E_s\epsilon_s$, and it also contains an oak section of modulus E_o, an area A_o, and a stress-strain relationship $\sigma_o = E_o\epsilon_o^{1/3}$. A rigid cap is laid across the top and loaded by a vertical load P as shown in Fig. 3.66. By virtual work, determine the value of the applied load P if the rigid cap is forced down a uniform distance Δ.

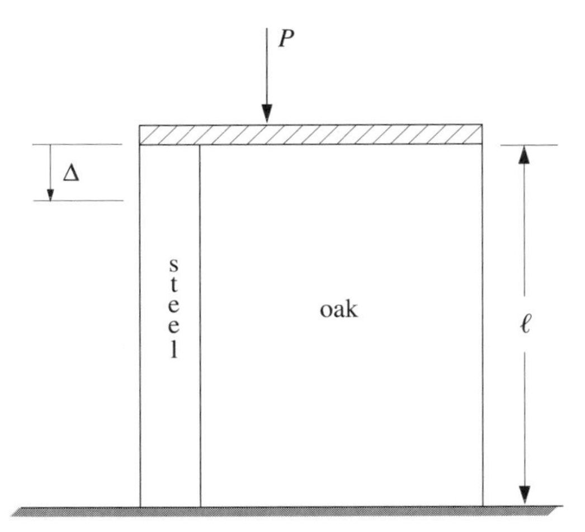

Figure 3.66 Problem 3.14.

3.15 Figure 3.67 shows a steel rod of area A_s passing through a copper tube with a net area A_c. The two are secured together by nuts and rigid washers. The steel has a Young's modulus of E_s and a coefficient of thermal expansion α_s. The copper has a Young's modulus of E_c and a coefficient of thermal expansion α_c. If the assembly is heated ΔT degrees, determine by virtual work the equilibrium equations and the stress in the steel rod.

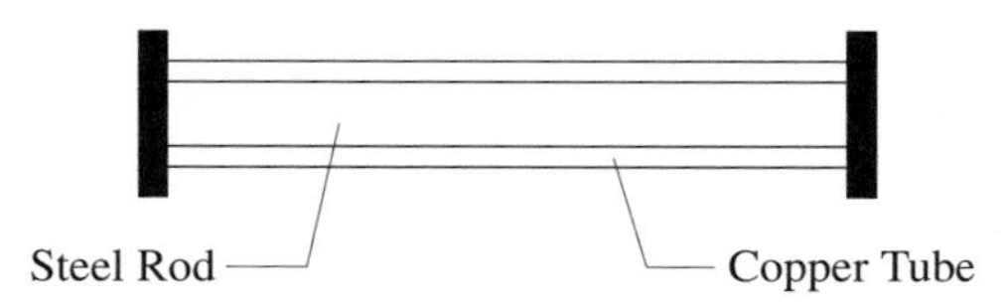

Figure 3.67 Problem 3.15.

3.16 A bar B has a length ℓ, an elastic modulus E_b, and a cross-sectional area A_b. Two cables with turnbuckles are attached to rigid caps over the ends of the bar as shown in Fig. 3.68. The cables each have a length ℓ, an elastic modulus E_c, and a cross-sectional area A_c. The pitch of the threads for the *double-acting* turnbuckles is p. By virtual work, derive a formula for the number of turns n of each turnbuckle required to prestress the bar B to a uniform compressive stress σ_o. (Hint: Use the fact that the extension in the cables plus the shortening of the bar is equal to the displacement of the caps caused by the turning of the turnbuckles.)

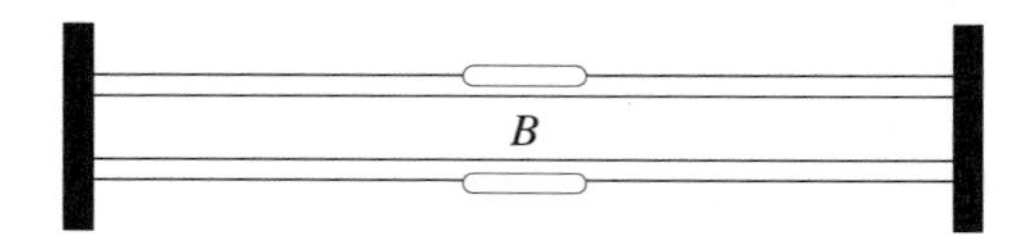

Figure 3.68 Problem 3.16.

3.17 For the structure shown in Fig. 3.69, determine by virtual work the tension T in the cable AB. Express the result in terms of the applied load W and the angles α and β.

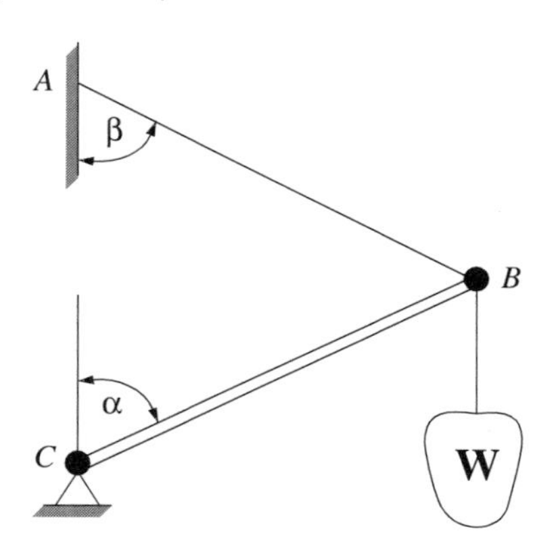

Figure 3.69 Problem 3.17.

3.18 A rigid bar of length ℓ and uniform weight W is leaned against a frictionless wall as shown in Fig. 3.70. The base of the bar is supported by a linear spring with modulus K. When the bar is at an angle ϕ_0, the spring is unstretched. After the bar is released, it settles into its equilibrium position ϕ. By virtual work, determine the expression for ϕ in terms of W, ℓ, K, and ϕ_0.

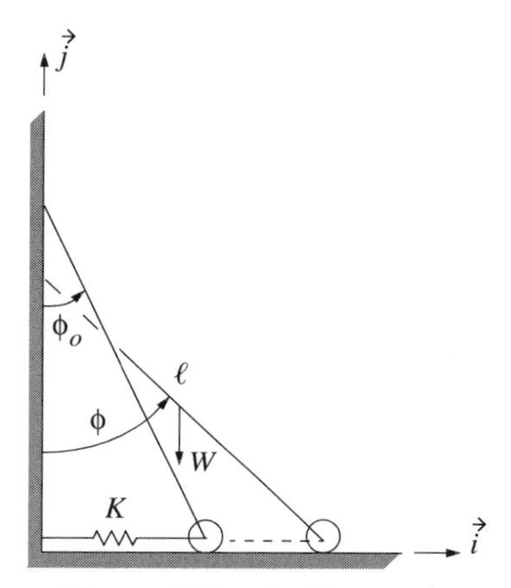

Figure 3.70 Problem 3.18.

3.19 The pin-jointed truss of Fig. 3.71 is loaded as shown. By virtual work, determine the force in member mp. This can be done with a single equation.

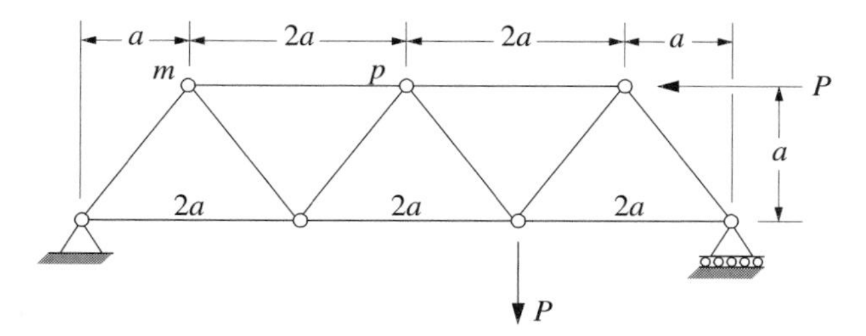

Figure 3.71 Problem 3.19.

3.20 The pin-jointed truss of Fig. 3.72 is loaded as shown. By virtual work, determine the force in member GF. This can be done with a single equation.

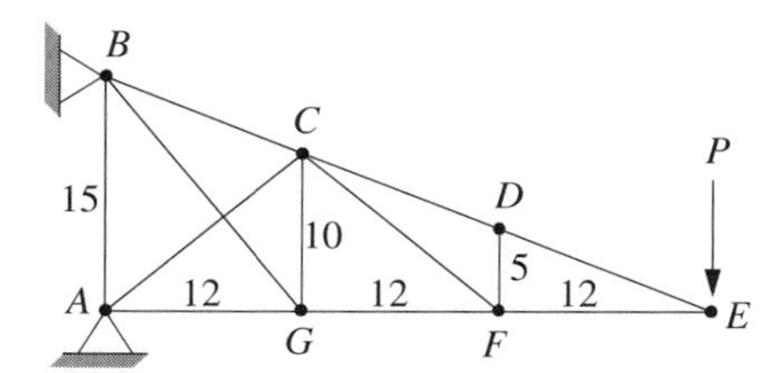

Figure 3.72 Problem 3.20.

3.21 The truss in Fig. 3.73 supports a force P that acts at an angle θ to the vertical. The cross-sectional area of the member AB is A_1; that of BC is A_2. Using virtual work, find the value of the angle θ for which the joint deflection B is in the same direction as the force P.

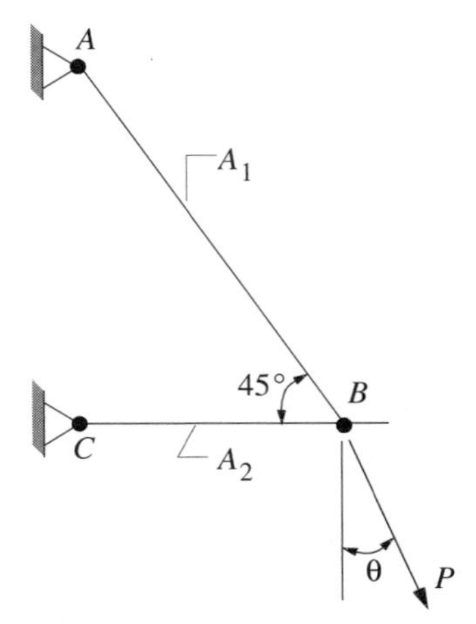

Figure 3.73 Problem 3.21.

3.22 The bar in Fig. 3.74 has a unit weight γ and follows the stress-strain relationship $\sigma^n = B\epsilon$, where B and n are constants. By virtual work, determine the elongation $u(\ell)$ of the end of the bar under its own weight.

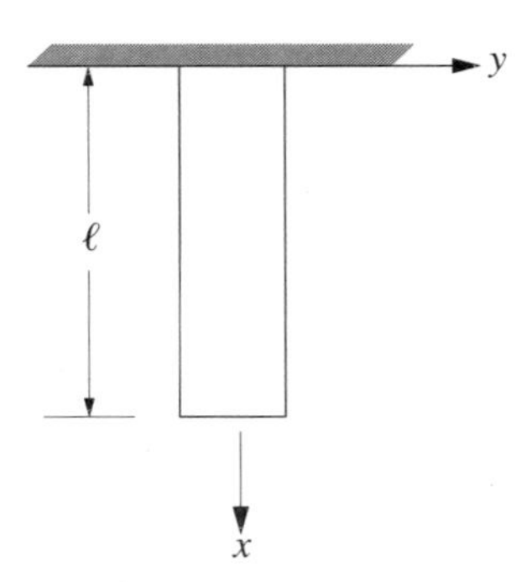

Figure 3.74 Problem 3.22.

3.23 For the element shown in Fig. 3.12:

(a) Expand the derivation of Eq. (3.42) to include thermal strain $\alpha\Delta T$.

(b) Use Eq. (3.48) to get the form of the thermal load vector in a global system.

3.24 For the pin-jointed truss shown in Fig. 3.75, AE/ℓ is the same for all members. Assume that member 1–2 on assembly is over the length by the amount $\Delta\ell$. By virtual work, compute the displacement of node 1. (Hint: Let $\alpha\Delta T = \Delta\ell/\ell$ as the thermal strain in member 1–2.)

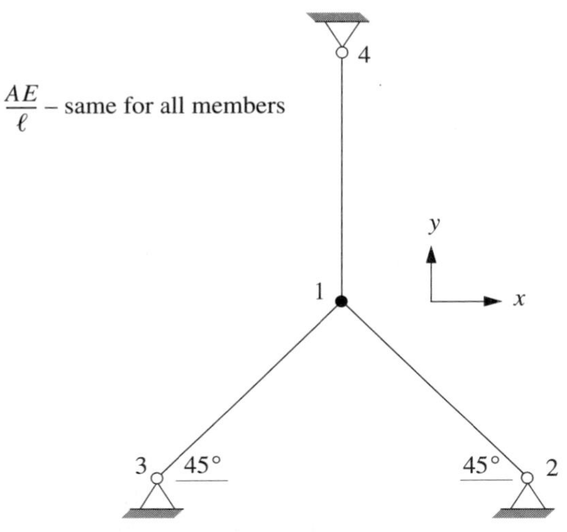

Figure 3.75 Problem 3.24.

3.25 In the truss shown in Fig. 3.76, member 1–2 is heated by $\Delta T = T_a$; member 1–3, by $\Delta T = T_b$. Both members have the same α, A, and E. By the unit-displacement method, determine the vertical displacement of joint 1 in terms of α, ℓ, T_a, and T_b.

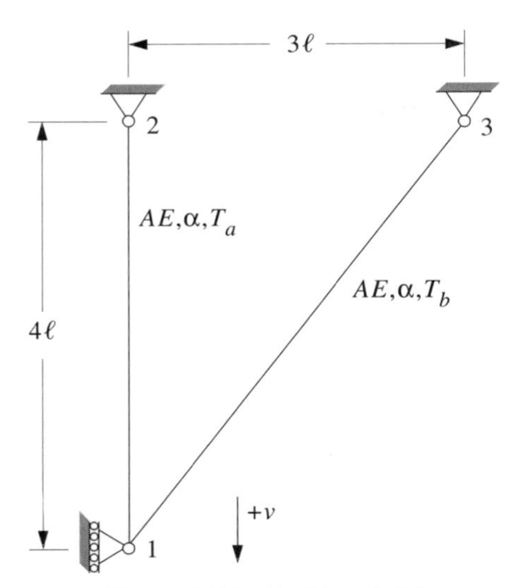

Figure 3.76 Problem 3.25.

3.26 The elastic beam shown in Fig. 3.77 has a flexural rigidity EI and a length ℓ, and it carries a uniform load p. The beam is clamped at its left end and has a vertical support at its right end. Also at the right end is a rotational spring of stiffness k. By virtual work, determine

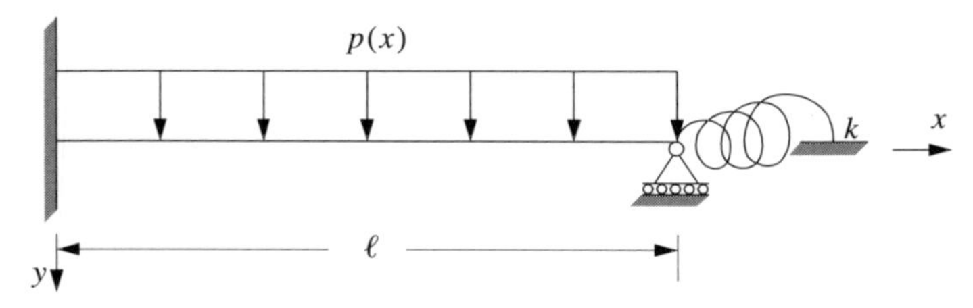

Figure 3.77 Problem 3.26.

(a) the governing fourth-order differential equation;

(b) the boundary conditions; and

(c) the rotation $v'(\ell)$.

3.27 The elastic beam shown in Fig. 3.78 has a flexural rigidity EI, a length ℓ, and an applied load P. It is clamped at the left end and is supported at its right end by the vertical spring of stiffness k_1 and the rotational spring of stiffness k_2. By virtual work, determine

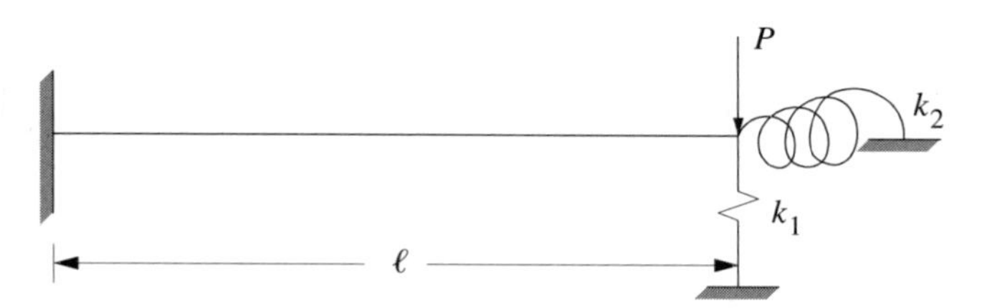

Figure 3.78 Problem 3.27.

(a) the governing fourth-order differential equation;

(b) the boundary conditions; and

(c) the displacement $v(\ell)$.

Show that your solution is reasonable for the following cases:

(a) $k_1 = 0, \quad k_2 = 0$

(b) $k_1 \rightarrow \infty, \quad k_2 \rightarrow \infty$

(c) $k_1 = 0, \quad k_2 \rightarrow \infty$

3.28 The beam shown in Fig. 3.79 is loaded with a triangular load distribution with p_0 the load intensity at the right end. The beam has a length ℓ, a height h, and a flexural rigidity EI, and it is subject to the temperature change $\Delta T = T_0 xy/(\ell h)$. The coefficient of thermal expansion is α. By virtual work, determine

(a) the equation of the elastic curve, and

(b) the bending stress at any section.

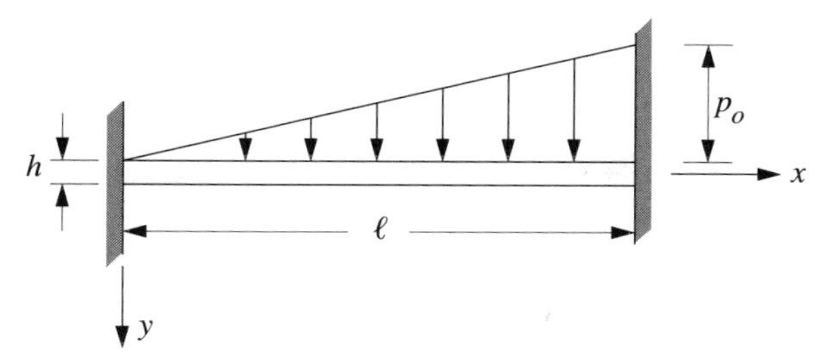

Figure 3.79 Problem 3.28.

3.29 The beam shown in Fig. 3.80 is free of any external loading but is subject to the temperature change $\Delta T = T_0(1 - 4y^2/h^2)$. The beam has a length ℓ, a height h, a flexural rigidity EI, and an axial rigidity AE. The coefficient of thermal expansion is α. By virtual work, determine the stress σ.

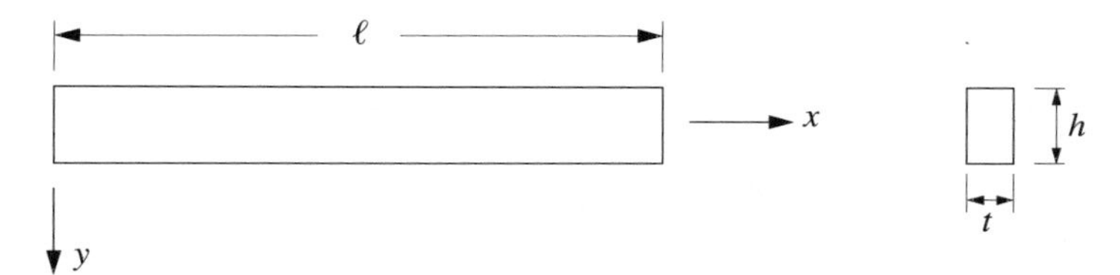

Figure 3.80 Problem 3.29.

3.30 At a given temperature, the free end of the cantilever beam shown in Fig. 3.81 just rests against the frictionless plane. The plane is oriented at an angle $\beta = 45°$ from the horizontal. The temperature of the beam is then raised by a uniform ΔT. The beam has a length ℓ, a flexural rigidity EI, and an axial rigidity AE. The coefficient of thermal expansion is α. By virtual work and neglecting the effect of the axial force on the *bending* deflection, determine

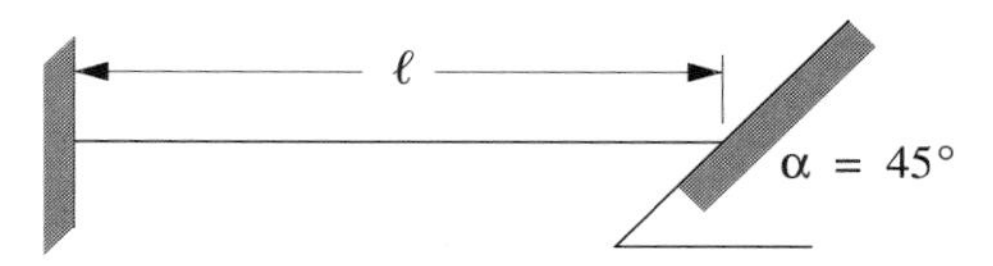

Figure 3.81 Problem 3.30.

(a) the tip deflection, and

(b) the tip reaction forces.

Then, sketch a free body showing all reactions on the beam.

3.31 The beam in Fig. 3.82 has a constant flexural rigidity EI. By virtual work, determine the fixed end moments.

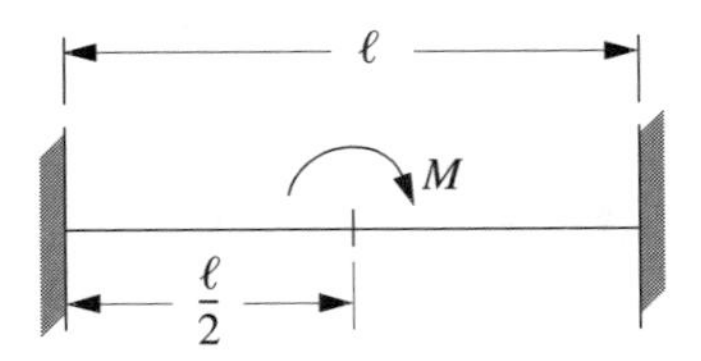

Figure 3.82 Problem 3.31.

3.32 The beam shown in Fig. 3.83 in an unloaded condition just touches a spring at midspan. The beam is then loaded across its entire span with a uniform load p_0. Determine by virtual work the spring force k so that the forces in all three supports are equal. To do so,

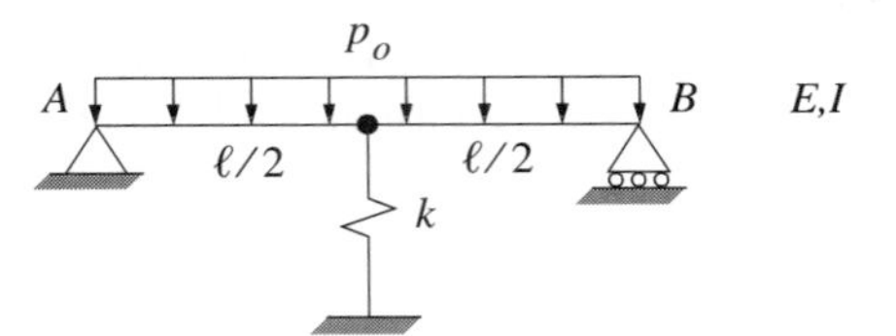

Figure 3.83 Problem 3.32.

(a) use symmetry and only the left half of the beam, and

(b) use the total beam.

3.33 The three bars—AD, BD, and CD—shown in Fig. 3.84 all have the same axial rigidity AE. By the unit-displacement method, determine the horizontal and vertical displacement of joint D in terms of P, H, AE, and $\sin \alpha$.

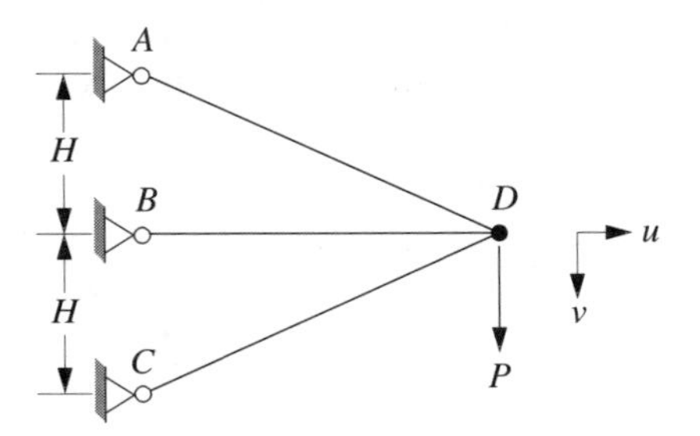

Figure 3.84 Problem 3.33.

3.34 The simply supported beam of Fig. 3.85 has a length ℓ, a height h, a flexural rigidity EI, and an axial rigidity AE. The coefficient of thermal expansion is α. The beam undergoes a temperature change $\Delta T = [(T_2 - T_1)/h]y + (T_1 + T_2 - 2T_0)/2$, where T_1 is the temperature at the top of the beam, T_2 is the temperature at the bottom of the beam, and T_0 is the initial temperature. By virtual work, determine the equation for the deflection curve of the beam.

Figure 3.85 Problem 3.34.

3.35 The cantilevered beam of Fig. 3.86 has a length ℓ, a height h, a flexural rigidity EI, and an axial rigidity AE. The coefficient of thermal expansion is α. The beam undergoes a temperature change $\Delta T = [(T_2 - T_1)/h]y + (T_1 + T_2 - 2T_0)/2$, where T_1 is the temperature at the top of the beam, T_2 is the temperature at the bottom of the beam, and T_0 is the initial temperature. By virtual work, determine

Figure 3.86 Problem 3.35.

(a) the tip deflection, and

(b) the tip rotation.

3.36 The simply supported beam of Fig. 3.87 has a length ℓ, a height h, and a flexural rigidity EI. The coefficient of thermal expansion is α. The beam undergoes a temperature change $\Delta T = T_0 xy/(\ell h)$. By virtual work, determine the maximum deflection of the beam.

Figure 3.87 Problem 3.36.

3.37 The cantilevered beam of Fig. 3.88 has a length ℓ, an area A, a flexural rigidity EI, and an axial rigidity AE. The coefficient of thermal expansion is α. The beam undergoes a temperature change $\Delta T = by^2$, where b is a specified constant. By virtual work, compute the stress σ in the beam.

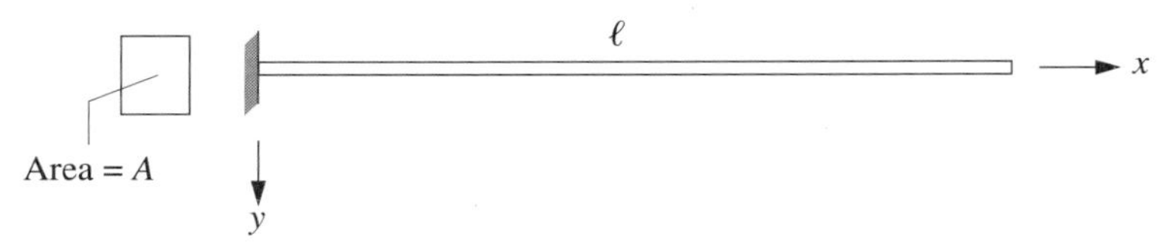

Figure 3.88 Problem 3.37.

3.38 The uniform bar in Fig. 3.89 has length ℓ with a constant axial rigidity AE and a coefficient of thermal expansion α. The end $x = \ell$ has a spring of modulus k, whereas the end $x = 0$ is fixed. The bar is raised a uniform ΔT degrees. By virtual work, determine

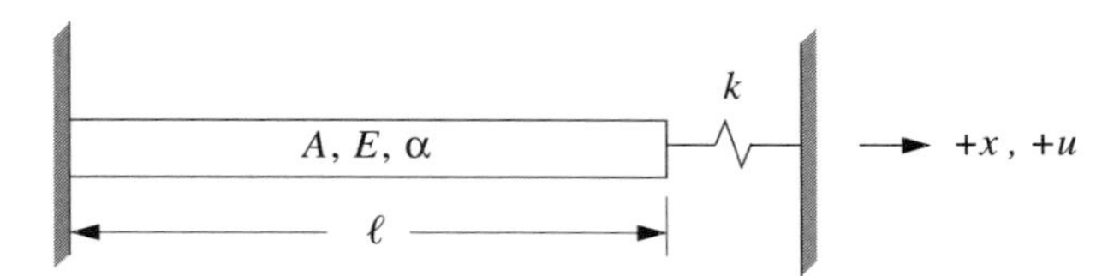

Figure 3.89 Problem 3.38.

(a) the boundary conditions;

(b) the governing differential equation;

(c) the elongation of the bar $u(\ell)$ at $x = \ell$; and

(d) the axial stress in the bar.

3.39 The uniform bar in Fig. 3.90 has a length ℓ with a constant axial rigidity AE and a coefficient of thermal expansion α. At the end $x = 0$ is a spring of modulus k_1; at the end $x = \ell$ is a spring of modulus k_2. The bar is raised a uniform ΔT degrees. By virtual work, determine

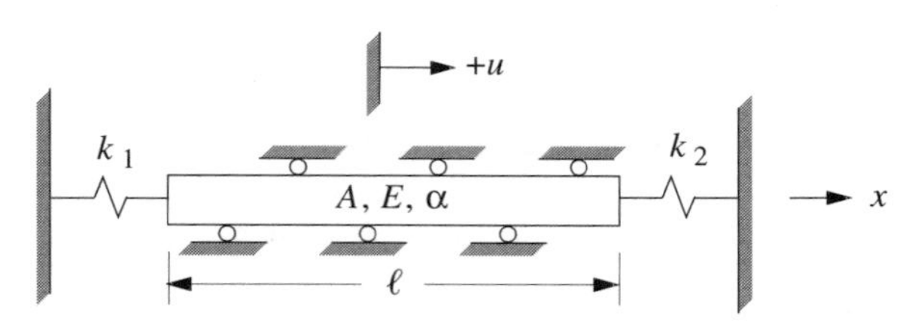

Figure 3.90 Problem 3.39.

(a) the boundary conditions;

(b) the governing differential equation;

(c) the elongation of the bar $u(0)$ at $x = 0$;

(d) the elongation of the bar $u(\ell)$ at $x = \ell$; and

(e) the axial stress in the bar.

3.40 The cantilevered beam of Fig. 3.91 has a length ℓ, a height h, a flexural rigidity EI, and an axial rigidity AE. The coefficient of thermal expansion

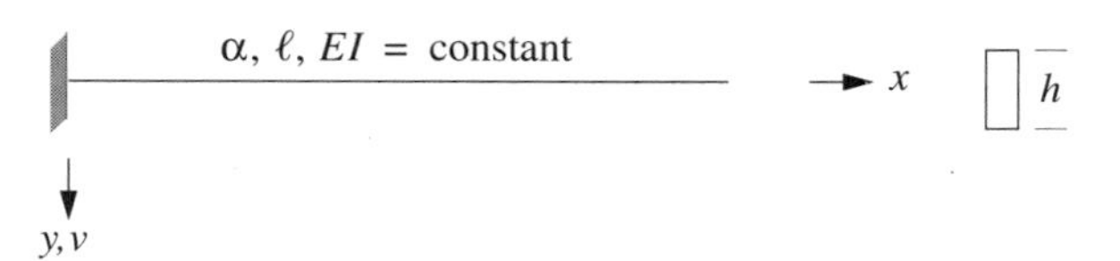

Figure 3.91 Problem 3.40.

is α. The beam undergoes a temperature change $\Delta T = T_0 xy/(\ell h)$. By virtual work, determine

(a) the tip deflection, and

(b) the tip rotation.

3.41 The uniform bar in Fig. 3.92 is of length ℓ with a constant axial rigidity AE. The bar is clamped at its left end and is free at its right end. It is subject to a uniform load per unit length of q. By virtual work, determine

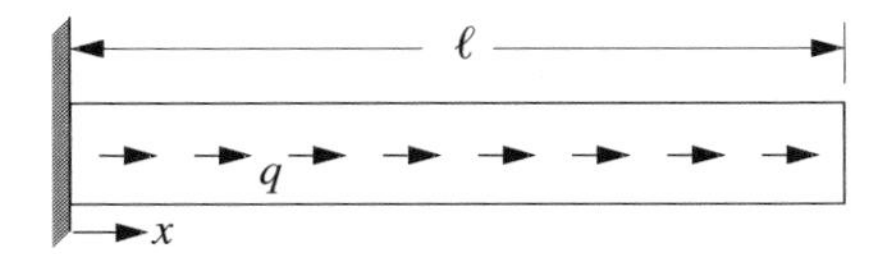

Figure 3.92 Problem 3.41.

(a) the boundary conditions;

(b) the governing differential equation; and

(c) the axial displacement $u(x)$ at any point.

3.42 The beam shown in Fig. 3.93 is built in at the left end and is supported by an elastic rod at the right end. The beam is of length ℓ and has a flexural rigidity EI. The rod is of length L and has an axial rigidity AE and a coefficient of thermal expansion α. In addition, the rod has its temperature raised a uniform ΔT degrees. By virtual work, determine

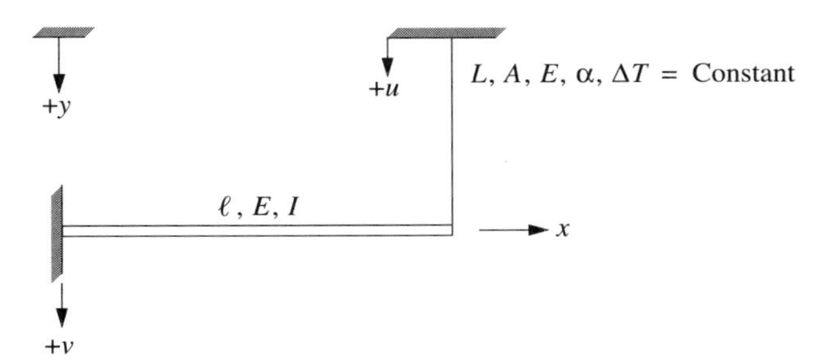

Figure 3.93 Problem 3.42.

(a) the governing differential equation for the beam;

(b) the governing differential equation for the rod;

(c) the natural and forced boundary conditions;

(d) the tip deflection in terms of A, E, I, L, ℓ, ΔT, and α;

(e) the beam shear at $x = \ell$ in terms of A, E, I, L, ℓ, ΔT, and α; and

(f) the axial load in the rod in terms of A, E, I, L, ℓ, ΔT, and α.

3.43 The beam shown in Fig. 3.94 is built in at the left end and is supported by an elastic spring at the right end. The beam is of length ℓ and depth h, and it has a flexural rigidity EI. The spring has a modulus $k = 3EI/\ell^3$.

The beam has a coefficient of thermal expansion α and is subject to a temperature change given by the expression $\Delta T = T_0 yx/(\ell h)$. By virtual work, determine

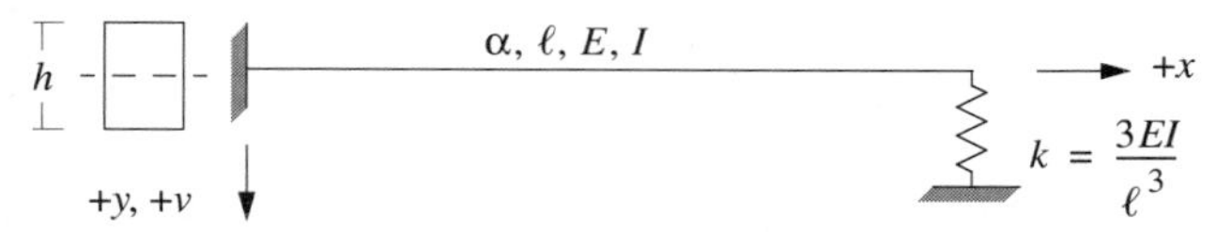

Figure 3.94 Problem 3.43.

(a) the governing fourth-order differential equation;

(b) the natural and forced boundary conditions; and

(c) the tip deflection in terms of E, I, h, ℓ, T_0, and α.

3.44 The beam shown in Fig. 3.95 is prevented from rotating at the left end but is allowed to displace vertically at that end. The vertical displacement at the left end is resisted by a spring of modulus k. The right end of the beam is simply supported. The beam is of length ℓ and depth h, and it has a flexural rigidity EI and a coefficient of thermal expansion α. The beam also undergoes a change in temperature distribution given by the relationship $\Delta T = T_0(y/h)^3$. Note: $\int y\, dA = \int y^3\, dA = 0$, $\int y^2\, dA = I$, and $\int y^4\, dA = 3h^2 I/20$. By virtual work, determine

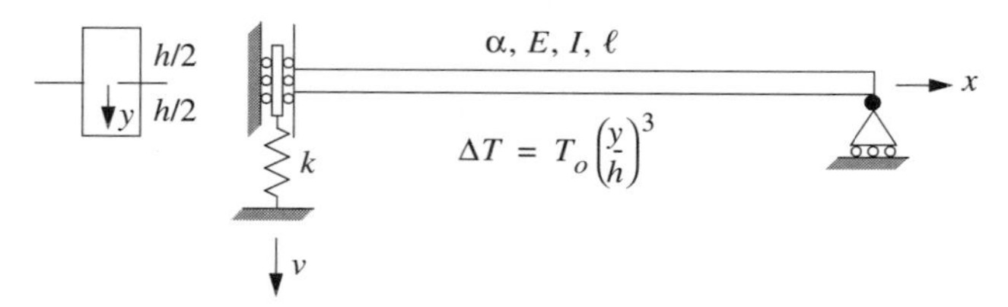

Figure 3.95 Problem 3.44.

(a) the governing fourth-order differential equation;

(b) the natural and forced boundary conditions;

(c) the deflection $v(0)$ at $x = 0$ in terms of α, EI, T_0, ℓ, h, and k; and

(d) the bending stress σ at any section in terms of α, EI, T_0, ℓ, h, k, $v(0)$, x, and y.

3.45 A simple rod of area A, length ℓ, and modulus E is shown in Fig. 3.96. Its left end is supported by a spring of modulus k and its right end is fixed. The rod is subject to a uniformly distributed axial load q_0 along its length

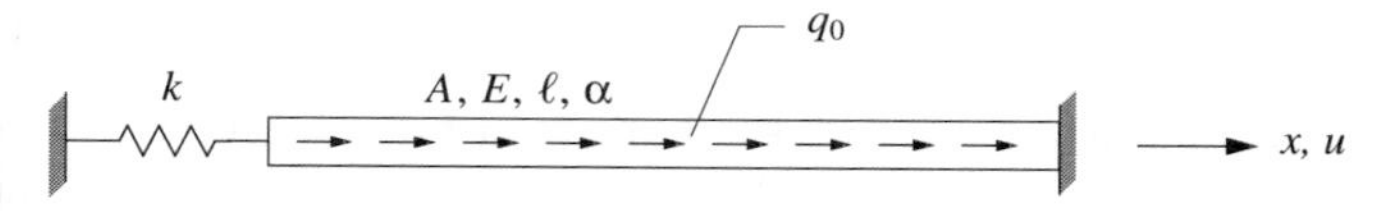

Figure 3.96 Problem 3.45.

and has a coefficient of thermal expansion α. The rod also undergoes a temperature change $\Delta T = T_0 x/\ell$. By virtual work, determine

(a) the governing second-order differential equation;

(b) the natural and forced boundary conditions;

(c) the horizontal displacement $u(0)$ at the left end; and

(d) the axial stress σ at any section in terms of A, k, q_0, $u(0)$, and x.

3.46 The beam shown in Fig. 3.97 is simply supported at both ends. The left end also has a rotational spring of modulus $k = EI/\ell$ to provide some rotational constraint. The beam is of length ℓ and has a flexural rigidity EI. A uniformly distributed load of magnitude p is applied along the span. By virtual work, determine

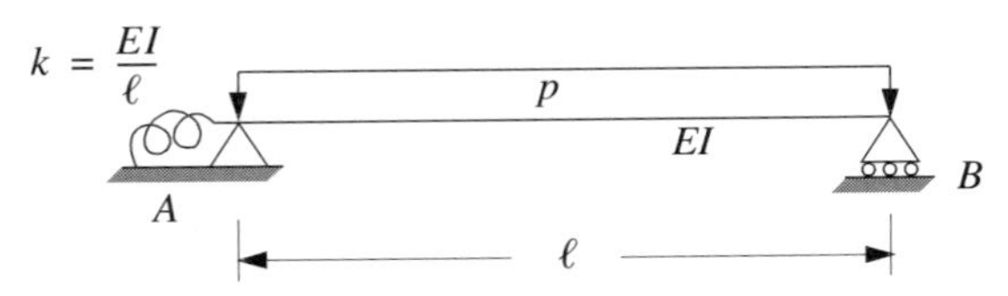

Figure 3.97 Problem 3.46.

(a) the governing fourth-order differential equation;

(b) the natural and forced boundary conditions; and forced boundary conditions; and

(c) the rotation at the left end.

3.47 A simple rod of area A, length ℓ, and modulus E is shown in Fig. 3.98. Its left end is fixed, and the right end is attached to an axial spring of modulus k, which is attached to a mechanism. This mechanism gives a *prescribed* horizontal displacement δ to the right. The rod is subject to a uniformly distributed axial load q_0 along its length and has a coefficient of thermal expansion α. The rod undergoes a temperature change $\Delta T = T_0 x/\ell$. By virtual work, determine

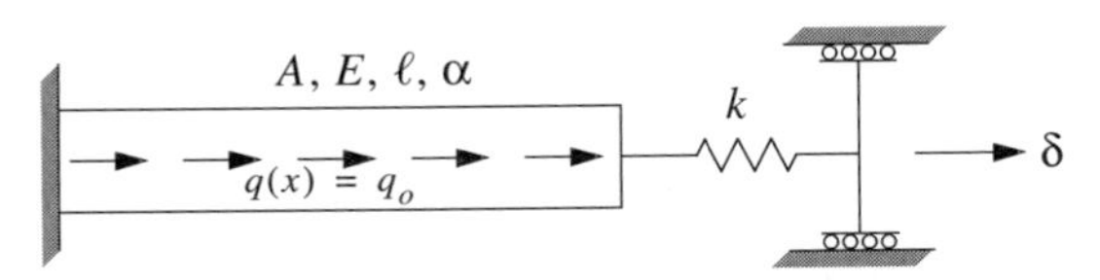

Figure 3.98 Problem 3.47.

(a) the governing second-order differential equation;

(b) the natural and forced boundary conditions;

(c) the horizontal displacement $u(0)$ at the right end of the bar; and

(d) the axial stress σ at any section in terms of A, k, q_0, $u(\ell)$, δ, and x.

3.48 The beam shown in Fig. 3.99 is of length ℓ and has a flexural rigidity EI and an axial rigidity AE. It has a coefficient of thermal expansion α. The left end is built in, and the right end is supported by a spring of modulus k inclined upward and to the left at a run of 3 units to a rise of 4 units. The beam experiences a temperature change of $\Delta T = T_0(1 + a_1 y)/h$ degrees, where a_1 is a specified constant. By virtual work, determine

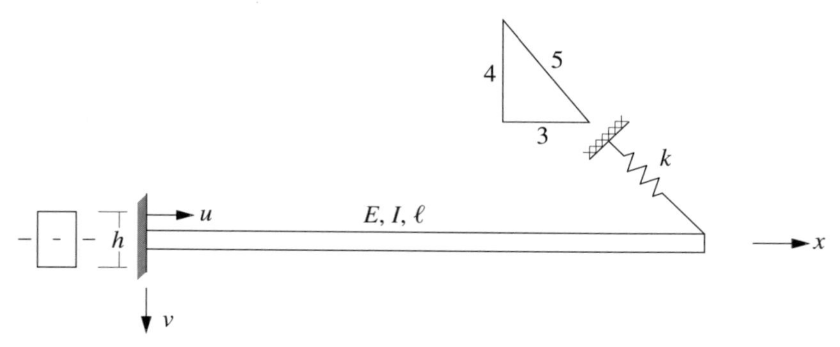

Figure 3.99 Problem 3.48.

(a) the governing differential equations, and

(b) the natural and forced boundary conditions.

3.49 The beam shown in Fig. (3.100) is of length ℓ and has a flexural rigidity EI and an axial rigidity AE. The left end is built in, and the right end is supported by a spring of modulus k. The spring is attached to a mechanism that rides on a inclined plane at an angle θ counterclockwise from the horizontal. The mechanism gives a *prescribed* displacement Δ along the inclined plane to tension the spring. By virtual work, determine

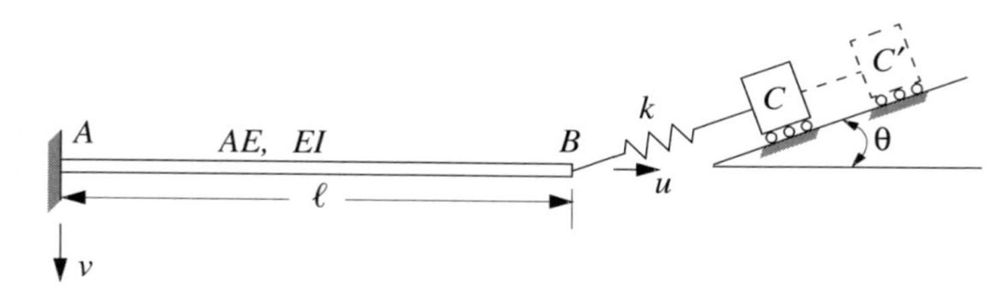

Figure 3.100 Problem 3.49.

(a) the governing differential equations, and

(b) the natural and forced boundary conditions.

3.50 For the beam cross section shown in Fig. 3.101, $I_x = 5a^3t/4$. Determine the following in terms of a, t, and the reacting shear force V:

(a) the shear distribution in the top flange;

(b) the resulting force in the top flange; and

(c) the location of the shear center e. (Use equilibrium considerations.)

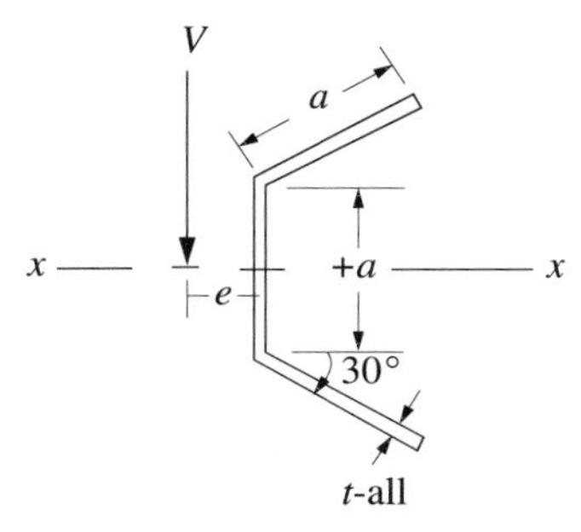

Figure 3.101 Problem 3.50.

3.51 For the section shown in Fig. 3.102, the moments of inertia of area about the xy axes are defined as $I_x = \int y^2\, dA$, $I_y = \int x^2\, dA$, and $I_x = \int xy\, dA$.

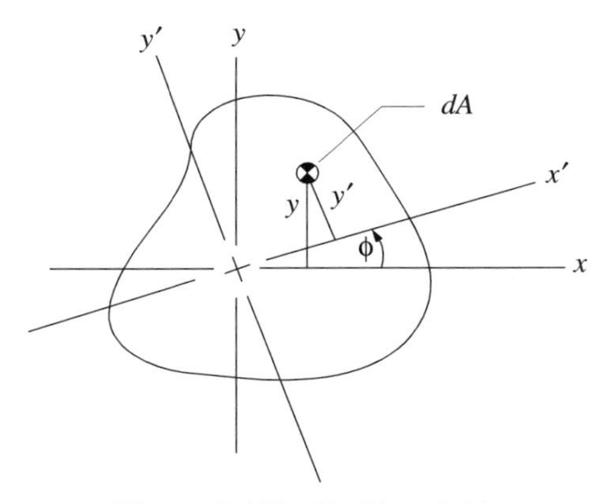

Figure 3.102 Problem 3.51.

(a) Show that

$$I_{x'} = \frac{I_x + I_y}{2} + \frac{I_x - I_y}{2} \cos 2\phi - I_{xy} \sin 2\phi$$

$$I_{x'y'} = \frac{I_x - I_y}{2} \sin 2\phi + I_{xy} \cos 2\phi$$

(b) Show that the expressions obtained above are the parametric equations of a circle.

The above parametric circle represents Mohr's circle for computation of inertia properties. The circle is constructed as follows:

(a) Lay out the abscissa as I_x and I_y and the ordinate as I_{xy}.

(b) Plot the two points (I_x, I_{xy}) and $(I_y, -I_{xy})$.

(c) Connect these two points by a straight line forming the circle diameter.

(d) The center of the circle is at the intersection of this diameter and the abscissa. Angles in the circle represent twice the actual angle but with the same sense.

3.52 For the section shown in Fig. 3.101, show that for $t \ll a$, $I_x = 5a^3t/4$.

3.53 The bar shown in Fig. 3.103 has a length ℓ and an axial rigidity AE. It is loaded by a axial load P at its right end and is fixed at its left end. By virtual work for the assumed displacement fields given in the list that follows, determine the value of the generalized coordinates.

Figure 3.103 Problem 3.53.

(a) $u = ax^2$.

(b) $u = ax^2 + bx^3$.

(c) $u = ax + bx^2$. (Discuss your results for this case.)

3.54 The simply supported beam in Fig. 3.104 is uniformly loaded with a load intensity per unit length of p_0. It is of length ℓ and has a flexural rigidity EI. By virtual work for the assumed displacement fields given in the list that follows, determine the value of the generalized coordinates.

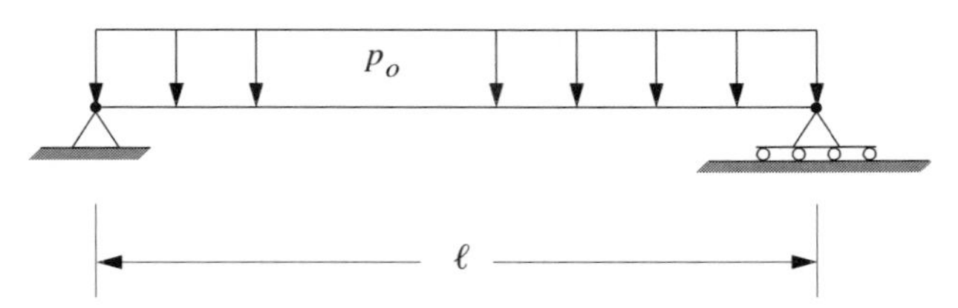

Figure 3.104 Problem 3.54.

(a) $v = a \sin(\pi x/\ell)$.

(b) $v = ax(\ell - x)$.

3.55 The simply supported beam in Fig. 3.105 has a vertical load P applied downward at the beam's center span. It is of length ℓ and has a flexural rigidity EI. By virtual work, determine the following:

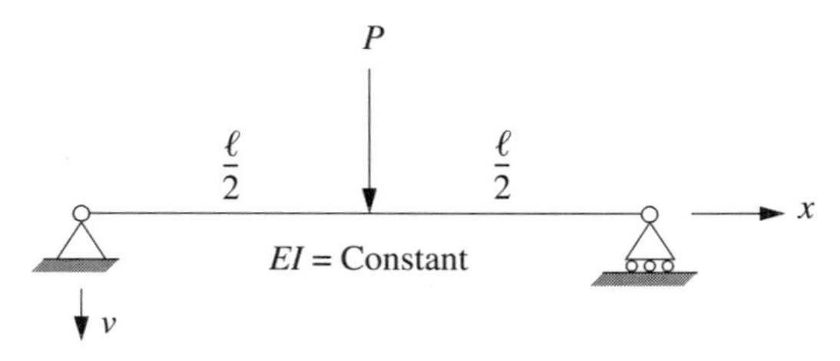

Figure 3.105 Problem 3.55.

(a) Using symmetry, find the undetermined constant a in the assumed displacement field:

$$v = a(3\ell^2 x - 4x^2); \qquad 0 \leq x \leq \ell/2$$

(b) Why is symmetry valid for this assumed field?

3.56 The beam in Fig. 3.106 is built in at the left end and is unsupported at the right end. It is of length ℓ and height h with a flexural rigidity EI and a coefficient of thermal expansion α. The beam experiences a temperature change $\Delta T = T_0 y/h$ through its depth. Using the assumed displacement function $v = a[1 - \cos \pi x/(2\ell)]$, determine the following by virtual work:

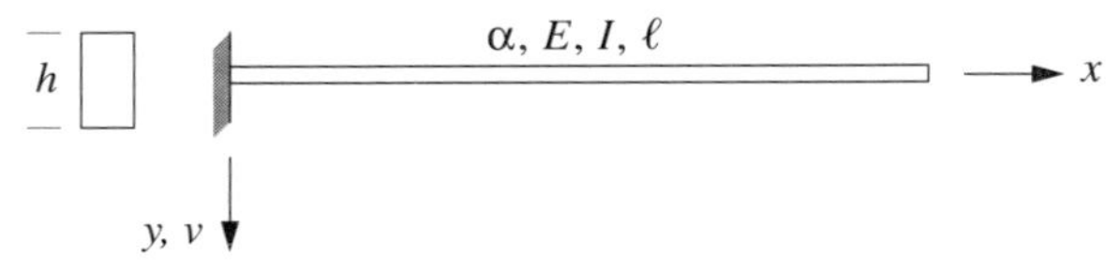

Figure 3.106 Problem 3.56.

(a) The parameter a in terms of h, α, T_0, and ℓ.

(b) The tip deflection $v(\ell)$ in terms of h, α, T_0, and ℓ.

3.57 The beam in Fig. 3.107 is built in at the left end and has a vertical spring support of modulus $k = 3EI/\ell$ at the right end. It is of length ℓ and height h with a flexural rigidity EI and a coefficient of thermal expansion α. The beam experiences a temperature change $\Delta T = T_0 xy/(\ell h)$ along its length and through its depth. Using the assumed displacement function $v = a[1 - \cos \pi x/(2\ell)]$, determine by virtual work the parameter a in terms of h, α, T_0, and ℓ.

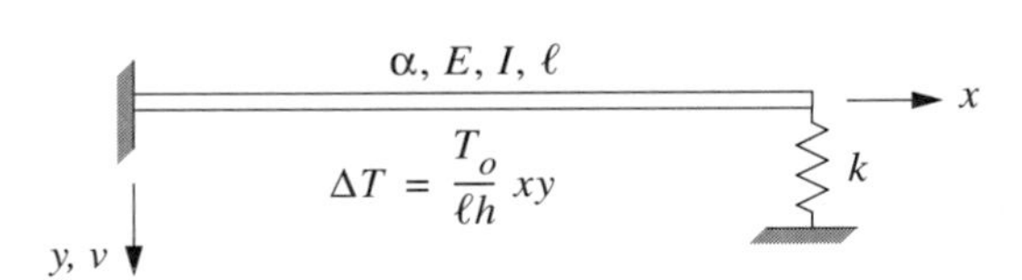

Figure 3.107 Problem 3.57.

3.58 The beam in Fig. 3.108 is simply supported at both ends and carries a uniformly distributed load p_0. The center of the beam is restrained with a vertical spring of modulus k. It is of length ℓ with a flexural rigidity

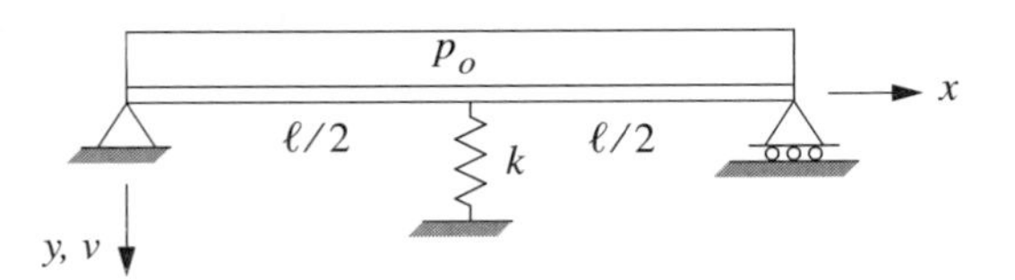

Figure 3.108 Problem 3.58.

EI. Using the assumed displacement function $v = a \sin \pi x/\ell$, determine by virtual work the parameter a in terms of k, ℓ, EI, and p_0.

3.59 The beam in Fig. 3.109 is built in at the left end and is unsupported at the right end. The beam carries a triangular load distribution, with p_0 the load intensity at the right end. It is of length ℓ with a flexural rigidity EI. Using the assumed displacement function $v = c_1x^2 + c_2x^3$, determine by virtual work the values of the generalized coordinates c_1 and c_2.

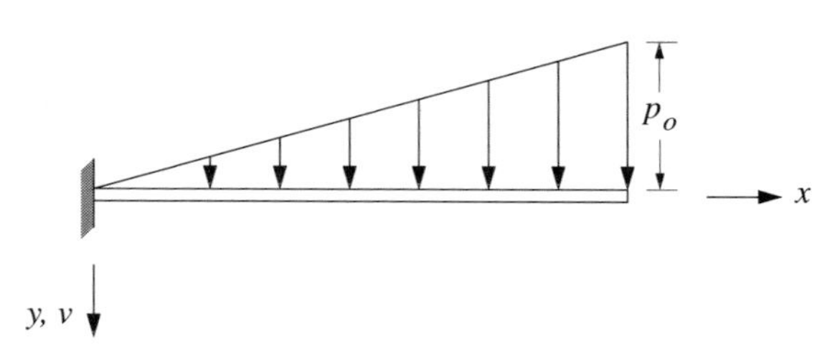

Figure 3.109 Problem 3.59.

3.60 Figure 3.110 shows a flat metal sheet of thickness, t, depth $2b$, and length ℓ fixed between supports AB and CD. The sheet carries shear load only. A stringer OO' is attached to the center of the sheet and has a cross-sectional area A_s. A tensile load P_0 is applied to the left end of the stringer. The stringer carries all of the tensile load. A displacement function is assumed that gives the finite displacement near point O but zero displacement at $\ell \to \infty$. The function is of the form

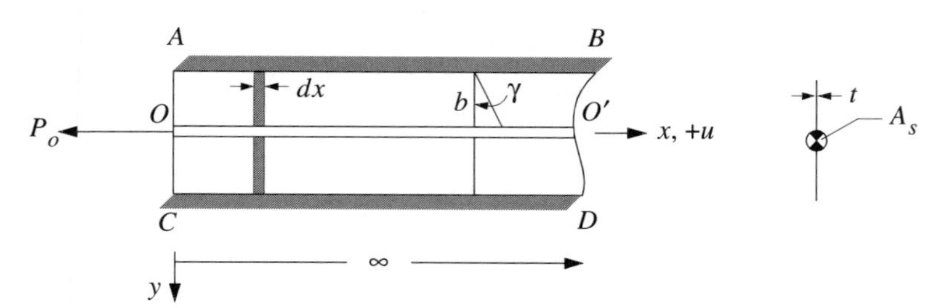

Figure 3.110 Problem 3.60.

$$u = u_0 e^{-qx}$$

where u_0 and q are unknown parameters to be determined by virtual work. Take positive x and u to the right. The strain relationships are of the form

$$\epsilon = \frac{du}{dx} = u'; \qquad \sigma = E\epsilon$$

and

$$\gamma = \frac{u}{b}; \qquad \tau = G\gamma$$

Define the internal virtual work as

$$\delta U = \int_0^\infty \sigma \;\, \delta\epsilon \; A_s \; dx + \int_0^\infty \tau \; \delta\gamma \; (2bt) \; dx$$

Pay particular attention to the differences in form between

$$\frac{d}{dx}\,(e^{-qx}) \quad \text{and} \quad \delta(e^{-qx})$$

(a) Determine q in terms of G, t, A_s, E, and b.
(b) Determine u_0 in terms of G, P_0, A_s, E, and q.

CHAPTER 4

COMPLEMENTARY VIRTUAL WORK

4.1 COMPLEMENTARY VIRTUAL WORK DEFINITION

In a similar fashion to virtual work, we may define the concept of complementary virtual work as a virtual force moving through a real displacement, or

$$\delta W^* = \delta \vec{P} \cdot \vec{u}$$

where δW^* stands for the complementary virtual work. Referring to Fig. 3.1 but considering the unshaded area between $\widehat{AB}$ and the ordinate axis, we can write for ΔW^* the following expression:

$$\Delta W^* = \frac{1}{2}\ [u + (u + \delta u)]\ \delta P = \delta P u + \frac{\delta P\ \delta u}{2}$$

In the above equation, the first-order term $\delta P u$ is defined as the complementary virtual work. The increment of displacement δu and the resulting actual incremental change δP does not enter into the definition of complementary virtual work.

Complementary work W^* is the dual or conjugate of the work W. The area under the curve of Fig. 3.1 represents the total work in moving the load P through a displacement u. If $d^2W/du^2 \neq 0$ (to ensure that P is not independent of u), then Fig. 3.1 implies the existence of a function

$$W = f(u)$$

where

$$P = \frac{dW}{du}$$

is the slope at any point. From Fig. 3.1, it is natural to consider the function $W^* = g(P)$, defined as

$$W^*(P) = Pu - W(u)$$

But this is identical to Eq. (2.20) with no passive variables, and we conclude that W^* is the Legendre transformation of W with the following result:

$$u = \frac{dW^*}{dP}$$

Also, if δ is considered an operator acting on the above expression for W^*, we get the following result:

$$\delta W^* = \delta Pu + P\,\delta u - \delta W$$

However, from the principle of virtual work, $\delta W = P\,\delta u$; hence

$$\delta W^* = \delta Pu$$

as previously stated.

4.2 COMPLEMENTARY VIRTUAL WORK OF A DEFORMABLE BODY

Consider a variational form of the equations of equilibrium represented as

$$\frac{\partial \delta\,\sigma_{xx}}{\partial x} + \frac{\partial \delta \tau_{xy}}{\partial y} + \frac{\partial \delta \tau_{xz}}{\partial z} + \delta X_b = 0 \tag{4.1}$$

$$\frac{\partial \delta\,\sigma_{yy}}{\partial y} + \frac{\partial \delta \tau_{yx}}{\partial x} + \frac{\partial \delta \tau_{yz}}{\partial z} + \delta Y_b = 0 \tag{4.2}$$

$$\frac{\partial \delta\,\sigma_{zz}}{\partial z} + \frac{\partial \delta \tau_{zx}}{\partial x} + \frac{\partial \delta \tau_{zy}}{\partial y} + \delta Z_b = 0 \tag{4.3}$$

with the following corresponding boundary equilibrium conditions:

$$\delta X_s = l\,\delta\sigma_{xx} + m\,\delta\tau_{xy} + n\,\delta\tau_{xz} \tag{4.4}$$

$$\delta Y_s = l\,\delta\tau_{yx} + m\,\delta\sigma_{yy} + n\,\delta\tau_{yz} \tag{4.5}$$

$$\delta Z_s = l\,\delta\tau_{zx} + m\,\delta\tau_{zy} + n\,\delta\sigma_{zz} \tag{4.6}$$

Since we started with a variational form of the equations of equilibrium, we conclude the requirement that variational forces must be self-equilibrating.

We now define the complementary virtual work of a deformable body as

$$\begin{aligned}\delta W^* \stackrel{\text{def}}{=} & \int_V \left(\frac{\partial\delta\sigma_{xx}}{\partial x} + \frac{\partial\delta\tau_{xy}}{\partial y} + \frac{\partial\delta\tau_{xz}}{\partial z} + \delta X_b \right) u\,dV \\ & + \int_V \left(\frac{\partial\delta\sigma_{yy}}{\partial y} + \frac{\partial\delta\tau_{yx}}{\partial x} + \frac{\partial\delta\tau_{yz}}{\partial z} + \delta Y_b \right) v\,dV \\ & + \int_V \left(\frac{\partial\delta\sigma_{zz}}{\partial z} + \frac{\partial\delta\tau_{zx}}{\partial x} + \frac{\partial\delta\tau_{zy}}{\partial y} + \delta Z_b \right) w\,dV = 0\end{aligned} \tag{4.7}$$

where u, v, w are the displacements compatible with the constraints.

Equation (4.7) defines the complementary virtual work for a deformable body. It is important to understand the following fine points of the definition:

- The deformable body is in equilibrium and the virtual forces are self-equilibrating.
- Associated with the deformed body in its equilibrium state is a unique set of values for strain compatible with the external displacements.
- Also associated with the deformed body in its equilibrium state is a unique set of values for strain and displacement that are consistent with the given boundary conditions.
- The virtual forces are superimposed on the true existing forces of the deformable body.
- Since it is complementary virtual work that is defined, it is important to remember that the internal and external deformation distributions remain unchanged by the virtual forces.

Applying Green's first theorem, we obtain

$$
\begin{aligned}
\delta W^* = &- \int_V \left[\delta \sigma_{xx} \frac{\partial u}{\partial x} + \delta \sigma_{yy} \frac{\partial v}{\partial y} + \delta \sigma_{zz} \frac{\partial w}{\partial z} \right. \\
&\left. + \delta \tau_{xy} \left(\frac{\partial u}{\partial y} + \frac{\partial v}{\partial x} \right) + \delta \tau_{yz} \left(\frac{\partial v}{\partial z} + \frac{\partial w}{\partial y} \right) + \delta \tau_{zx} \left(\frac{\partial w}{\partial x} + \frac{\partial u}{\partial z} \right) \right] dV \\
&+ \int_V (\delta X_b u + \delta Y_b v + \delta Z_b w) \, dV + \oint_{\partial V} (\delta X_s u + \delta Y_s v + \delta Z_s w) \, dS
\end{aligned}
\tag{4.8}
$$

The first integral of Eq. (4.8) still needs to be clarified. We have explicitly defined $\delta \sigma_{xx}$ and so on as a set of self-equilibrating internal loads and the u, v, w as the set of true displacements of the deformable body. Therefore, terms such as $\partial u / \partial x$ are related to the strain. Up to this point, the principle is entirely general and valid for large strain[1]; however, we will now restrict the principle to classical strain theory by relating the displacement gradients to the form $\partial u / \partial x = \epsilon_x$ (etc.).

The principle of complementary virtual work then takes the following form:

$$
\begin{aligned}
\delta W^* = &- \int_V (\delta \sigma_{xx} \epsilon_{xx} + \delta \sigma_{yy} \epsilon_{yy} + \delta \sigma_{zz} \epsilon_{zz} \\
&+ \delta \tau_{xy} \gamma_{xy} + \delta \tau_{yz} \gamma_{yz} + \delta \tau_{zx} \gamma_{zx}) \, dV \\
&+ \int_V (\delta X_b u + \delta Y_b v + \delta Z_b w) \, dV + \oint_{\partial V} (\delta X_s u + \delta Y_s v + \delta Z_s w) \, dS
\end{aligned}
\tag{4.9}
$$

The first integral in Eq. (4.9) then represents the total internal complementary virtual work δU^*. The remaining two integrals represent the complementary virtual work of the applied body forces and surface forces denoted by δW_e^*. Then, Eq. (4.9) may be expressed as

$$
\delta W^* = -\delta U^* + \delta W_e^* = 0
$$

or

[1]While the principle of complementary virtual work and the principle of stationary enengy (to be discussed later) are valid for the generalized nonlinear case, their application under nonlinear circumstances is difficult because of the coupling between strain and stress components. An article by Koiter (ref. [42]) presents the Zubov-Koiter formulation in terms of the second Piola stress tensor, allowing for the solution of the strain tensor in terms of a symmetric tensor formed from the Piola tensor. As noted by the article, solutions are then extremely difficult.

$$\delta W_e^* = \delta U^* \tag{4.10}$$

Thus we may state the following:

The strains and displacements are compatible and consistent with the constraints if the sum of the total external complementary virtual work and the total internal complementary virtual work is zero for every system of virtual forces and stresses that satisfy equilibrium.

We have proved the necessary condition for the principle of complementary virtual work. To prove sufficiency (for small-strain theory), we start with the identities

$$\epsilon_x - \frac{\partial u}{\partial x} = 0$$

$$\epsilon_y - \frac{\partial v}{\partial y} = 0$$

$$\epsilon_z - \frac{\partial w}{\partial z} = 0$$

$$\gamma_{xy} - \frac{\partial u}{\partial y} - \frac{\partial v}{\partial x} = 0$$

$$\gamma_{yz} - \frac{\partial v}{\partial z} - \frac{\partial w}{\partial y} = 0$$

$$\gamma_{zx} - \frac{\partial w}{\partial x} - \frac{\partial u}{\partial z} = 0$$

in the interior, and

$$u - \tilde{u} = 0$$
$$v - \tilde{v} = 0$$
$$w - \tilde{w} = 0$$

on the boundary, where the tildes represent prescribed boundary displacements, and then form the following:

$$
\begin{aligned}
\delta W^* = -\int_V \Bigg[& \left(\epsilon_x - \frac{\partial u}{\partial x}\right) \delta\sigma_x + \left(\epsilon_y - \frac{\partial v}{\partial y}\right) \delta\sigma_y + \left(\epsilon_z - \frac{\partial w}{\partial z}\right) \delta\sigma_z \\
& + \left(\gamma_{xy} - \frac{\partial u}{\partial y} - \frac{\partial v}{\partial x}\right) \delta\tau_{xy} + \left(\gamma_{yz} - \frac{\partial v}{\partial z} - \frac{\partial w}{\partial y}\right) \delta\tau_{yz} \\
& + \left(\gamma_{zx} - \frac{\partial w}{\partial x} - \frac{\partial u}{\partial z}\right) \delta\tau_{zx} \Bigg] \, dV \\
& + \int_V (\delta X_b u + \delta Y_b v + \delta Z_b w) \, dV \\
& + \oint_{\partial V} [\delta X_s(u - \tilde{u}) + \delta Y_s(v - \tilde{v}) + \delta Z_s(w - \tilde{w})] \, dS
\end{aligned}
$$

After integrating by parts and using the requirement that the variational equilibrium equations be satisfied identically in the interior and on the boundary, the above equation reduces to

$$\delta W^* = -\delta U^* + \delta W_e^* = 0$$

as before.

Several points about the principle of complementary virtual work should be noted.

- The principle holds irrespective of any material stress-strain relationships.
- The principle is an alternative way of expressing compatibility conditions. Thus, once a self-equilibrating load set has been assumed, a set of compatibility equations results.
- The principle has nothing to do with the conservation of energy.

The point is often made that in applications, because true strains depend upon material properties, unlike the principle of virtual work the principle of complementary virtual work requires that material properties be known in advance of its use. Such a statement implies that the principle of complementary virtual work is less general than the principle of virtual work. It should be pointed out, however, that with the exception of statically determinate structures, the principle of virtual work also requires that the material properties be known in advance.

As an example, consider the beam shown in Fig. 4.1. The beam is deformed under a system of loads not shown to yield the given displacements in the *YX* plane. Then, a system of self-equilibrating virtual loads is applied. Considering the stress resultants at the boundary as external loading, we have for the external complementary virtual work

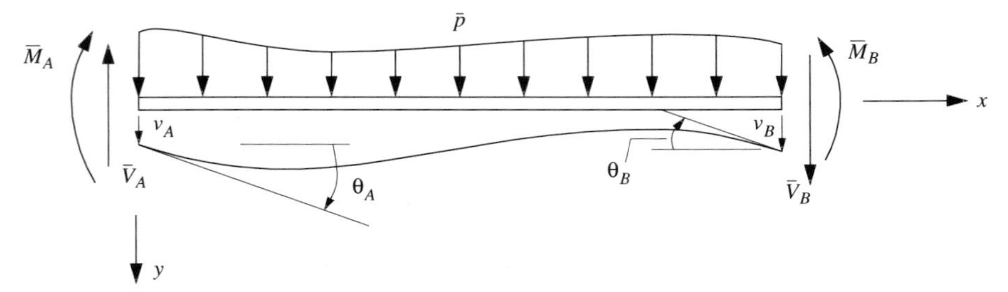

Figure 4.1 Beam with virtual loads.

$$\delta W_e^* = \int_A^B \bar{p}v\, dx + (\overline{V}_B v_B - \overline{V}_A v_A - \overline{M}_B \theta_B + \overline{M}_A \theta_A) \tag{4.11}$$

The complementary internal virtual work is

$$\delta U^* = \int_V \epsilon_x \overline{\sigma}_x\, dV = \int_V (-yv'')\overline{\sigma}_x\, dA\, dx \tag{4.12}$$

where we have used $\epsilon_x = -yv''$ for the strain, and where dA represents the beam's differential cross-sectional area and dx represents a differential beam length. In the second integral of δU^* is the term

$$\int_A y\overline{\sigma}_x\, dA = \int_A d\overline{M} = \overline{M}$$

which is the virtual internal moment.

Assuming principal axes and bending only about the z axis and using the second term of Eq. (3.63) under the assumption of a linear-elastic beam, we get the following result:

$$\delta U^* = \int_A^B \frac{M + M_T}{EI}\, \overline{M}\, dx \tag{4.13}$$

A clear understanding of the terms in Eq. (4.13) is vital. The M_T is due to any prescribed thermal loading. The M is caused by the actual loading of the beam and is the true internal moment. The $\overline{M}$ is the internal virtual bending moment from any convenient self-equilibrating virtual loading system. Such a virtual loading system is chosen in any manner that facilitates a solution.

Example 4.1 As an example, consider the beam shown in Fig. 3.22 under the

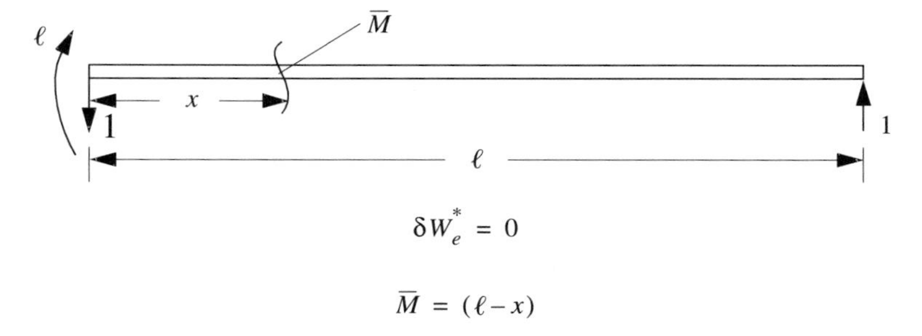

$$\delta W_e^* = 0$$

$$\overline{M} = (\ell - x)$$

Figure 4.2 A virtual load.

temperature distribution $\Delta T = T_0[1 + y/(2h)]$ and let us determine directly the vertical reaction R under the right support. We use Eq. (4.10) in the following form:

$$\delta U^* = \int \frac{M + M_T}{EI} \overline{M}\, dx = \delta W_e^*$$

The actual bending moment at any section is given by

$$M = R(\ell - x) - p\, \frac{(\ell - x)^2}{2}$$

For the given temperature distribution, we have

$$\begin{aligned} M_T &= \alpha E \int \Delta T y\, dA \\ &= \alpha E T_0 \int \left(1 + \frac{y}{2h}\right) y\, dA \\ &= \frac{\alpha E I T_0}{2h} \end{aligned}$$

All that remains is determining a virtual loading system that is in equilibrium and suitable as an aid in solving the problem. To this end, consider the beam of length ℓ shown in Fig. 4.2.

The virtual loading system shown is in equilibrium. The complementary virtual work of the applied loads is zero; hence, $\delta W_e^* = 0$. This is because the complementary virtual work of the right support is a virtual load of $1 \times v_B$, where v_B is the true or actual vertical displacement at the right end of the beam shown in Fig. 3.22. Also, the left end of the beam in Fig. 3.22 is clamped; hence the complementary virtual work of not only the virtual moment ℓ but also the left vertical reaction 1 is zero. Therefore, the expression for complementary virtual work becomes

$$\delta U^* = \int_0^\ell \left[\frac{R(\ell - x)}{EI} - \frac{p(\ell - x)^2}{2EI} + \frac{\alpha T_0}{2h} \right] (\ell - x)\, dx = 0$$

where from Fig. 4.2 we have determined $\overline{M}$ to be

$$\overline{M} = (\ell - x)$$

Expanding the integral for δU^*, we obtain the reaction R as

$$R = \frac{3p\ell}{8} - \frac{3\alpha EIT_0}{4h\ell}$$

For simple straight beams, we have the following general expression for δU^* for bending only:

$$\begin{aligned} \delta U^* &= \int \epsilon_x \overline{\sigma}_x \, dV \\ &= \int (u_0' - yv'' - zw'') \overline{\sigma}_x \, dA \, dx \end{aligned}$$

From Eq. (3.61) with σ replaced by $\overline{\sigma}$ and by the use of Eq. (3.63), we get the following result:

$$\begin{aligned} \delta U^* = \int \bigg[& \frac{N + N_T}{AE} \overline{N} + \frac{(M_z + M_{Tz})I_y - (M_y + M_{Ty})I_{yz}}{E(I_y I_z - I_{yz}^2)} \overline{M}_z \\ & + \frac{(M_y + M_{Ty})I_z - (M_z + M_{Tz})I_{yz}}{E(I_y I_z - I_{yz}^2)} \overline{M}_y \bigg] \, dx \end{aligned} \tag{4.14}$$

For an initially curved beam, we have the following general expression for δU^* for bending only:

$$\delta U^* = \int \epsilon_x \overline{\sigma}_x \, dV$$

$$= \left[\left(u_0' - \frac{v}{R} \right) - \left(\frac{v}{R^2} + v'' \right) \frac{y}{1 - y/R} - \frac{zw''}{1 - y/R} - \frac{\eta}{R} \frac{1}{1 - y/R} \right] \overline{\sigma}_s \, dA(1 - y/R) \, ds$$

From Eq. (3.61) with σ replaced by $\overline{\sigma}$ and by the use of Eq. (3.124), we get the following result:

$$\begin{aligned} \delta U^* = \int \Bigg[& \left(\frac{N + \hat{N}_T}{AE} - \frac{M_z + \hat{M}_{Tz}}{RAE} \right) \overline{N} \\ & + \Bigg(\frac{(M_z + \hat{M}_{Tz})\hat{I}_y - (M_y + \hat{M}_{Ty})\hat{I}_{yz}}{E(\hat{I}_y \hat{I}_z - \hat{I}_{yz}^2)} \\ & + \frac{M_z + \hat{M}_{Tz}}{R^2 AE} - \frac{N + \hat{N}_T}{RAE} \Bigg) \overline{M}_z \\ & + \frac{(M_y + \hat{M}_{Ty})\hat{I}_z - (M_z + \hat{M}_{Tz})\hat{I}_{yz}}{E(\hat{I}_y \hat{I}_z - \hat{I}_{yz}^2)} \overline{M}_y \Bigg] ds \end{aligned} \tag{4.15}$$

In obtaining the above equation, we assumed that

$$\frac{1}{R} \int_A \left(\int_0^y \alpha \Delta T \, dy \right) \overline{\sigma} \, dA \ll \frac{\overline{M}_z}{R}$$

For thin walled beams of open section, we have the following expression for δU^* for bending only:

$$\delta U^* = \int \epsilon_x \overline{\sigma}_x \, dV$$

$$= \int (u_0' - yv'' - zw'' - \omega\phi'')\overline{\sigma}_x \, dA \, dx$$

From Eqs. (3.61) and (3.163) with σ replaced by $\overline{\sigma}$ and by the use of Eq. (3.177), we get the following result:

$$\delta U^* = \int \left[\frac{N + N_T}{AE} \overline{N} + \frac{(M_z + M_{Tz})I_y - (M_y + M_{Ty})I_{yz}}{E(I_y I_z - I_{yz}^2)} \overline{M}_z \right.$$

$$+ \frac{(M_y + M_{Ty})I_z - (M_z + M_{Tz})I_{yz}}{E(I_y I_z - I_{yz}^2)} \overline{M}_y$$

$$\left. + \frac{M_\omega + M_{T\omega}}{EI_\omega} \overline{M}_\omega \right] dx \tag{4.16}$$

for the case of principal sectorial centroid and principal pole. From Eqs. (3.61) and (3.185) with σ replaced by $\overline{\sigma}$ and by the use of Eq. (3.187), we get the following result:

$$\delta U^* = \int \left[\frac{N + N_T}{AE} \overline{N} + \frac{(M_z + M_{Tz})I_y - (M_y + M_{Ty})I_{yz}}{E(I_y I_z - I_{yz}^2)} \overline{M}_z \right.$$

$$+ \frac{(M_y + M_{Ty})I_z - (M_z + M_{Tz})I_{yz}}{E(I_y I_z - I_{yz}^2)} \overline{M}_y$$

$$\left. + \frac{M_{\omega 0} + M_{T\omega 0}}{EI_{\omega 0}} \overline{M}_{\omega 0} \right] dx \tag{4.17}$$

for the case of arbitrary sectorial origin and principal pole.

The complementary virtual work for beams of circular or hollow cross section is obtained under the Coulomb assumption that the shear stress τ_{SV} in the plane of the cross section is everywhere perpendicular to the radius drawn from the center of the cross section to the point of interest. If ϕ represents the angle of twist of a circular cross section, then the Coulomb assumption, which implies that there is no cross-sectional warping, results in the following expression for the shear strain γ_{xs}:

$$\gamma_{xs} = r\phi' \tag{4.18}$$

where

- r is the radius to a point on the cross section
- x is measured along the beam axis
- s is measured along the arc of the cross section

If G is the shear modulus, for elastic deformation we get the result for shear stress τ_{SV} as

$$\tau_{SV} = Gr\phi' \tag{4.19}$$

If we define the torsional stress resultant T_{SV} as

$$T_{SV} = \int r\tau_{SV}\, dA$$

and compute the virtual work of the above integral for a virtual rotation $\overline{\phi}$, we get the well-known result

$$T_{SV} = I_p G\phi' \tag{4.20}$$

where I_p is the polar second moment of area defined as

$$I_p = \int r^2\, dA$$

For a circular cross section, $\phi' = d\phi/dx = \theta$ = constant. The complementary virtual work is then defined as

$$\begin{aligned}\delta U^* &= \int \gamma_{xs}\overline{\tau}_{SV}\, dV \\ &= \int r\phi'\overline{\tau}_{SV}\, dA\, dx\end{aligned}$$

or

$$\delta U^* = \int \frac{T_{SV}}{GI_p}\, \overline{T}_{SV}\, dx \tag{4.21}$$

Equation (4.21) is also reasonably accurate for curved beams of circular cross section. For curved beams with a deep, circular cross section, the above expression tends to underpredict the stresses on the side nearest the radius of curvature of the beam.

For thin walled beams of open section, the contribution of Saint Venant's torsional shear τ_{SV} to the complementary virtual work δU^* is

$$\begin{aligned}\delta U^* &= \int \gamma_{xs}\overline{\tau}_{SV}\, dV \\ &= \int 2n\phi'\overline{\tau}_{SV}\, dA\, dx\end{aligned}$$

where we have used Eq. (3.194). From Eq. (3.214b) with τ_{SV} replaced by $\overline{\tau}_{SV}$ in the above expression and Eq. (3.214d), we get the following result:

$$\delta U^* = \int \frac{T_{SV}}{GJ} \overline{T}_{SV} \, dx \tag{4.22}$$

Reference [43] shows that Eq. (4.20) with I_p replaced by $\mathcal{J}$, a generalized torsion constant dependent on the cross-sectional shape, is a fundamental relationship for torsion under the condition of unrestrained warping. Thus the complementary virtual work of a beam of any cross-sectional shape under Saint Venant's torsion can be expressed as

$$\delta U^* = \int \frac{T_{SV}}{G\mathcal{J}} \overline{T}_{SV} \, dx \tag{4.23}$$

where $\mathcal{J}$ for various sections is given by Roark and Young in *Formulas for Stress and Strain* (pp. 290–294). Saint Venant arrived at the following approximate expression for solid sections with no reentrant corners:

$$\mathcal{J} = \frac{A^4}{40I_p} \tag{4.24}$$

where A is the area of the cross section and I_p is the polar second moment of area. The constant is $4\pi^2 \approx 40$.

In the previous discussions of shear in beams, it was noted that the assumption of plane sections precluded the existence of a compatible virtual strain. Thus, under the assumption of plane sections, the virtual work principle does not yield a result for transverse shear deformation. Complementary virtual work, however, provides a simple approximate formulation for transverse shear deformation that is acceptably accurate under many design conditions.

For simplicity, consider a symmetrical straight beam with bending in the yx plane and let τ be the average shear in the yx plane. Then, from Section 3.12,

$$\tau = \frac{V_y Q_z}{b I_z}$$

Assuming a linear-elastic material, we may write

$$\gamma = \frac{\tau}{G}$$

The complementary virtual work δU^* is now given as

$$
\begin{aligned}
\delta U^* &= \int \gamma \overline{\tau}\, dV \\
&= \int \frac{V_y Q_z}{G b I_z} \frac{\overline{V}_y Q_z}{b I_z}\, dV \\
&= \int \frac{V_y}{G} \overline{V}_y \left(\frac{1}{I_z^2} \int \frac{Q_z^2}{b^2}\, dA \right) dx
\end{aligned}
$$

If we now define a nondimensional form factor f_s as

$$
f_s = \frac{A}{I_z^2} \int \frac{Q_z^2}{b^2}\, dA \tag{4.25}
$$

then we have

$$
\delta U^* = \int \frac{f_s V_y}{AG} \overline{V}_y\, dx \tag{4.26}
$$

When used to compute beam shear deflections and when compared against elasticity solutions for beam deflections, Eq. (4.26) gives reasonable results. However, these results are sensitive to boundary conditions: for example, whether a beam cantilevered at one end and loaded with a concentrated load at the other will or will not undergo transverse shear deformation depends on how warping is constrained at the built-in end. The solution obtained using Eq. (4.26) matches elasticity solutions that assume that at the built-in end warping occurs freely, and also that the vertical side of a differential element located at the neutral axis remains vertical.

For solid cross sections, Eq. (4.25) is used directly. For example, consider a rectangular beam of height h and width b. Equation (4.25) then yields

$$
f_s = \frac{144}{bh^5} \int_{-h/2}^{h/2} \frac{1}{4} \left(\frac{h^2}{4} - y^2 \right) b\, dy = \frac{6}{5}
$$

For beams of open section or for beams with cutouts, the assumption is usually made that the shear is uniformly distributed over the height of the web and that

$$
f_s = \frac{A}{A_{\text{web}}}
$$

As mentioned in the discussion of the virtual work of curved beams, when the cross section is in the shape of an **I** or **T** section, values of deflection are too small because of the distortion of the flanges. In this case, a modified area

(A_{modified}) and a modified second moment of area ($I_{z\,\text{modified}}$) are determined by computing an effective flange width, as developed by Bleich and listed in Table 6 of ref. [30]. Then, for curved beams of **I** or **T** cross section, terms such as

$$\int \frac{N}{AE}\,\overline{N}\,ds$$

and

$$\int \frac{M_z}{EI_z}\,\overline{M}_z\,ds$$

become

$$\int \frac{N}{A_{\text{modified}}E}\,\overline{N}\,ds$$

and

$$\int \frac{M_z}{EI_{z\,\text{modified}}}\,\overline{M}_z\,ds$$

For curved beams in which the depth of the section is small relative to the radius of curvature (say, a ratio of <0.2), Eq. (4.15) is replaced by Eq. (4.14).

4.3 SYMMETRY

Many structures are made up of nearly identical segments that are symmetrically arranged with respect to an axis. Structures of this type fall into the category of simple rotational symmetry, in which the segments may in general be double-curved surfaces, and of dihedral symmetry, in which each segment has a plane of reflective symmetry and the boundaries between segments are planar. Principles of cyclic symmetry can reduce the analysis region to a single segment in the case of dihedral symmetry and to a pair of segments in the case of simple rotational symmetry (see refs. [44] and [45]). If u^n is any component of displacement, force, stress, and so on in the nth segment, and if $\alpha = 2\pi/N$ is the angle between segments (with N the number of segments), then the periodicity of u^n can be expressed as a finite Fourier transform [46]. While this is an important form of structural symmetry, it is not pursued here any further because its efficient application is best conducted through computer techniques.

Another form of symmetry is reflective symmetry. This form often permits the reduction of a structure to a half- or quarter-model about the plane or planes of symmetry. In reflective symmetry, the properties of the structure are considered mirror images about the plane of symmetry. Antisymmetry can also be considered where the properties of the structure are mirrored in the reverse sense.

The loads on the structure need not be symmetrical. A general loading is called asymmetric. Such a loading may be broken into a symmetric portion and an antisymmetric portion. When this is done, the symmetric loads are applied with symmetric boundary conditions enforced on the plane of symmetry, and the antisymmetric loads are applied with antisymmetric boundary conditions enforced on the plane of antisymmetry. If $P_R(x, y, z)$ is the portion of the asymmetric load acting on the right side of a plane of symmetry, and if $P_L(x, -y, z)$ is the portion of the asymmetric load acting on the left side of the plane of symmetry where the reflection is assumed to be about the y axis, then the symmetric load is defined as

$$P_S = \frac{P_R + P_L}{2}$$

If the right side is modeled, then the antisymmetric load is defined as

$$P_A = \frac{P_R - P_L}{2}$$

If the left side is modeled, then the antisymmetric load is defined as

$$P_A = \frac{P_L - P_R}{2}$$

The general solution is obtained by the addition of the results of the symmetric model (consisting of its symmetric loads and symmetric boundary conditions) and the antisymmetric model (consisting of its antisymmetric loads and antisymmetric boundary conditions). In general, this requires that the behavior of the structure be linear.

Figures 4.3–4.6 show the required boundary conditions, applied load, and boundary reaction forces for symmetry and antisymmetry planes. From Fig. 4.3, note that there can be no direct applied load component normal to and acting directly on the plane of symmetry or applied moment component parallel to and acting directly on the plane of symmetry—thus the requirement that $P_y = 0$, $M_x = 0$, and $M_z = 0$ on the plane of symmetry. Since any motion out of the plane of symmetry would destroy the symmetry plane, the symmetry plane must be constrained so that $u_y = 0$, $\theta_x = 0$, and $\theta_z = 0$.

From Fig. 4.4, note that there can be no direct applied load component parallel to and acting directly on the plane of antisymmetry or an applied moment component normal to and acting directly on the plane of antisymmetry—thus the requirement that $P_x = 0$, $P_z = 0$, and $M_y = 0$ on the plane of antisymmetry.

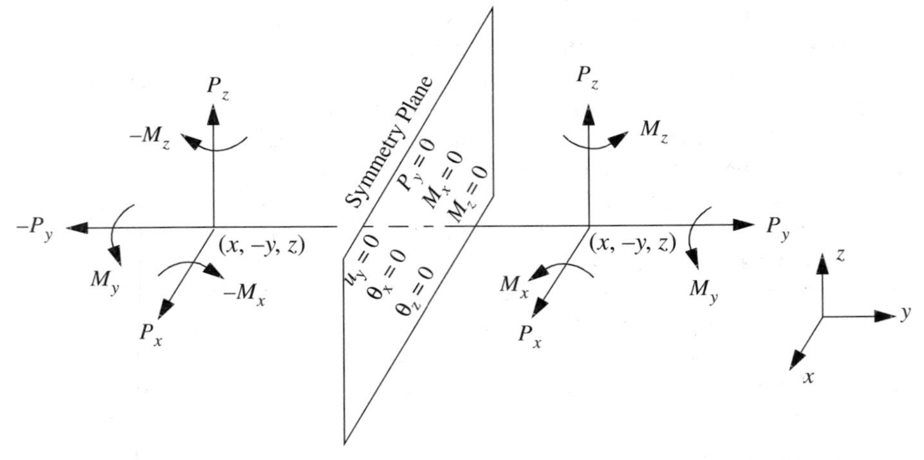

Figure 4.3 Symmetry plane with boundary conditions and applied loads.

Since any motion in the plane of antisymmetry would destroy the antisymmetry plane, the antisymmetry plane must be constrained so that $u_x = 0$, $u_z = 0$, and $\theta_y = 0$.

In Figs. 4.5 and 4.6, $\mathcal{N}_y$ is the direct normal force; $\mathcal{S}_x$ and $\mathcal{S}_z$ are the two in-plane boundary shear forces; and $\mathcal{M}_x$, $\mathcal{M}_y$, and $\mathcal{M}_z$ are the three boundary moments. From Fig. 4.5, note that boundary displacements required for symmetry dictate that there be nonzero reactions $\mathcal{N}_y$, $\mathcal{M}_x$, $\mathcal{M}_z$; from Fig. 4.6, note that boundary displacements required for antisymmetry dictate that there be nonzero reactions $\mathcal{M}_y$, $\mathcal{S}_x$, $\mathcal{S}_z$.

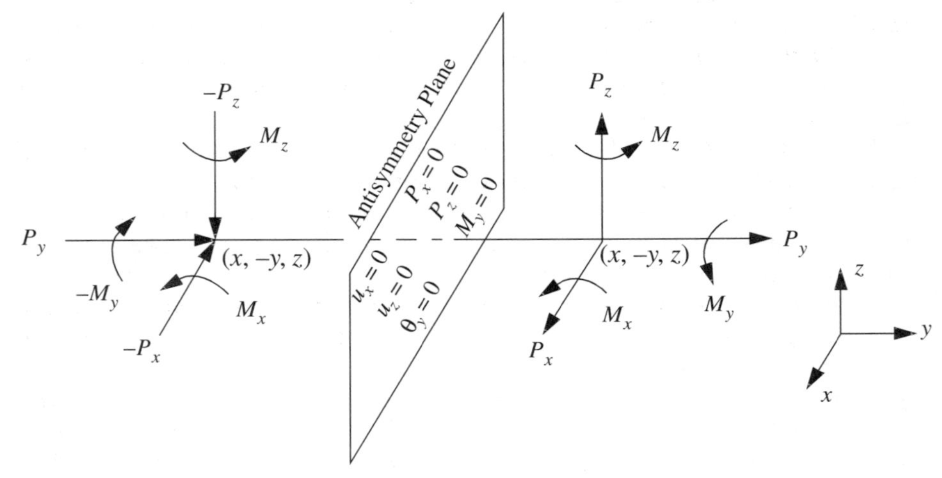

Figure 4.4 Antisymmetry plane with boundary conditions and applied loads.

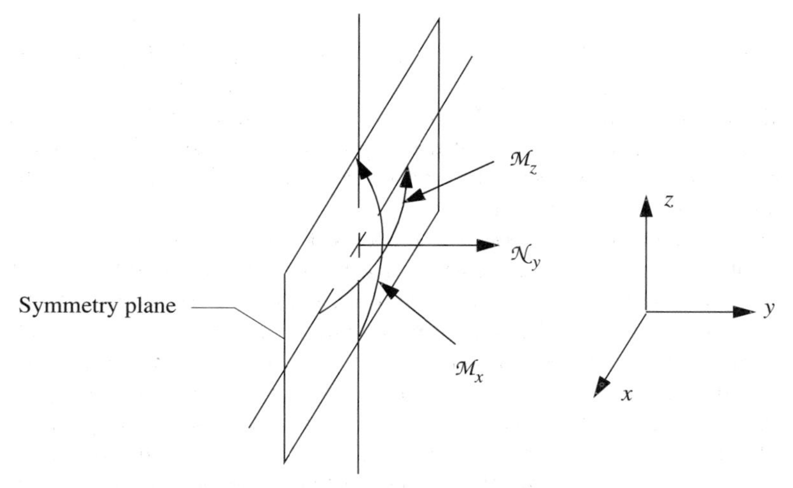

Figure 4.5 Symmetry plane with boundary reactions.

Example 4.2 To illustrate the method, consider the frame shown in Fig. 4.7(a) under the action of *EI* bending only. Let the bending stiffness *EI* be the same in all members. The vertical members have length 2ℓ and the horizontal member has length ℓ. The vertical members are fully restrained at *A* and *B*. Figure 4.7(b) shows the asymmetrical load broken down into its symmetrical and antisymmetrical components. Figure 4.7(c) shows a free body of the right portion of the frame under the condition of the symmetrical load and symmetrical boundary

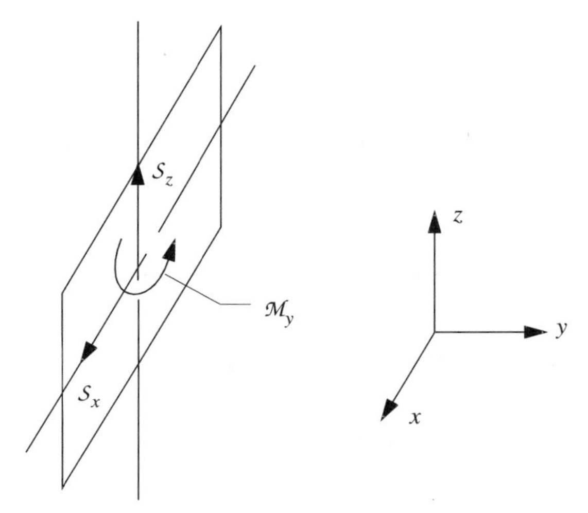

Figure 4.6 Antisymmetry plane with boundary reactions.

condition. Figures 4.7(d) and (e) show sets of self-equilibrating virtual loads corresponding to the virtual moment at the symmetry boundary and a virtual direct force at the symmetry boundary, respectively. Figure 4.7(f) shows a free body of the right portion of the frame under the condition of the antisymmetrical load and the antisymmetrical boundary condition. Figure 4.7(g) shows the corresponding virtual load system for a virtual shear acting under the condition of the antisymmetrical boundary.

All the virtual load systems shown have been chosen so that the external complementary virtual work δW_e^* is zero. By definition, δW_e^* is a virtual load acting through a real displacement (or rotation). Since the frame is fully restrained at B, all the real displacements and rotations there are zero. Hence the complementary virtual work of all the virtual forces at B is zero. Since on the symmetry boundary there is the constraint requirement that the symmetry plane neither rotates about an axis normal to the plane of the page nor translates left or right in the plane of the page, these real symmetry displacements are required to be zero. Thus the complementary virtual work of the moment $\overline{M}_C = 1$ and $\overline{N}_C = 1$ is zero. The antisymmetry condition requires that the plane of antisymmetry not translate up or down in the plane of the page. Thus in this direction the antisymmetry displacement is zero, and the complementary virtual work of the virtual shear force $\overline{V}_C = 1$ is zero.

From Figs. 4.7(c) and (d) as well as 4.7(c) and (e), we get for the complementary virtual work

$$\frac{1}{EI}\int_0^{\ell/2}\left(M_C - \frac{p_0 x^2}{4}\right)(1)\,dx + \frac{1}{EI}\int_0^{2\ell}\left(M_C + N_C y - \frac{p_0\ell^2}{16}\right)(1)\,dy = 0 \tag{4.27}$$

and

$$\frac{1}{EI}\int_0^{2\ell}\left(M_C + N_C y - \frac{p_0\ell^2}{16}\right)(y)\,dy = 0 \tag{4.28}$$

respectively.

Expanding, we get the following two expressions:

$$M_C + \frac{4}{5}\,\ell N_C = \frac{13}{240}\,p_0\ell^2$$

$$M_C + \frac{4}{3}\,\ell N_C = \frac{1}{16}\,p_0\ell^2 \tag{4.29}$$

Solving, we get

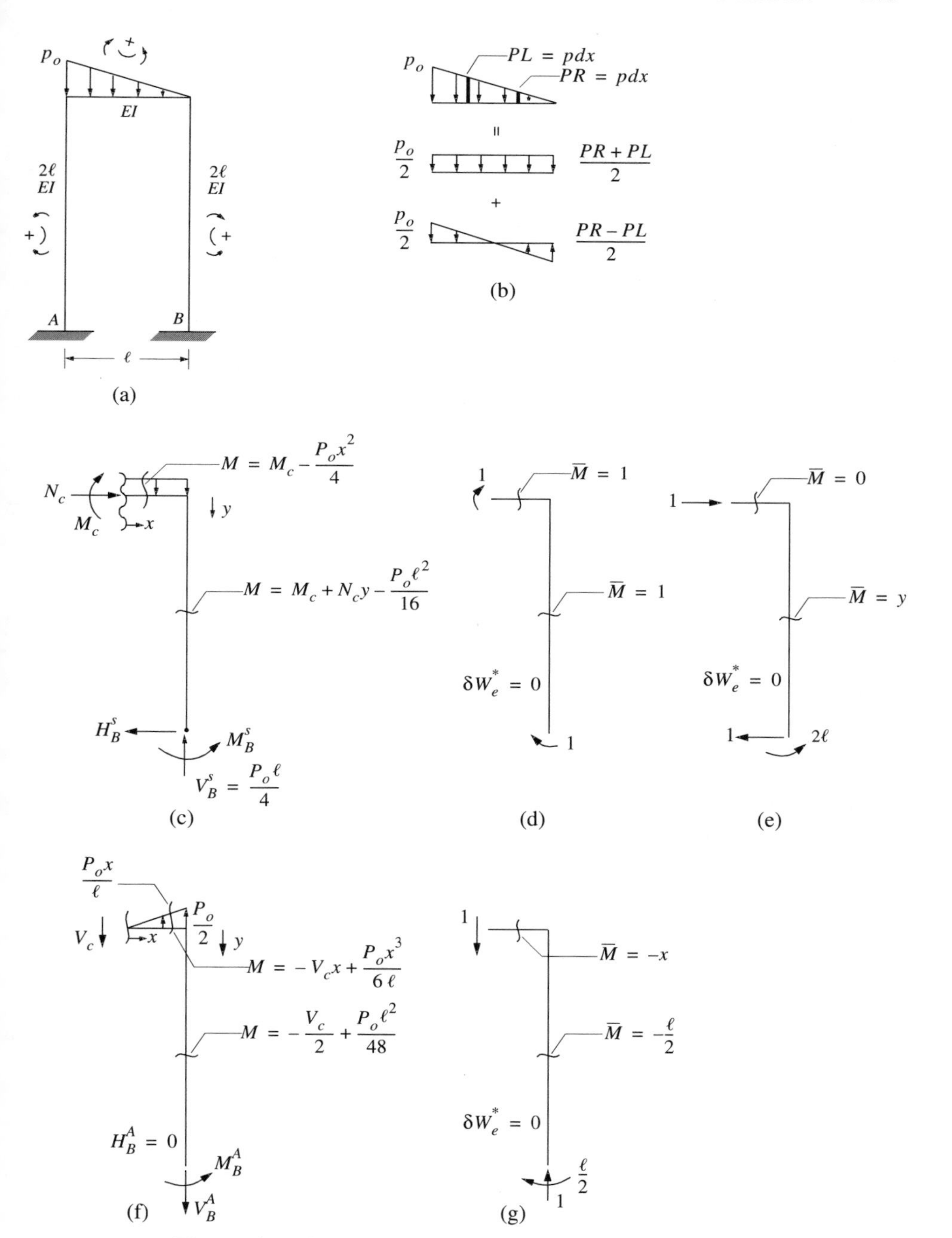

Figure 4.7 Symmetric structure with asymmetrical loading.

$$M_C = \frac{1}{24}\, p_0 \ell^2$$

$$N_C = \frac{1}{64}\, p_0 \ell \tag{4.30}$$

From Fig. 4.7(c), static equilibrium yields the following for the reactions at B:

$$M_B^S = \frac{1}{96}\, p_0 \ell^2$$

$$H_B^S = \frac{1}{64}\, p_0 \ell$$

$$V_B^S = \frac{1}{4}\, p_0 \ell \tag{4.31}$$

where the superscript S represents the symmetrical components.

Figures 4.7(f) and (g) yield the complementary virtual work for the antisymmetry condition as

$$\frac{1}{EI}\int_0^{\ell/2}\left(-V_C x + \frac{p_0 x^3}{6\ell}\right)(-x)\,dx + \frac{1}{EI}\int_0^{2\ell}\left(-\frac{V_C \ell}{2} + \frac{p_0 \ell^2}{48}\right)\left(-\frac{\ell}{2}\right)\,dy = 0 \tag{4.32}$$

which yields

$$V_C = \frac{21}{520}\, p_0 \ell \tag{4.33}$$

From Fig. 4.7(f), static equilibrium yields the following for the reactions at B:

$$M_B^A = \frac{1}{1560}\, p_0 \ell^2$$

$$H_B^A = 0$$

$$V_B^A = \frac{11}{130}\, p_0 \ell \tag{4.34}$$

where the superscript A represents the antisymmetrical components.

Adding Eqs. (4.31) and (4.34), we get the following for the total reaction at B:

$$M_B = M_B^S + M_B^A = \frac{1}{96} p_0 \ell^2 + \frac{1}{1560} p_0 \ell^2 = \frac{23}{2080} p_0 \ell^2$$

$$H_B = H_B^S + H_B^A = \frac{1}{64} p_0 \ell + 0 = \frac{1}{64} p_0 \ell$$

$$V_B = V_B^S + V_B^A = \frac{1}{4} p_0 \ell - \frac{11}{130} p_0 \ell = \frac{43}{260} p_0 \ell \tag{4.35}$$

4.4 THE UNIT LOAD METHOD

The examples that we have thus far discussed were in actuality solved by the unit load method. The method isolates a single displacement in a structure by applying a unit virtual load to the structure at the point and in the direction of the required displacement.

In symbolic form, we can state

(Applied unit load) (Actual displacement at point of application of unit load)
= (Internal forces statically equivalent to applied unit load)
· (Actual internal deformation)

In other words, the true displacement at a given point and direction of a deformable body is equal to the integral over the volume of the product of the true strains ϵ_x, $\epsilon_y, \ldots, \gamma_{yz}$ and the virtual stresses $\delta\sigma_x^1$, $\delta\sigma_y^1, \ldots, \delta\tau_{yz}^1$ produced by a unit virtual load $\delta P = 1$ applied at the same point and direction as the required displacement. The superscript 1 is to indicate that the virtual stresses arise because of the applied unit load. To formalize, consider a virtual load $\overline{P}_u = 1$ applied in the direction u and its corresponding internal virtual stress $\{\overline{\sigma}\}_u$. The subscript used is to indicate that the stress arises because of the virtual load. Next, consider a true load $P_v = 1$ applied in the direction v. This will generate a true internal strain pattern $\lfloor \epsilon \rfloor_v$, where the subscript is used to indicate that the strain is a true strain because of the real load P_v. Define the flexibility g_{uv} as the displacement produced in the direction of u from the application of a unit load in the direction of v. Then, complementary virtual work gives

$$(\overline{P}_u = 1) \cdot g_{uv} = \int_V \lfloor \epsilon \rfloor_v \{\overline{\sigma}\}_u \, dV \tag{4.36}$$

The right-hand side of Eq. (4.36) is composed of the various terms represented by Eqs. (4.14)–(4.17) for bending contribution, Eqs. (4.21)–(4.23) for torsion contributions, and Eq. (4.26) for transverse shear contribution. For example, considering a beam with *EI* stiffness only, $\overline{\sigma} \equiv \overline{M}$ and $\epsilon \equiv d\theta = [M/(EI)]\, dx$, where $d\theta$ is the rotation caused by the bending and the $\equiv$ symbol stands for "is equivalent to."

By using arguments similar to those used to show that for virtual displacements the Maxwell-Betti reciprocity law takes the form $k_{uv} = k_{vu}$, we can show that for virtual loads this reciprocity law takes the following form:

$$g_{uv} = g_{vu}$$

However, the Maxwell-Betti reciprocity law can be shown to be a property of the differential operator. In Section 2.6, we discussed self-adjoint operators. The operator

$$L = \frac{d^2}{dx^2} EI \frac{d^2}{dx^2}$$

can, by the methods of Section 2.6, be shown to be self-adjoint. The self-adjointness is represented by the expression

$$\int vLu \, dx = \int uLv \, dx$$

For a simple illustration, consider a symmetrically simple elastic beam under bending in the xy plane with $M_z = M$ and so on. Its equation of equilibrium is

$$\frac{d^2}{dx^2} EI \frac{d^2 v}{dx^2} = p(x)$$

In operator form, it is

$$Lv = p$$

Consider a system of loads I with associated moments M_I and displacements v_I. Form the inner product of L_{v_I} and v_{II} (the displacement associated with another system of loads II with corresponding moments M_{II}) to get the expression

$$\int v_{II} \frac{d^2}{dx^2} EI \frac{d^2 v_I}{dx^2} \, dx$$

Integrate twice by parts to get

$$\left(v_{II} \frac{d}{dx} EI \frac{d^2 v_I}{dx^2} - \frac{dv_{II}}{dx} EI \frac{d^2 v_I}{dx^2} \right) \bigg| + \int \frac{d^2 v_{II}}{dx^2} EI \frac{d^2 v_I}{dx^2} \, dx \bigg|$$

Next, consider the inner product

$$\int v_I \frac{d^2}{dx^2} EI \frac{d^2 v_{II}}{dx^2} dx$$

and integrate twice by parts to get

$$\left(v_I \frac{d}{dx} EI \frac{d^2 v_{II}}{dx^2} - \frac{dv_I}{dx} EI \frac{d^2 v_{II}}{dx^2} \right) \bigg| + \int \frac{d^2 v_I}{dx^2} EI \frac{d^2 v_{II}}{dx^2} dx \bigg|$$

The self-adjointness implies that

$$(\theta_{II} M_I - v_{II} V_I)| = (\theta_I M_{II} - v_I V_{II})|$$

where we have used identities such as $\theta = dv/dx$ for the beam rotation, $d(EId^2v/dx^2)/dx = -V$ for the beam shear force, and so forth.

The above expression states that the work done by a system of forces *I*, through displacements caused by a second system of forces *II*, equals the work done by the second system of forces *II* through displacements caused by the first system of forces *I*. This is the Maxwell-Betti reciprocity law.

Approximate methods, such as the finite element method, take operators such as L and replace them with expressions that are often written in matrix form. In the complementary virtual work method, we have

$$u_i = \sum_{j=1}^{N} g_{ij} P_j$$

where g_{ij} is the structural flexibility matrix for the constrained structure, P_j is the applied loading, and u_i is the vector of nodal displacements. The operator L is

$$L = \sum_{j=1}^{N} g_{ij}$$

and we have

$$L[\{P\}] = [G]\{P\} = \{u\}$$

where $[G]$ is the assembled matrix of g_{ij}s

In the virtual work method, we have

$$P_i = \sum_{j=1}^{N} k_{ij} u_j$$

where k_{ij} is the structural stiffness matrix for the constrained structure. Then L is defined as

$$L = \sum_{j=1}^{N} k_{ij}$$

and we have

$$L[\{u\}] = [K]\{u\} = \{P\}$$

Since the original differential operators were self-adjoint, the requirement on the matrix operators is that

$$g_{ij} = g_{ji}$$

and

$$k_{ij} = k_{ji}$$

To see this, consider the definition of self-adjointness for $[K]$ and two displacement systems $\{u^I\}$ and $\{u^{II}\}$

$$\begin{aligned}(\{u^{II}\}, L\{u^I\}) &= \sum_i u_i^{II} \sum_j k_{ij} u_j^I \\ &= \sum_j \sum_i u_j^{II} k_{ji} u_i^I \\ &= \sum_i u_i^I \sum_j k_{ij} u_j^{II} \\ &= (\{u^I\}, L\{u^{II}\})\end{aligned}$$

which can hold only if $k_{ij} = k_{ji}$.

Example 4.3 Figure 4.8(a) shows a beam of length ℓ loaded with a uniform load p_0 and simply supported at its right end, clamped at its left end. Using the reaction at the simple support as the unknown, the moment at any section is given as

$$M = Rx - \frac{p_0 x^2}{2}$$

From Fig. 4.8(b), the corresponding moment from a unit load at the left end is

$$\overline{M} = x$$

Considering EI stiffness only, we have

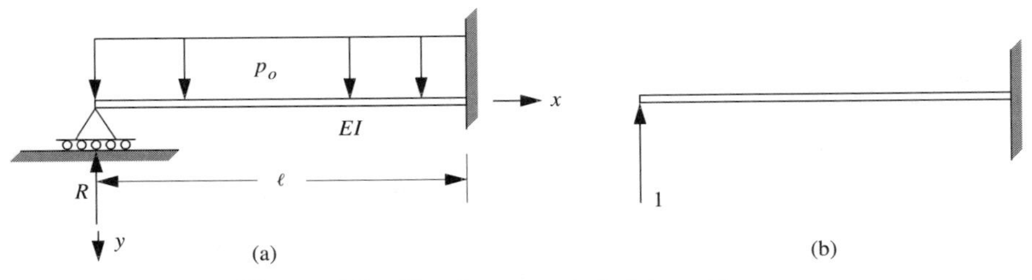

Figure 4.8 Simply supported clamped beam.

$$v(0) = \int_0^\ell \frac{M\overline{M}}{EI}\, dx = \frac{1}{EI}\int_0^\ell \left(Rx - \frac{p_0 x^2}{2}\right)(x)\, dx$$

or

$$v(0) = \frac{R\ell^3}{3EI} - \frac{p_0 \ell^4}{8EI}$$

Since R is the reaction of the fixed support, $v(0) = 0$. Therefore,

$$R = \frac{3}{8}\, p_0 \ell$$

Example 4.4 Figure 4.9(a) shows a beam of length ℓ loaded with a concentrated load P at its center and clamped at both ends. From symmetry, the vertical reactions at the wall supports are $P/2$ upward. Using the moment reaction M_0 at the left wall as the unknown, the moment at any section between 0 and $\ell/2$ is given as

$$M = \frac{Px}{2} - M_0$$

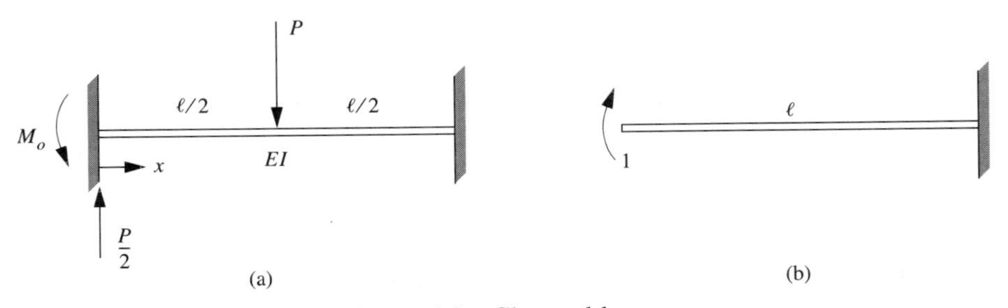

Figure 4.9 Clamped beam.

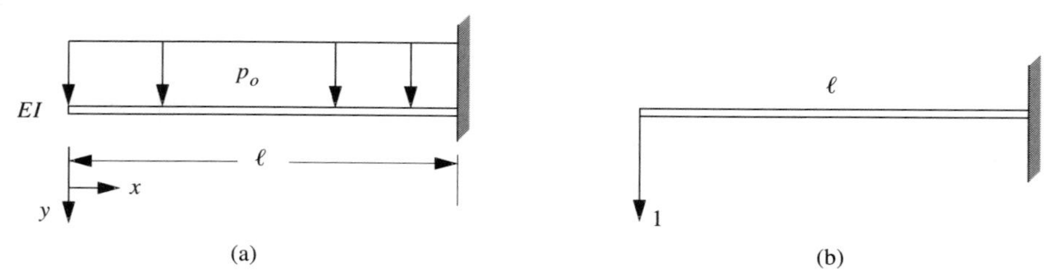

Figure 4.10 Clamped free beam.

From Fig. 4.9(b), the corresponding moment from a unit moment at the left wall is

$$\overline{M} = 1$$

Considering EI stiffness only, we have

$$v'(0) = \int_0^{\ell} \frac{M\overline{M}}{EI}\, dx = \frac{2}{EI} \int_0^{\ell/2} \left(\frac{Px}{2} - M_0\right)(1)\, dx$$

or

$$v'(0) = \frac{P\ell^2}{8EI} - \frac{M_0 \ell}{EI}$$

Since M_0 is the moment reaction of the fixed wall, $v'(0) = 0$. Therefore,

$$M_0 = \frac{1}{8} P\ell$$

Example 4.5 Figure 4.10(a) shows a beam of length ℓ loaded with a uniform load p_0 and free at its right end, clamped at its left end. The moment at any section is given as

$$M = -\frac{p_0 x^2}{2}$$

From Fig. 4.10(b), the corresponding moment from a unit load at the left end is

$$\overline{M} = -x$$

Considering bending only but with a nonlinear-elastic parabolic stress-strain relationship given as

$$\sigma = B\ \text{sgn}(\epsilon)|\epsilon|^{1/2}$$

where B is a constant and sgn(ϵ) is defined as +1 if $\epsilon > 0$, -1 if $\epsilon < 0$, and 0 if $\epsilon = 0$.

The strain ϵ is still given as $-yv''$; hence, δU^* is given as

$$\delta U^* = -\int v''\overline{M}\,dx$$

The internal moment is given as

$$\begin{aligned} M &= \int \sigma y\,dA \\ &= B\ \text{sgn}(\epsilon)\int |\epsilon|^{1/2} y\,dA \\ &= B\breve{I}\ \text{sgn}(\epsilon)(v'')^{1/2} \end{aligned}$$

where

$$\breve{I} = \int_A y^{3/2}\,dA$$

Then,

$$v'' = \left(\frac{M}{B\breve{I}}\right)^2$$

and δU^* becomes

$$\delta U^* = -\int \left(\frac{M}{B\breve{I}}\right)^2 \overline{M}\,dx$$

Then the vertical displacement at the left end is

$$v(0) = -\int_0^\ell \left(\frac{-p_0 x^2}{2B\breve{I}}\right)^2 (-x)\,dx = \frac{p_0^2 \ell^6}{24B^2\breve{I}^2}$$

Example 4.6 Figure 4.11(a) shows an aircraft wing idealized as a beam of length ℓ clamped at the left end, free at the right end, and subject to a uniformly distributed lifting force p_0 per unit of length. There is also an active control problem represented by a concentrated load $P = kv(\ell)$, where k is a constant of

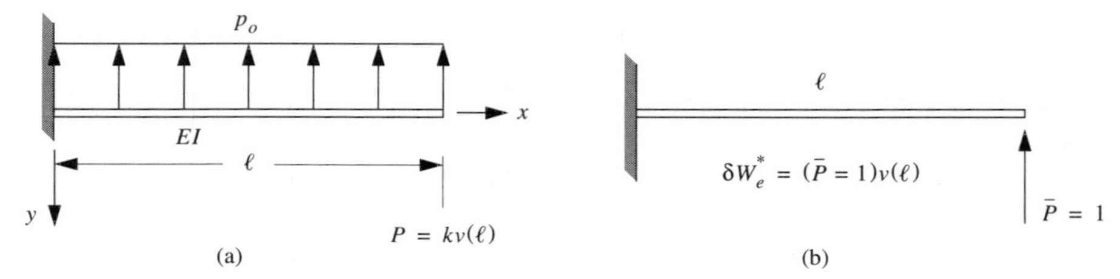

Figure 4.11 Idealized aircraft wing.

proportionality. Thus the load is unknown because it depends on the deflection. We can consider the load as parameterized, with the unknown parameter $P = kv(\ell)$.

The moment at any section is given as

$$M = kv(\ell)(\ell - x) + \frac{p_0}{2}(\ell - x)^2$$

From Fig. 4.11(b), the corresponding moment from a unit load at the right end is

$$\overline{M} = (\ell - x)$$

Considering EI stiffness only, we have

$$v(\ell) = \int_0^\ell \frac{M\overline{M}}{EI}\,dx = \frac{1}{EI}\int_0^\ell \left[kv(\ell)(\ell - x) - \frac{p_0}{2}(\ell - x)^2\right](\ell - x)\,dx$$

or

$$v(\ell) = \frac{kv(\ell)\ell^3}{3EI} + \frac{p_0\ell^4}{8EI}$$

Define

$$\beta = \frac{k\ell^3}{3EI}$$

and solve to get

$$v(\ell) = \frac{p_0\ell^4}{8EI}\,\frac{1}{1 - \beta}$$

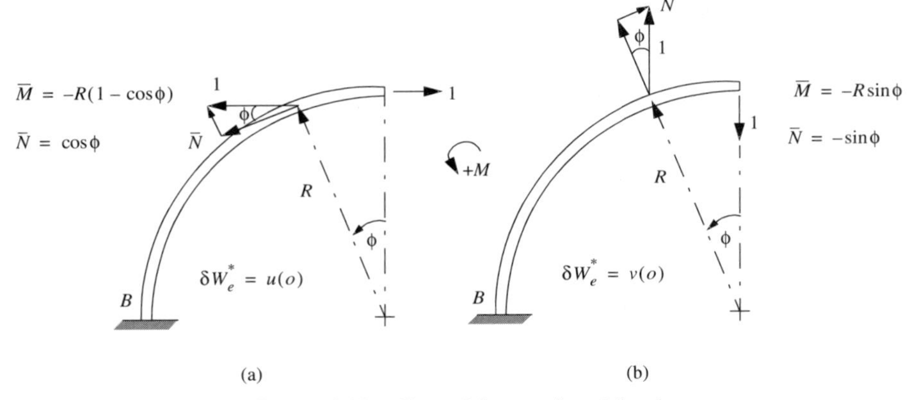

Figure 4.12 Curved beam virtual loads.

The tip load P is therefore

$$P = \frac{p_0 \ell^4}{8EI} \frac{k}{1-\beta}$$

Thus, within the limits of small-deflection theory, the wing is stable for $\beta < 1$.

Example 4.7 As a curved beam example, consider the curved beam shown in Fig. 3.29. Figure 4.12 shows the virtual loading to determine the $u(0)$ and $v(0)$ tip displacement.

Equation (4.15) takes the form

$$\delta U^* = \int \left(\frac{R\hat{N}_T}{AE} - \frac{\hat{M}_T}{AE} \right) \overline{N}\, d\phi + \int \left(\frac{R\hat{M}_T}{E\hat{I}} + \frac{\hat{M}_T}{RAE} - \frac{\hat{N}_T}{AE} \right) \overline{M}\, d\phi$$

From Fig. 4.12(a), we get

$$\overline{M} = -R(1 - \cos\phi)$$

and

$$\overline{N} = \cos\phi$$

and also

$$\delta W_e^* = u(0)$$

Then equating the internal complementary virtual work, to external complementary virtual work, we obtain

$$u(0) = \frac{R^2\hat{M}_T}{E\hat{I}}\left(1 - \frac{\pi}{2}\right) + \frac{\pi}{2}\,\frac{R\hat{N}_T}{AE} - \frac{\pi}{2}\,\frac{\hat{M}_T}{AE}$$

Similarly, from Fig. 4.12(b) we get

$$\overline{M} = -R\sin\phi$$

and

$$\overline{N} = -\sin\phi$$

with

$$\delta W_e^* = v(0)$$

which yields the vertical tip deflection

$$v(0) = -\frac{R^2\hat{M}_T}{E\hat{I}}$$

Example 4.8 Figure 4.13(a) shows a truss structure with one redundant member that, for the purpose of definiteness, is chosen as member BD. The dimensions and properties of the members are shown in the figure.

The true strain in any member (i) is

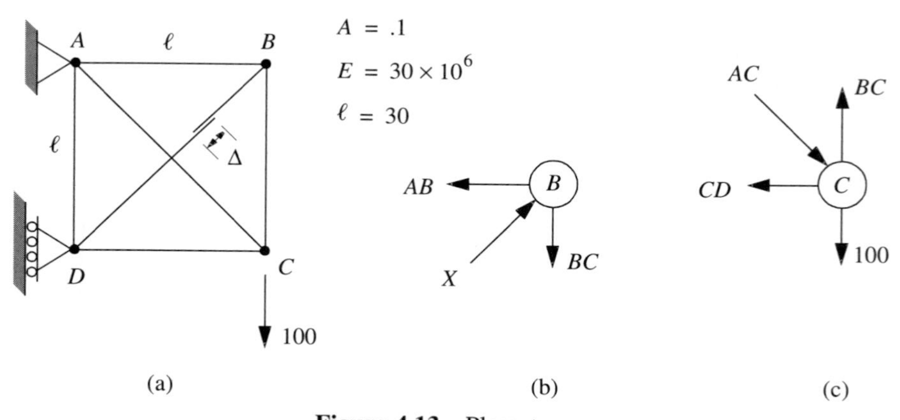

Figure 4.13 Plane truss.

$$\epsilon_i = \frac{N_i}{A_i E_i}$$

where N_i is the internal load in the (ith) member and $A_i E_i$ is its axial stiffness. The virtual stress in the (ith) member is

$$\overline{\sigma}_i = \frac{\overline{N}_i}{A_i}$$

Then, by the unit load method, the displacement Δ at the point of application of the unit load and in the same direction is

$$\Delta = \sum_{i=1}^{\substack{\text{all} \\ \text{members}}} \frac{\overline{N}_i}{A_i} \frac{N_i}{A_i E_i} A_i \ell_i$$

$$= \sum_i \overline{N}_i N_i f_i$$

where ℓ_i is the length of the (ith) member and $f_i = \ell_i / A_i E_i$ is the member flexibility.

Figures 4.13(b) and (c) show free bodies of joints B and C, respectively, and the unknown load in BD is represented by X. Equilibrium of the free bodies yields

$$AB = \frac{\sqrt{2}}{2} X$$

$$BC = \frac{\sqrt{2}}{2} X$$

$$AC = -100\sqrt{2} + X$$

$$CD = -100 + \frac{\sqrt{2}}{2} X$$

In general, for a linear structure, the force in any member is

$$N_i = N_{i0} + N_{i1} X$$

where N_{i0} is the contribution from the applied load obtained by cutting BD ($X = 0$) and by solving the determinate system, and $N_{i1} X$ is the contribution from the redundant load obtained by letting $X = 1$ with no applied loads. Let the self-equilibrating system be

$$\overline{N}_i = N_{i1}\overline{X}$$

Then,

$$\begin{aligned}\Delta &= \sum_i N_i \overline{N}_i f_i \\ &= \sum_i (N_{i0} + N_{i1}X) f_i N_{i1}\overline{X} \\ &= \left(\sum_i N_{i0} N_{i1} f_i + X \sum_i N_{i1} N_{i1} f_i \right) \overline{X}\end{aligned}$$

If we let Δ represent the relative displacement between a cut in member BD, then $\Delta = 0$ in the true structure. Thus we obtain the solution

$$X = -\frac{\sum_i N_{i0} N_{i1} f_i}{\sum_i N_{i1}^2 f_i}$$

For Fig. 4.13(a), we can construct the first six columns of the following table:

Member	f	N_0	N_1	$N_0 N_1 f$	$N_1^2 f$	XN_1	N
AB	10^{-5}	0	−0.707	0	0.5×10^{-5}	39.7	39.7
BC	10^{-5}	0	−0.707	0	0.5×10^{-5}	39.7	39.7
CD	10^{-5}	−100.0	−0.707	70.7×10^{-5}	0.5×10^{-5}	39.7	−60.3
AD	10^{-5}	0	−0.707	0	0.5×10^{-5}	39.7	39.7
AC	1.41×10^{-5}	141.0	1.0	200×10^{-5}	1.41×10^{-5}	−56.2	84.8
BD	1.41×10^{-5}	0	1.0	0	1.41×10^{-5}	−56.2	−56.2
				270.7×10^{-5}	4.82×10^{-5}		

From the sum of columns (5) and (6) of the table, we get by using the relationship

$$X = -\frac{\text{column (5)}}{\text{column (6)}}$$

the value of X as

$$X = -56.2$$

with the final member loads shown in column (8).

For the structure shown in Fig. 4.13(a), the foregoing table used to obtain a solution is already quite extensive. To mechanize the process even further, we introduce the force method of analysis, the first step of which is the formulation of the equilibrium equations. These take the form

$$[E]\{F\} = \{P\} \tag{4.37}$$

where $[E]$ is, in general, an $m \times n$ matrix, but it is $m < n$ after rigid-body modes have been removed. The coefficients of $[E]$ are the direction cosines used in resolving the assembled vector of independent generalized element forces $\{F\}$ to the same coordinate system as the external applied loads $\{P\}$. Since $[E]$ cannot generally be inverted (except for determinate structures), we can partition $\{F\}$ into static force variables $\{S\}$ and $n-m$ redundant force variables $\{X\}$. This is represented by the following relationship:

$$\{F\} = [T_S | T_X] \begin{Bmatrix} S \\ X \end{Bmatrix} \tag{4.38}$$

where $[T_S]$ and $[T_X]$ are Boolean matrices that simply rearrange the variables into the desired sets. The form of Eq. (4.38) implies a linear relationship. Thus, all that follows represents a linear solution.

Substituting Eq. (4.38) into Eq. (4.37), we get

$$[E_S]\{S\} + [E_X]\{X\} = \{P\} \tag{4.39}$$

where

$$[E_S] = [E][T_S]$$
$$[E_X] = [E][T_X]$$

Since $\{S\}$ is a static set, $[E_S]$ can be inverted. Hence,

$$\{S\} = [E_S]^{-1}(\{P\} - [E_X]\{X\}) \tag{4.40}$$

Substituting Eq. (4.40) into Eq. (4.38), we get

$$\{F\} = [T_S \quad T_X] \begin{bmatrix} E_S^{-1} & -E_S^{-1}E_X \\ 0 & I \end{bmatrix} \begin{Bmatrix} P \\ X \end{Bmatrix} \tag{4.41}$$

or

$$\{F\} = [B_S]\{P\} + [B_X]\{X\} \tag{4.42}$$

where

$$[B_S] = [T_S][E_S]^{-1} \tag{4.43}$$

and

$$[B_X] = [T_X] - [T_S][E_S]^{-1}[E_X] \tag{4.44}$$

The matrix $[B_S]$ represents the load distribution in the determinate structure from the applied loads, and the matrix $[B_X]$ represents the load distribution in the structure from unit loads in the redundants.

Substituting Eq. (4.42) back into Eq. (4.37), we get

$$[E][B_S]\{P\} + [E][B_X]\{X\} = \{P\}$$

which holds when

$$[E][B_S] = [I] \tag{4.45a}$$

$$[E][B_X] = [0] \tag{4.45b}$$

with the Eq. (4.45b) following from the fact that the redundant forces must be self-equilibrating.

If $\{d\}$ represents the assembled vector of independent generalized element displacement vectors and $\{u\}$ represents the external structural displacement vector, then structural continuity can be represented by the following relationship:

$$\{d\} = [a]\{u\} \tag{4.46}$$

From the principle of complementary virtual work, we have

$$\lceil \overline{P} \rfloor \{u\} = [\overline{F}]^{\mathrm{T}}\{d\}$$

where $\lceil \overline{P} \rfloor$ is a diagonal matrix of ones for a unit virtual load applied to the structure at each degree of freedom one at a time. The columns of $[\overline{F}]^{\mathrm{T}}$ represent the corresponding internal virtual loads for each external virtual load. Then, using Eqs. (4.37) and (4.46), we have

$$[\overline{F}]^{\mathrm{T}}([E]^{\mathrm{T}}\{u\} - \{d\}) = 0$$

Since the columns of $[\overline{F}]^{\mathrm{T}}$ are formed from an arbitrary set of virtual forces, we have

$$[a] = [E]^{\mathrm{T}} \tag{4.47}$$

and Eq. (4.46) becomes

$$\{d\} = [E]^{\mathrm{T}}\{u\} \tag{4.48}$$

Consider $\{d\}$ to consist of two parts, namely, an elastic part $\{d_e\}$ and an initial deformation part $\{d_0\}$, or

$$\{d\} = \{d_e\} + \{d_0\} \tag{4.49}$$

The elastic part $\{d_e\}$ is related to $\{F\}$ by the nonsingular flexibility matrix $[\mathcal{F}]$ by the relationship

$$\{d_e\} = [\mathcal{F}]\{F\} \tag{4.50}$$

Substituting Eqs. (4.49) and (4.50) into Eq. (4.47) and rearranging, we get

$$[\mathcal{F}]\{F\} + \{d_0\} - [E]^{\mathrm{T}}\{u\} = \{0\} \tag{4.51}$$

If Eq. (4.51) is multiplied by Eq. (4.45b), the last term is identically zero and the equation becomes

$$[B_X]^{\mathrm{T}}[\mathcal{F}]\{F\} + [B_X]^{\mathrm{T}}\{d_0\} = \{0\} \tag{4.52}$$

Substituting Eq. (4.42) for $\{F\}$ into the above equation and expanding, we get

$$[V]\{X\} + [B_X]^{\mathrm{T}}[\mathcal{F}][B_S]\{P\} + [B_X]^{\mathrm{T}}\{d_0\} - \{0\} \tag{4.53}$$

where

$$[V] = [B_X]^{\mathrm{T}}[\mathcal{F}][B_X] \tag{4.54}$$

is the redundant flexibility matrix. Because $[V]$ can be inverted, we can solve for $\{X\}$ to obtain

$$\{X\} = -[V]^{-1}[B_X]^{\mathrm{T}}([\mathcal{F}][B_S]\{P\} + \{d_0\}) \tag{4.55}$$

Substituting Eq. (4.55) into Eq. (4.42) yields the complete set of independent force variables as

$$\{F\} = ([I] - [B_X][V]^{-1}[B_X]^{\mathrm{T}}[\mathcal{F}])[B_S]\{P\} - [B_X][V]^{-1}[B_X]^{\mathrm{T}}\{d_0\} \tag{4.56}$$

The structural displacements can be obtained by multiplying Eq. (4.51) by $[B_S]^{\mathrm{T}}$, noting Eq. (4.45a), and rearranging the result to get

$$\{u\} = [B_S]^{\mathrm{T}}([\mathcal{F}]\{F\} + \{d_0\}) \tag{4.57}$$

Finally, substituting Eq. (4.56) into Eq. (4.57), we get

$$\{u\} = [G]\{P\} + [G_0]\{d_0\} \tag{4.58}$$

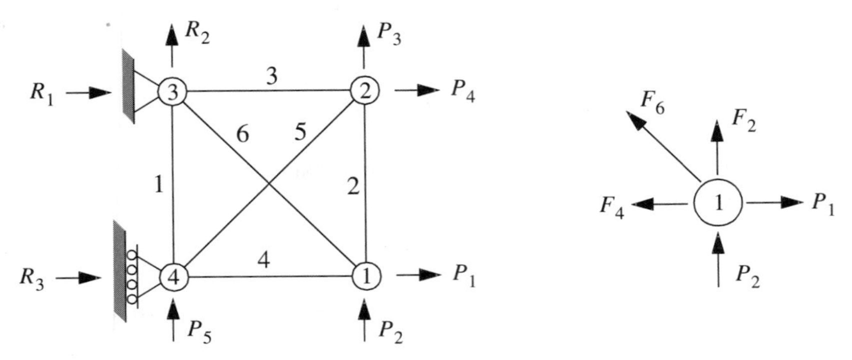

Figure 4.14 Plane truss with generalized loading.

where

$$[G] = [B_S]^{\mathrm{T}}[\mathcal{F}]([I] - [B_X][V]^{-1}[B_X]^{\mathrm{T}}[\mathcal{F}])[B_S] \tag{4.59}$$

is the constrained nonsingular structural flexibility matrix that we have shown to be symmetric, and

$$[G_0] = [B_S]^{\mathrm{T}}([I] - [\mathcal{F}][B_X][V]^{-1}[B_X]^{\mathrm{T}}) \tag{4.60}$$

is the flexibility matrix for initial deformations.

Equations (4.37)–(4.60) represent the matrix forms of the force method. As presented, the method requires the engineer to select the redundancies. Reference [47] presents a detailed discussion of the automated selection of redundancies.

Example 4.9 Figure 4.14(a) represents the same structure as Fig. 4.13(a) but with a general set of external loads. Figure 4.14(b) represents a general free body of node 1 under the influence of the generalized internal forces $\{F\}$ and the external loads $\{P\}$ acting at that node. Equilibrium in the P_1 direction yields

$$P_1 = F_4 + \frac{\sqrt{2}}{2} F_6$$

For all nodes, we can collect each equilibrium equation and form the relationship $[E_T]\{F\} = \{P\}$ as

$$
\begin{bmatrix}
0 & 0 & 0 & 1 & 0 & \sqrt{2}/2 \\
0 & -1 & 0 & 0 & 0 & -\sqrt{2}/2 \\
0 & 0 & 1 & 0 & \sqrt{2}/2 & 0 \\
0 & 1 & 0 & 0 & \sqrt{2}/2 & 0 \\
-1 & 0 & 0 & 0 & -\sqrt{2}/2 & 0 \\
--- & --- & --- & --- & --- & --- \\
0 & 0 & -1 & 0 & 0 & -\sqrt{2}/2 \\
1 & 0 & 0 & 0 & 0 & \sqrt{2}/2 \\
0 & 0 & 0 & -1 & -\sqrt{2}/2 & 0
\end{bmatrix}
\begin{Bmatrix} F_1 \\ F_2 \\ F_3 \\ F_4 \\ F_5 \\ F_6 \end{Bmatrix}
=
\begin{Bmatrix} P_1 \\ P_2 \\ P_3 \\ P_4 \\ P_5 \\ --- \\ R_1 \\ R_2 \\ R_3 \end{Bmatrix}
$$

where the $\{R\}$ represents the reaction loads at the external supports and $[E_T]$ is the total equilibrium matrix that, for convenience, can be partitioned as shown above into

$$
\begin{bmatrix} E \\ - \\ E_R \end{bmatrix} \{F\} = \begin{Bmatrix} P \\ - \\ R \end{Bmatrix}
$$

The set of internal loads $\{F\}$ corresponding to the static set $\{S\}$ and the redundant set $\{X\}$ can be selected by a Boolean matrix as follows:

$$
\begin{Bmatrix} F_1 \\ F_2 \\ F_3 \\ F_4 \\ F_5 \\ F_6 \end{Bmatrix}
=
\left[\begin{array}{ccccc|c}
1 & 0 & 0 & 0 & 0 & 0 \\
0 & 1 & 0 & 0 & 0 & 0 \\
0 & 0 & 1 & 0 & 0 & 0 \\
0 & 0 & 0 & 1 & 0 & 0 \\
0 & 0 & 0 & 0 & 0 & 1 \\
0 & 0 & 0 & 0 & 1 & 0
\end{array}\right]
\begin{Bmatrix} S_1 \\ S_2 \\ S_3 \\ S_4 \\ S_5 \\ - \\ X_1 \end{Bmatrix}
$$

or

$$
\{F\} = [T_S \quad | \quad T_X] \begin{Bmatrix} S \\ - \\ X \end{Bmatrix}
$$

The matrices $[E_S] = [E][T_S]$ and $[E_X] = [E][T_X]$ each take the forms

$$[E_S] = \begin{bmatrix} 0 & 0 & 0 & 1 & \sqrt{2}/2 \\ 0 & -1 & 0 & 0 & -\sqrt{2}/2 \\ 0 & 0 & 1 & 0 & 0 \\ 0 & 1 & 0 & 0 & 0 \\ -1 & 0 & 0 & 0 & 0 \end{bmatrix}$$

and

$$[E_X] = \begin{bmatrix} 0 \\ 0 \\ \sqrt{2}/2 \\ \sqrt{2}/2 \\ -\sqrt{2}/2 \end{bmatrix}$$

The inverse of $[E_S]$ is

$$[E_S]^{-1} = \begin{bmatrix} 0 & 0 & 0 & 0 & -1 \\ 0 & 0 & 0 & 1 & 0 \\ 0 & 0 & 1 & 0 & 0 \\ 1 & 1 & 0 & 1 & 0 \\ 0 & -\sqrt{2} & 0 & -\sqrt{2} & 0 \end{bmatrix}$$

Then, the load distribution $[B_S] = [T_S][E_S]^{-1}$ in the determinate structure from the applied load $\{P\}$ is determined as

$$[B_S] = \begin{bmatrix} 0 & 0 & 0 & 0 & -1 \\ 0 & 0 & 0 & 1 & 0 \\ 0 & 0 & 1 & 0 & 0 \\ 1 & 1 & 0 & 1 & 0 \\ 0 & 0 & 0 & 0 & 0 \\ 0 & -\sqrt{2} & 0 & -\sqrt{2} & 0 \end{bmatrix}$$

The load distribution $[B_X] = [T_X] - [T_S][E_S]^{-1}[E_X]$ in the structure from the unit load in the redundant is determined as

$$[B_X] = \begin{bmatrix} -\sqrt{2}/2 \\ -\sqrt{2}/2 \\ -\sqrt{2}/2 \\ -\sqrt{2}/2 \\ 1 \\ 1 \end{bmatrix}$$

The assembled element flexibility matrix $[\mathcal{F}]$ is

$$[\mathcal{F}] = \frac{\ell}{AE} \begin{bmatrix} 1 & 0 & 0 & 0 & 0 & 0 \\ 0 & 1 & 0 & 0 & 0 & 0 \\ 0 & 0 & 1 & 0 & 0 & 0 \\ 0 & 0 & 0 & 1 & 0 & 0 \\ 0 & 0 & 0 & 0 & \sqrt{2} & 0 \\ 0 & 0 & 0 & 0 & 0 & \sqrt{2} \end{bmatrix}$$

The redundant flexibility matrix $[V] = [B_X]^{\mathrm{T}}[\mathcal{F}][B_X]$ is

$$V = 2(1 + \sqrt{2}) \frac{\ell}{AE} = 4.82 \times 10^{-5}$$

The applied load vector is

$$\{P\} = \begin{Bmatrix} 0 \\ -100 \\ 0 \\ 0 \\ 0 \end{Bmatrix}$$

and the redundant $\{X\} = -[V]^{-1}[B_X]^{\mathrm{T}}[\mathcal{F}][B_S]\{P\}$ is again obtained as

$$X = -56.2$$

4.5 FORCE ELEMENTS

Unfortunately, the matrix form of the force method of analysis introduced in the previous section does not lend itself to modern computational methods. For example, adding, deleting, and changing element type in a complex structure all require recalculating the appropriate redundant loads. Also, to ensure a well-behaved structural flexibility matrix, the solution algorithms often require full-pivot selection, which can be time-consuming. Finally, the method does not lend itself to a direct natural formulation when dealing with geometric nonlinearities.

However, individual finite elements based on the force method of analysis can be readily included in a displacement-type finite element formulation. These elements perform as well as and often better than conventional displacement-based elements.

Force elements are based on stress functions that, in the absence of body forces, identically satisfy the equations of equilibrium. For two dimensions, they are called the Airy stress functions and are defined as

$$\sigma_x = \frac{\partial^2 \phi}{\partial y^2}, \qquad \sigma_y = \frac{\partial^2 \phi}{\partial x^2}, \qquad \tau_{xy} = -\frac{\partial^2 \phi}{\partial x\,\partial y} \tag{4.61}$$

For three dimensions, there are the Maxwell stress functions, defined as

$$\begin{aligned}
\sigma_x &= \frac{\partial^2 \chi_3}{\partial y^2} + \frac{\partial^2 \chi_2}{\partial z^2}, \qquad & \tau_{yz} &= -\frac{\partial^2 \chi_1}{\partial y\,\partial z} \\
\sigma_y &= \frac{\partial^2 \chi_1}{\partial z^2} + \frac{\partial^2 \chi_3}{\partial x^2}, \qquad & \tau_{zx} &= -\frac{\partial^2 \chi_2}{\partial z\,\partial x} \\
\sigma_z &= \frac{\partial^2 \chi_2}{\partial x^2} + \frac{\partial^2 \chi_1}{\partial y^2}, \qquad & \tau_{xy} &= -\frac{\partial^2 \chi_3}{\partial x\,\partial y}
\end{aligned} \tag{4.62}$$

as well as the Morera stress functions, defined as

$$\begin{aligned}
\sigma_x &= \frac{\partial^2 \psi_1}{\partial y\,\partial z}, \qquad & \tau_{yz} &= -\frac{1}{2}\frac{\partial}{\partial x}\left(-\frac{\partial \psi_1}{\partial x} + \frac{\partial \psi_2}{\partial y} + \frac{\partial \psi_3}{\partial z}\right) \\
\sigma_y &= \frac{\partial^2 \psi_2}{\partial z\,\partial x}, \qquad & \tau_{zx} &= -\frac{1}{2}\frac{\partial}{\partial y}\left(+\frac{\partial \psi_1}{\partial x} - \frac{\partial \psi_2}{\partial y} + \frac{\partial \psi_3}{\partial z}\right) \\
\sigma_z &= \frac{\partial^2 \psi_3}{\partial x\,\partial y}, \qquad & \tau_{xy} &= -\frac{1}{2}\frac{\partial}{\partial z}\left(+\frac{\partial \psi_1}{\partial x} + \frac{\partial \psi_2}{\partial y} - \frac{\partial \psi_3}{\partial z}\right)
\end{aligned} \tag{4.63}$$

For special structures, such as plate bending, stress-resultant functions are used and take various forms, such as the following:

$$M_x = \frac{\partial \phi_2}{\partial y}, \qquad M_y = \frac{\partial \phi_1}{\partial x}, \qquad M_{xy} = \frac{1}{2}\left(\frac{\partial \phi_1}{\partial y} + \frac{\partial \phi_2}{\partial x}\right) \tag{4.64}$$

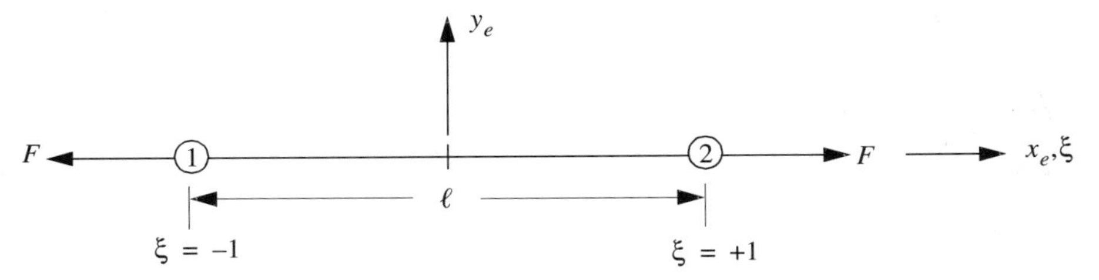

Figure 4.15 Axial-rod force element in an element system.

To develop a force element, stress functions of appropriate complexity are assumed, and the generalized stress field is obtained by differentiation and substitution into the desired set of stress-function relationships defined above. The results are then placed in the form

$$\{\sigma\} = [T_{\sigma F}]\{F\} \tag{4.65}$$

where $\{F\}$ is a vector of generalized independent-element force variables free of the zero-stress state and satisfying the stress-equilibrium conditions. The jth column of $[T_{\sigma F}]$ gives the stress distribution required for equilibrium for a unit value of F_j.

From Eq. (4.65), the element natural (or nonsingular) flexibility matrix may be obtained. To see how, consider the axial-rod force element of Fig. 3.12, shown again as Fig. 4.15. Note that in the figure, the load system represents a self-equilibrating system, and we can write

$$\begin{aligned} \sigma &= \frac{1}{A} F \\ &= [T_{\sigma F}]\,\{F\} \end{aligned} \tag{4.66}$$

The strain in the rod is determined as

$$\begin{aligned} \{\epsilon\} &= \frac{1}{E}\,\sigma \\ &= [D]^{-1}\{\sigma\} \end{aligned} \tag{4.67}$$

Then, applying the principle of complementary virtual work,

$$\delta U^* = \delta W_e^*$$

we get

$$
\begin{aligned}
\int \lfloor \bar{\sigma} \rfloor \{\epsilon\}\, dV
&= \int \bar{F} \frac{1}{A} \frac{1}{E} \frac{1}{A} F(A\, dx_e) \\
&= \bar{F} \frac{1}{AE} F \int_{-1}^{1} \frac{\ell}{2}\, d\xi \\
&= \bar{F} \frac{\ell}{AE} F \\
&= \bar{F} d
\end{aligned} \tag{4.68}
$$

Since the virtual force $\bar{F}$ is arbitrary, we arrive at the relationship

$$
\begin{aligned}
\frac{\ell}{AE} F &= d \\
[\mathcal{F}^e]\{F\} &= \{d\}
\end{aligned} \tag{4.69}
$$

where d is the independent generalized element displacement, and

$$
\begin{aligned}
[\mathcal{F}^e] &= \frac{\ell}{AE} \\
&= \int [T_{\sigma F}]^{\mathrm{T}} [D]^{-1} [T_{\sigma F}] |J|\, dV_\xi
\end{aligned} \tag{4.70}
$$

is the element-natural-flexibility matrix, and dV_ξ is the volume of the element in natural coordinates.

The element-natural-stiffness matrix is given as

$$
[\mathcal{F}^e]^{-1} = \frac{AE}{\ell} \tag{4.71}
$$

From this equation, we may readily form the standard singular element-stiffness matrix as follows: From a comparison of the load sets in Figs. 3.12 and 4.15, we see that the generalized element nodal forces $\{P^e\}$ are related to the generalized independent force variables $\{F\}$ by the relationship

$$
\begin{aligned}
\begin{Bmatrix} P_1^e \\ P_2^e \end{Bmatrix} &= \begin{bmatrix} -1 \\ 1 \end{bmatrix} \{F\} \\
\{P^e\} &= [E_{PF}]\{F\}
\end{aligned} \tag{4.72}
$$

Equation (4.72), however, is nothing more than an element form of Eq. (4.37). Following the same line of reasoning as Eqs. (4.46)–(4.48), we can write

$$\{d\} = [E_{PF}]^{\mathrm{T}}\{u^e\} \tag{4.73}$$

Since the element must be equilibrium, we can apply the virtual work expression

$$\delta U = \delta W_e$$

to get

$$\lceil \overline{u}^e \rfloor \{P^e\} = [\overline{d}]^{\mathrm{T}}\{F\}$$

where $\lceil \overline{u}^e \rfloor$ is a diagonal matrix of ones for a unit virtual displacement applied to the element at each degree of freedom one at a time. The columns of $[\overline{d}]^{\mathrm{T}}$ represent the corresponding independent generalized element displacement for each virtual nodal displacement. Using the relationship

$$\{F\} = [\mathcal{F}^e]^{-1}\{d\} \tag{4.74}$$

and Eq. (4.73), we get

$$\{P^e\} = [K^e]\{u^e\}$$

where the element-stiffness matrix is defined as

$$\begin{aligned} [K^e] &= [E_{PF}][\mathcal{F}^e]^{-1}[E_{PF}]^{\mathrm{T}} \\ &= \frac{AE}{\ell}\begin{bmatrix} 1 & -1 \\ -1 & 1 \end{bmatrix} \end{aligned} \tag{4.75}$$

Use of Eq. (4.75) makes it easy to include any force-type element into a displacement-type code [48].

4.6 GENERALIZED FORCE-DISPLACEMENT TRANSFORMATIONS

The transformation described by Eq. (4.75) is representative of a general class of transformations. Often, transformations are needed because experimental flexibility coefficients can be obtained by direct measurement of displacements, which can then be used to obtain experimental stiffness coefficients, or by nonsingular structural flexibilities that exist from older substructure models no longer in existence and must be coupled to current displacement models.

Inverting the nonsingular flexibility matrix yields the nonsingular stiffness matrix. If the structural displacements have been measured relative to some statically determinate support system, we may write

$$[K_{ff}]\{u_{f/s}\} = \{P_f\} \tag{4.76}$$

where $[K_{ff}]$ is the nonsingular stiffness matrix of the structure free to deflect and the notation $\{u_{f/s}\}$ is used to indicate that the deflections are measured relative to a static support system.

Since the static support reactions $\{P_s\}$ must be in equilibrium with the applied loads, we may write

$$\{P_s\} = [E_{sf}]\{P_f\} \tag{4.77}$$

where the coefficients of $[E_{sf}]$ relate the support reactions to the applied loads.

The static support displacements are, of course, zero, and we can write

$$\{u_{s/s}\} = \{0\} \tag{4.78}$$

Let us now define a vector $\{P_g\}$ of generalized forces as

$$\begin{aligned}\{P_g\} &= \begin{Bmatrix} P_f \\ P_s \end{Bmatrix} \\ &= \begin{bmatrix} I_{ff} \\ E_{sf} \end{bmatrix} \{P_f\} \\ &= [E_{gf}]\{P_f\}\end{aligned} \tag{4.79}$$

where $[I_{ff}]$ is a unit matrix.

The corresponding set of displacement vectors is $\{u_g\}$. For an unconstrained stiffness matrix, this set of displacement vectors must contain rigid-body motion. To this end, define

$$\{u_f\} = \{u_{f/s}\} + \{u_{f/r}\} \tag{4.80}$$

where $\{u_{f/r}\}$ is the displacement of the f degrees of freedom relative to a rigid-body motion. Also, define

$$\{u_s\} = \{u_{s/s}\} + \{u_{s/r}\} = \{u_{s/r}\} \tag{4.81}$$

because of Eq. (4.78). Collecting the results, we have

$$\{u_g\} = \begin{Bmatrix} u_f \\ u_s \end{Bmatrix}$$

$$= \begin{Bmatrix} u_{f/s} \\ 0 \end{Bmatrix} + \{u_{g/r}\} \tag{4.82}$$

where

$$\{u_{g/r}\} = \begin{Bmatrix} u_{f/r} \\ u_{s/r} \end{Bmatrix} \tag{4.83}$$

The vector $\{u_{g/r}\}$ represents rigid-body displacements consistent with equilibrium.

Complementary virtual work states that

$$\delta W^* = 0$$

This requires that

$$\lceil \overline{P}_g \rfloor \{u_{g/r}\} = 0$$

for equilibrium. Using Eq. (4.79), we get

$$[\overline{P}_f]^{\mathrm{T}}[E_{gf}]^{\mathrm{T}}\{u_{g/r}\} = \{0\}$$

or

$$[I_{ff} \quad E_{sf}^{\mathrm{T}}] \begin{Bmatrix} u_{f/r} \\ u_s \end{Bmatrix} = \{0\}$$

from which we get

$$\{u_{f/r}\} = -[E_{sf}]^{\mathrm{T}}\{u_s\} \tag{4.84}$$

From Eq. (4.80),

$$\{u_{f/s}\} = \{u_f\} + [E_{sf}]^{\mathrm{T}}\{u_s\}$$

$$= [I_{ff} \quad E_{sf}^{\mathrm{T}}] \begin{Bmatrix} u_f \\ u_s \end{Bmatrix}$$

$$= [E_{gf}]^{\mathrm{T}}\{u_g\} \tag{4.85}$$

From Eq. (4.76), we obtain

$$\{P_f\} = [K_{ff}]\{u_f\} + [K_{ff}][E_{sf}]^{\mathrm{T}}\{u_s\}$$

and from Eq. (4.77), we obtain

$$\{P_s\} = [E_{sf}][K_{ff}]\{u_f\} + [E_{sf}][K_{ff}][E_{sf}]^{\mathrm{T}}\{u_s\}$$

These two expressions can be arranged as

$$\{P_g\} = \begin{Bmatrix} P_f \\ P_s \end{Bmatrix} = \begin{bmatrix} K_{ff} & K_{ff}E_{sf}^{\mathrm{T}} \\ E_{sf}K_{ff} & E_{sf}K_{ff}E_{sf}^{\mathrm{T}} \end{bmatrix} \begin{Bmatrix} u_f \\ u_s \end{Bmatrix} \tag{4.86}$$

which can be written as

$$\{P_g\} = [K_{gg}]\{u_g\}$$

Using the last of Eq. (4.79), we note that $[K_{gg}]$ can be placed into the form

$$[K_{gg}] = [E_{gf}][K_{ff}][E_{gf}]^{\mathrm{T}} \tag{4.87}$$

and we see the similarity in form with Eq. (4.75).

PROBLEMS

4.1 The beam shown in Fig. 4.16 is simply supported as shown and has an applied moment M_0 at its left end. Its flexural rigidity is *EI*. By complementary virtual work, determine the slope of the overhanging left end.

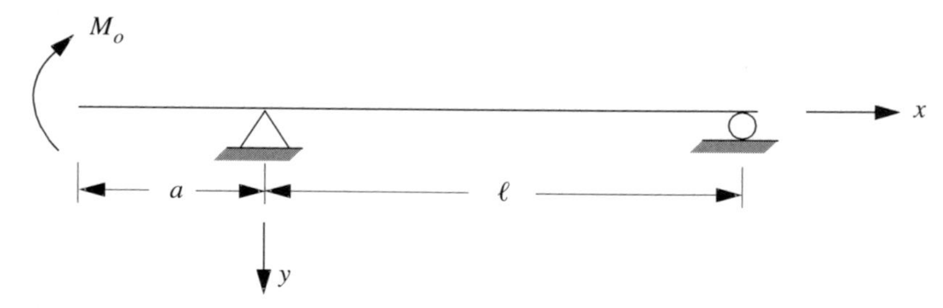

Figure 4.16 Problem 4.1.

4.2 The beam shown in Fig. 4.17 has simple supports at its ends *A* and *C* and its midspan at *B*, and it is loaded across its entire span with a uniform load *p*. The beam has a length 2ℓ and a flexural rigidity *EI*. By complementary virtual work, determine the reaction R_B of the center support.

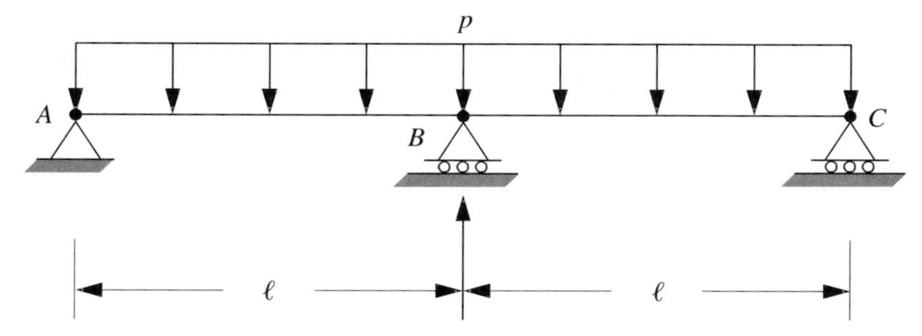

Figure 4.17 Problem 4.2.

4.3 The beam shown in Fig. 4.18 carries a uniformly distributed load $\omega = 16.0$ lb/in., is of length $\ell = 100.0$ in., and has for the section shown a moment of inertia of area of $I = 0.2$ in.^{4}. In final calculations, note the shape of the cross section. Young's modulus for the beam is $E = 10^7$ lb/in.^{2}. The beam is designed to be simply supported at the left end and fixed at the right end. The supports were designed to be at the same elevation; however, upon assembly, it is found that the end over the simple support has to be forced upward a distance of $\Delta = 3.0$ in. to have it rest on the support. The manufacturing code requires that the maximum tensile stress not exceed 60,000.0 lb/in.^{2}. By complementary virtual work, determine if the beam meets the code requirements. (First set up the problem in terms of symbols; only use values at the end.) If the beam does not meet the requirements, should it be rebuilt at an assembly cost of \$3,800 and a government-contract administration and late-delivery tax liability cost of \$296,200? Or is it OK as is?

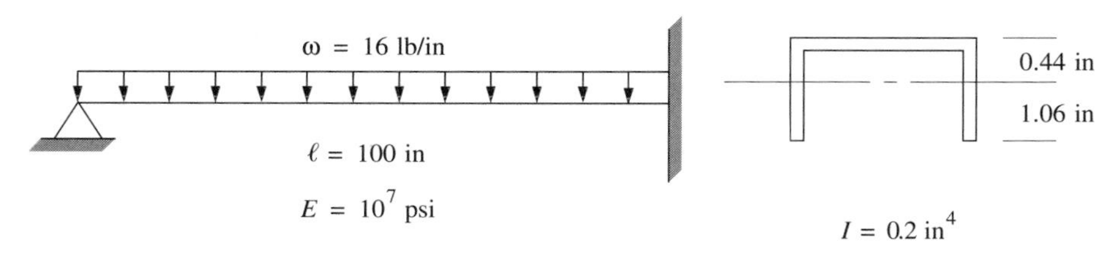

Figure 4.18 Problem 4.3.

4.4 The frame shown in Fig. 4.19 is simply supported at A and C and has a horizontal load P applied at the rigid joint B. Both members have a length ℓ and flexural rigidity EI. By complementary virtual work, determine the horizontal deflection at C.

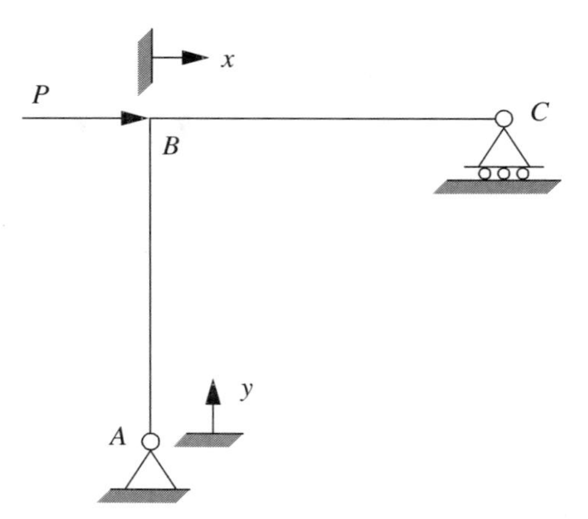

Figure 4.19 Problem 4.4.

4.5 A thin, circular ring with flexural rigidity EI and radius R is cut and spread open at a specified distance e by the application of two equal but opposite vertical forces N applied at the cut as shown in Fig. 4.20. By complementary virtual work, determine the following:

(a) The load N in terms of e, EI, and R.

(b) The maximum bending moment in terms of e, EI, and R.

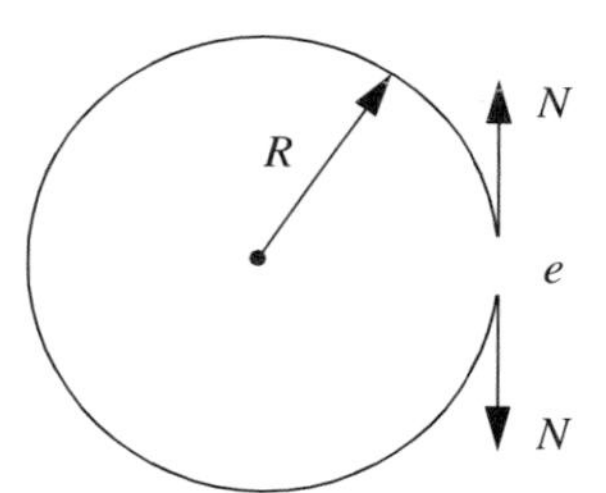

Figure 4.20 Problem 4.5.

4.6 A circular frame of radius R is subjected to equal and opposite vertical loads P as shown in Fig. 4.21. Assume the radius is very much greater than the cross-sectional depth and that the flexural rigidity is EI. Use symmetry considerations and only the upper right quadrant of the ring. by the unit-dummy-load method, determine

(a) the expression for the bending moment at any section of the ring;

(b) the decrease in radius at point A; and

(c) the value of the angle where the internal moment changes sign.

Then, sketch the moment distribution for the quadrant.

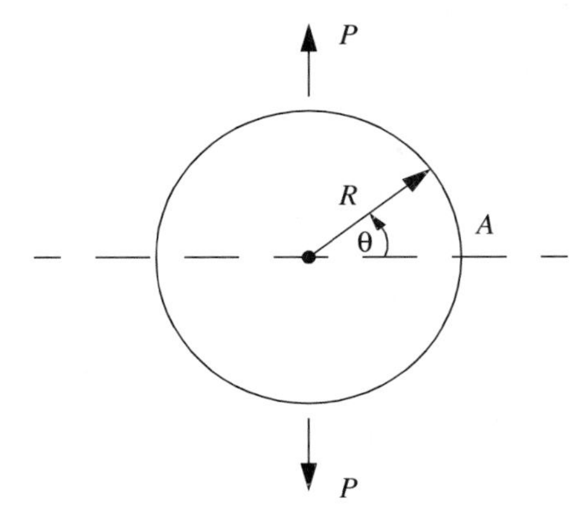

Figure 4.21 Problem 4.6.

4.7 The symmetrical two-hinged frame has a Young's modulus E, dimensions ℓ and h, and moments of inertia of area I_1 and I_2 as indicated in Fig. 4.22. The frame carries a load P as shown. Use symmetry considerations and only the left half of the frame. By the unit-dummy-load method, determine the horizontal support reaction.

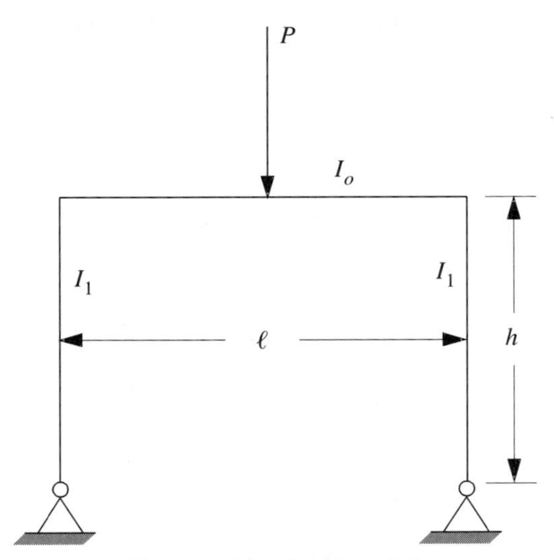

Figure 4.22 Problem 4.7.

4.8 Each member of the square frame shown in Fig. 4.23 has a length ℓ and a flexural rigidity EI, and it is fixed at A, pinned at D. Member AB carries a uniformly distributed load of intensity p_0 that pushes the frame to the right. By complementary virtual work, determine the horizontal and vertical reactions at the pin D.

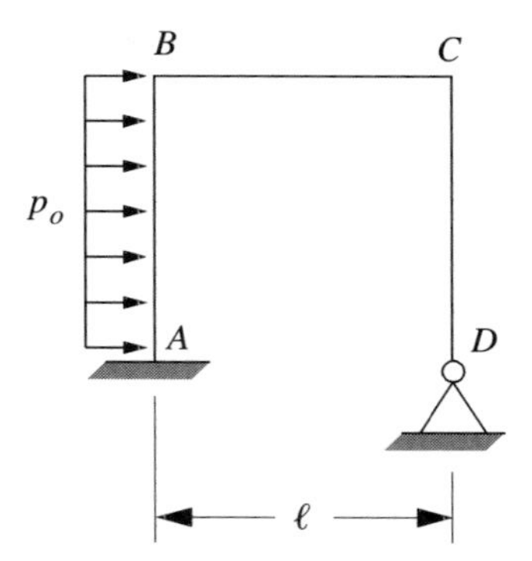

Figure 4.23 Problem 4.8.

4.9 The two-cell fuselage consists of a circular frame with radius r, with a rigidly attached straight member across the middle. The top circular frame and the floor both have a flexural rigidity EI, and the bottom circular frame has a flexural rigidity of $2EI$. The loading as shown in Fig. 4.24 consists of two antisymmetric moments M balanced by the antisymmetric loads M/r. Using antisymmetry and the left half of the structure, determine by complementary virtual work

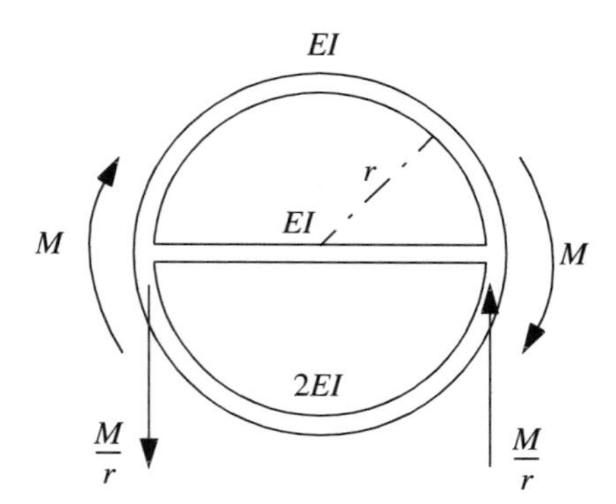

Figure 4.24 Problem 4.9.

(a) the bending moment distribution in each part of the frame, and

(b) on a sketch, show a free-body diagram of the left portion of the frame.

4.10 The circular fuselage frame of radius r shown in Fig. 4.25 supports a load P applied at the bottom of the frame. The load is reacted by a shear flow $q = P \sin \alpha/(\pi r)$ distributed around the circumference of the frame by the fuselage skin. The frame has a flexural rigidity EI. Using symmetry and the right half of the structure, determine by complementary virtual work

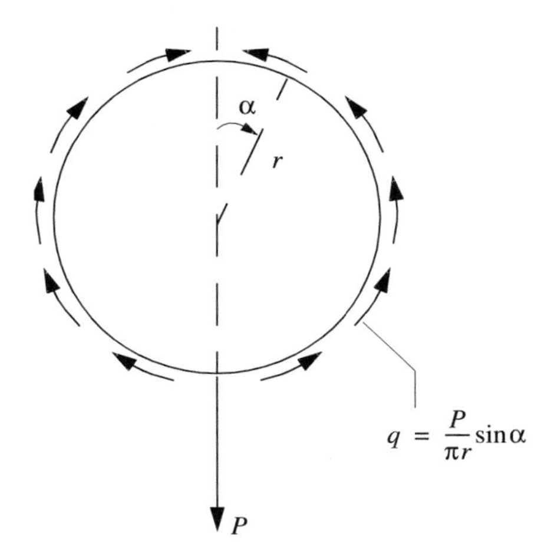

Figure 4.25 Problem 4.10.

(a) the bending moment distribution at any section, and

(b) a free-body diagram of the right portion of the frame.

4.11 The frame shown in Fig. 4.26 in a circular section fuel tank has a radius

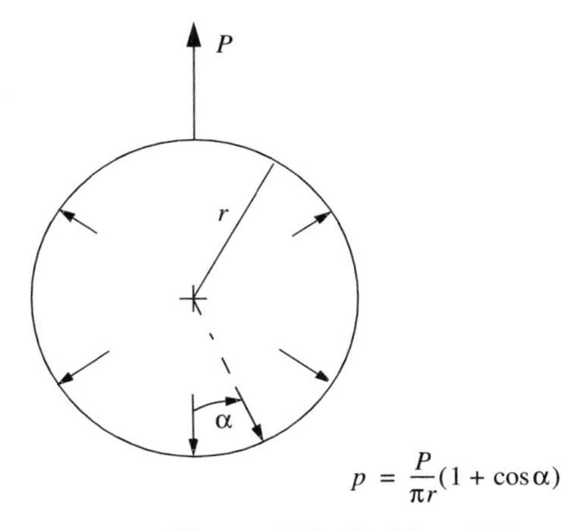

Figure 4.26 Problem 4.11.

r and a flexural rigidity EI. The loading consists of a vertical load P at the top of the frame from the fuel load and the fuel hydrostatic pressure $p = P(1 + \cos\alpha)/(\pi r)$. Using symmetry and the right half of the structure, determine by complementary virtual work the bending moment distribution at any section.

4.12 A curved bar shown in Fig. 4.27 bends through a 90° arc with one end clamped and the other end free. Its radius $r \gg h$ its depth. Its axial rigidity is AE and its flexural rigidity is EI. The bar undergoes a temperature change $\Delta T = T_0(1 + a_1 y/h + a_2 y^2/h^2)$, where a_1 and a_2 are known constants and T_0 is a given temperature. $I/A = h^2/12$. By complementary virtual work, determine the horizontal deflection of the free end.

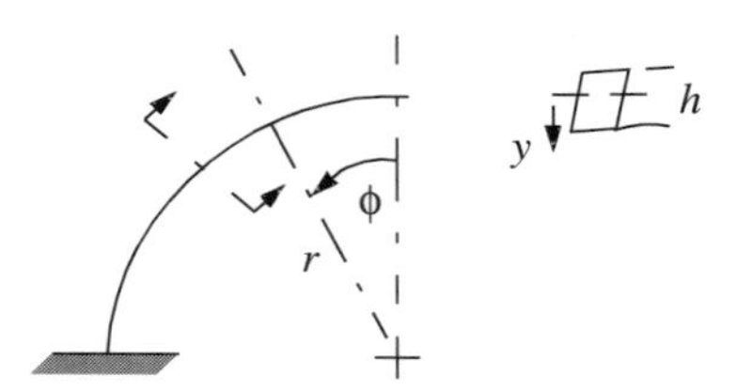

Figure 4.27 Problem 4.12.

4.13 For the frame shown in Fig. 4.28, the horizontal member is of length ℓ and has a flexural rigidity EI. The vertical members are of height h and have flexural rigidity EI_1, and the horizontal member carries a triangular

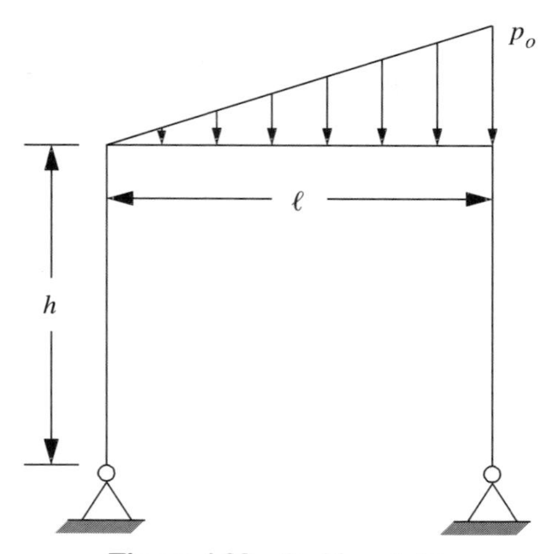

Figure 4.28 Problem 4.13.

load distribution with the load intensity p_0 at the right end. The vertical members are pinned at the ground. By complementary virtual work, determine the horizontal reactions at the pinned supports.

4.14 The beam shown in Fig. 4.29 is clamped at both ends. It is of length ℓ with a flexural rigidity EI. A counterclockwise applied moment M_0 is applied at $\ell/3$ from the left end. By complementary virtual work, determine the reactions at the left end.

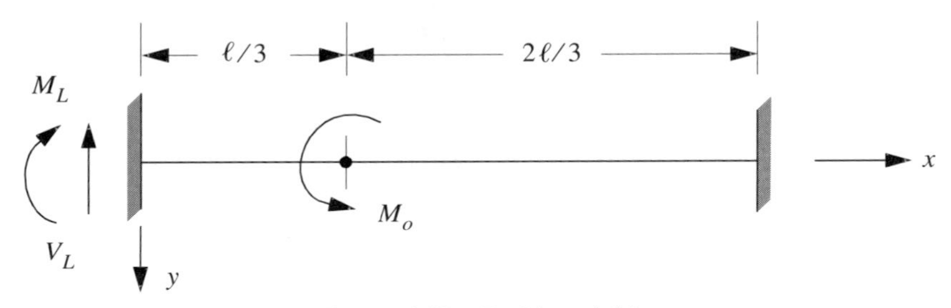

Figure 4.29 Problem 4.14.

4.15 A two-span beam shown in Fig. 4.30 rests on simple supports A and C before it is loaded. There is a small gap Δ between the beam and the simple center support at B. When the uniform load p_0 is applied, the gap is closed and reactions develop in all three supports. The beam is of length 2ℓ and has a flexural rigidity EI. By complementary virtual work, determine the magnitude of the gap Δ so that all three reactions are equal.

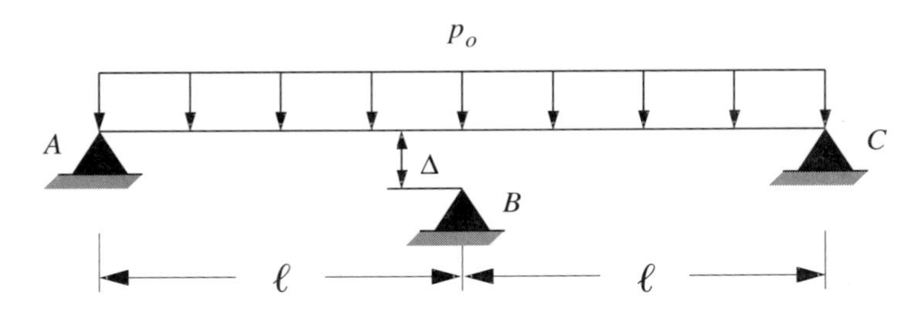

Figure 4.30 Problem 4.15.

4.16 Two cantilever beams AB and CD are supported at their left ends as shown in Fig. 4.31. A rigid roller lies between the two beams at D. The upper beam is of length ℓ with a flexural rigidity EI_1, and the lower beam is of length $\ell/2$ with a flexural rigidity EI_2. A vertical downward load P

is applied at B. By complementary virtual work, determine the reaction at D.

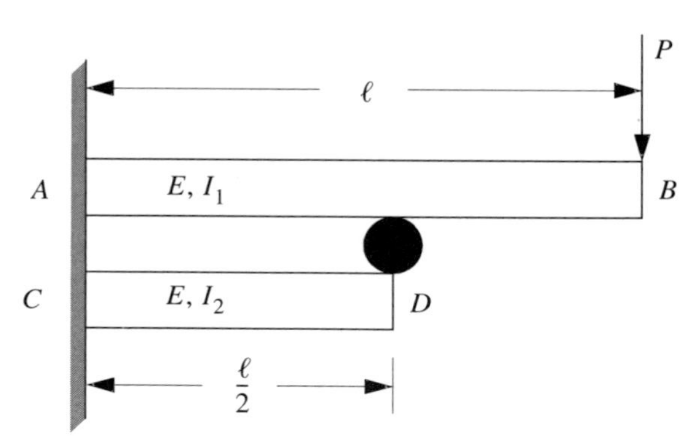

Figure 4.31 Problem 4.16.

4.17 Two cantilever beams AB and CD are supported at their left ends as shown in Fig. 4.32. A rigid roller lies between the two beams at D. The upper beam is of length ℓ with a flexural rigidity EI_1 and coefficient of thermal expansion α. The lower beam is of length ℓ with a flexural rigidity EI_2. The upper beam undergoes a temperature change $\Delta T = (T_2 - T_1)y/h$, where $T_2 < T_1$, T_1 is the temperature at the upper surface of beam AB, and T_2 is the temperature at the lower surface of beam AB. By complementary virtual work, determine the reaction of the rigid roller on the upper beam at B.

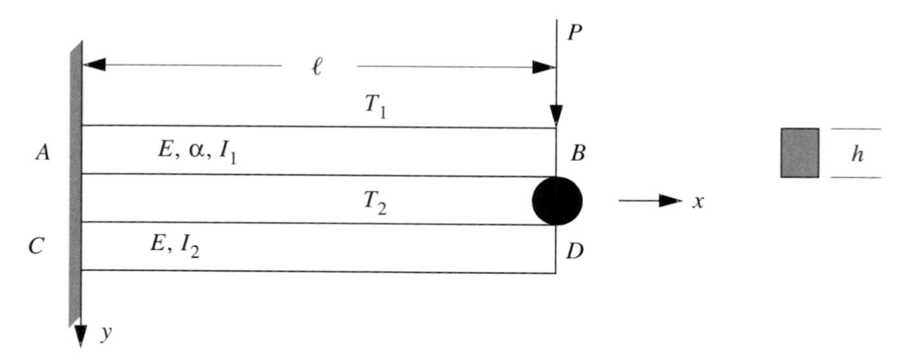

Figure 4.32 Problem 4.17.

4.18 The beam shown in Fig. 4.33 has a flexural rigidity EI and a length ℓ. Both ends A and B are fixed, but end B is displaced laterally through a distance Δ with respect to end A. By complementary virtual work, determine the moment and shear reactions at end A.

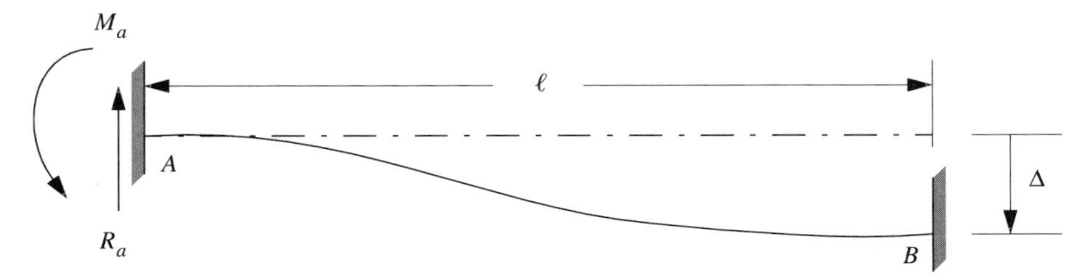

Figure 4.33 Problem 4.18.

4.19 For the frame shown in Fig. 4.34, the fixed support C sinks vertically downward at a distance Δ. The support at A allows for horizontal movement but restrains both vertical motion and rotation. Both members are of length ℓ and have flexural rigidity EI. By complementary virtual work, determine the moment and vertical reactions at end A.

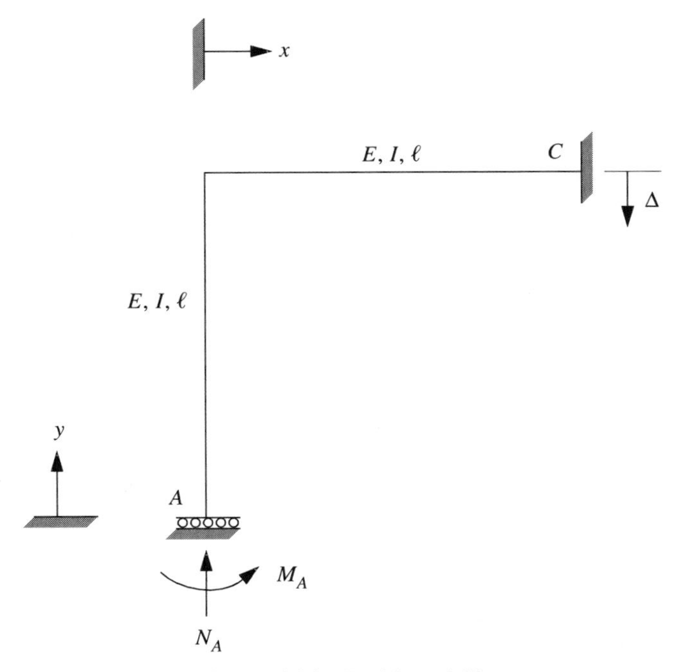

Figure 4.34 Problem 4.19.

4.20 The simple bar shown in Fig. 4.35 has an axial rigidity AE, is of length ℓ, and carries an axial load per unit length of q. Its left end is fixed and its right end is free. By complementary virtual work, determine its axial displacement $u(x)$ at any section.

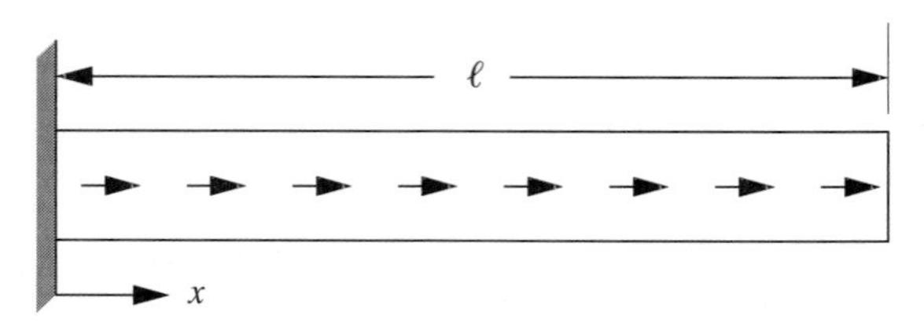

Figure 4.35 Problem 4.20.

4.21 For the tow frame shown in Fig. 4.36, the horizontal members have a flexural rigidity EI. The frame has a height h and a width ℓ. Equal but opposite vertical loads P are applied at $\ell/2$ of the horizontal members. Considering only the right section ABC, determine by complementary virtual work

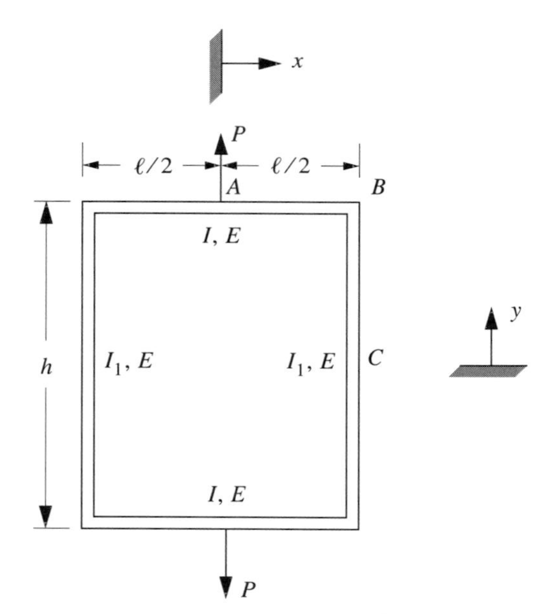

Figure 4.36 Problem 4.21.

(a) the bending moment at C, and

(b) the horizontal displacement at C.

4.22 The link shown in Fig. 4.37 is composed of two semicircles of radius r and two straight portions each of length 2ℓ. The members have flexural rigidity EI. Equal but opposite horizontal loads P are applied at $\ell/2$ of the vertical members. Considering only the right section ABC, determine by complementary virtual work the bending moment at A.

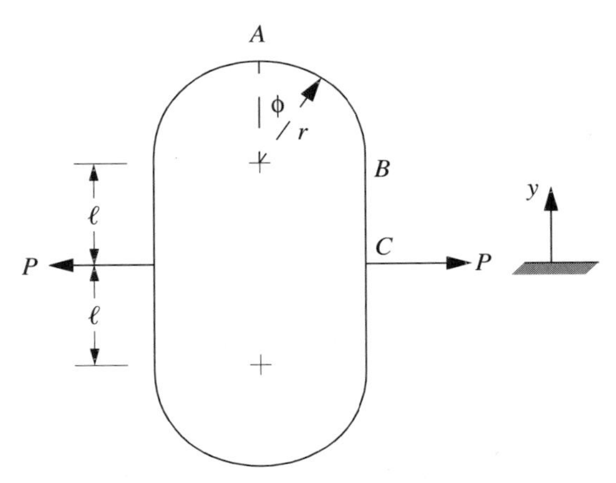

Figure 4.37 Problem 4.22.

4.23 The fixed, semicircular arch of radius R has a constant flexural rigidity EI and carries a uniform horizontal load of intensity p as shown in Fig. 4.38. Considering only the left section AB, by complementary virtual-work determine

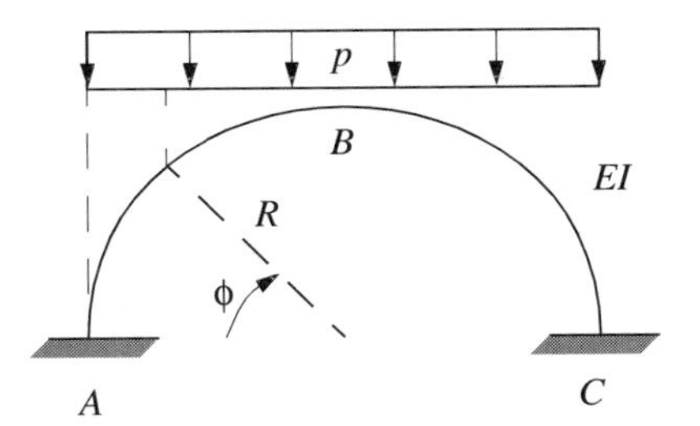

Figure 4.38 Problem 4.23.

(a) the horizontal reaction at A;

(b) the moment at A;

(c) the moment at any section;

(d) the moment at B; and

(e) the angle where the moment goes to zero.

4.24 The beam shown in Fig. 4.39 is simply supported at its right end B. Its left end A is restrained from vertical and horizontal motion by a simple support. A rotational spring of modulus $k = EI/\ell$ provides rotational constraint at the left end. The beam is of length ℓ with constant flexural rigidity EI and carries a uniform load of intensity p. By complementary virtual work, determine the vertical reaction at B.

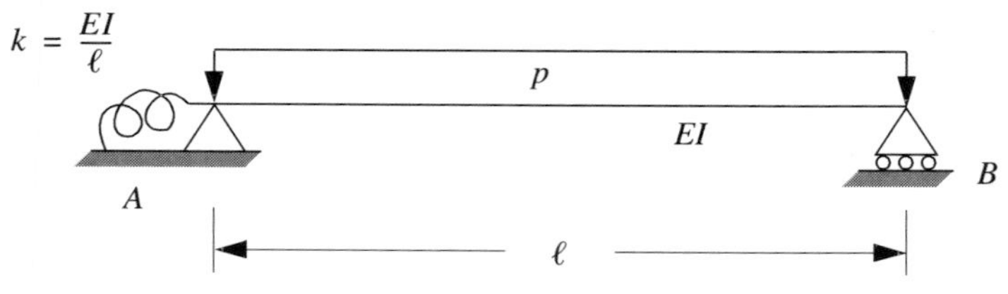

Figure 4.39 Problem 4.24.

4.25 The beam shown in Fig. 4.40 in an unloaded condition, just touching a spring at midspan, and is then loaded across its entire span with a uniform load p_0. Determine by complementary virtual work the spring force k so that the forces in all three supports are equal.

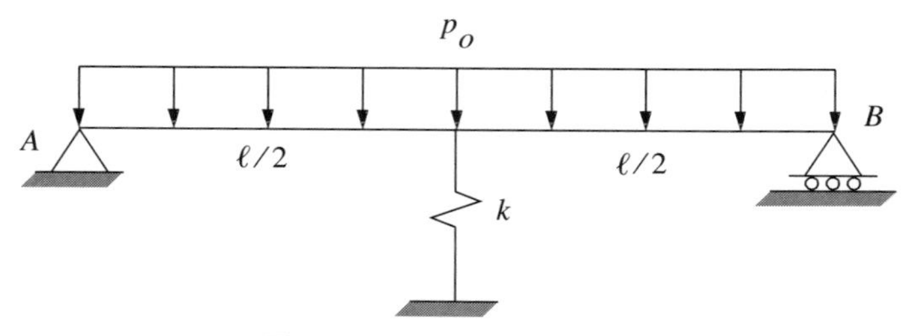

Figure 4.40 Problem 4.25.

(a) Use symmetry and only the left half of the beam.

(b) Use the total beam.

4.26 The beam shown in Fig. 4.41 is built in at its left end and is supported by an elastic rod at its right end. The beam is of length ℓ and has a flexural rigidity EI. The rod is of length L, has an axial rigidity of AE, and a coefficient of thermal expansion α. The rod is heated a uniform ΔT degrees. By complementary virtual work, determine the axial load in the rod.

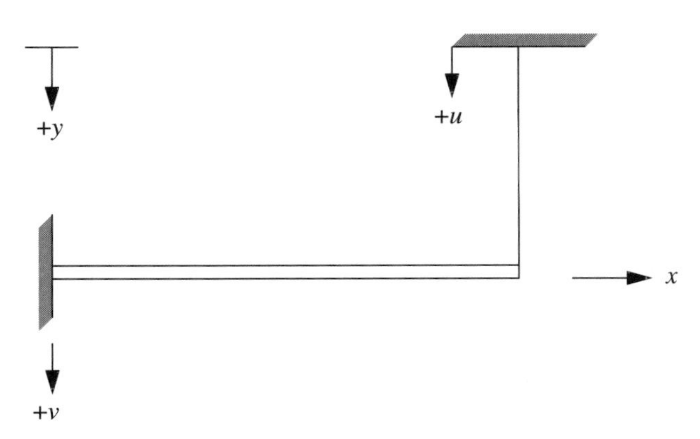

Figure 4.41 Problem 4.26.

4.27 The beam shown in Fig. 4.42 is of length ℓ and is clamped against translation and rotation at the left end and is subject to the temperature distribution along its length of $\Delta T = by^2$, where b is a specified constant. The beam has a symmetric cross section of area A and a second moment of area I. The elastic modulus is E and the coefficient of thermal expansion is α. By complementary virtual work, determine the horizontal displacement u of the right end of the beam in terms of α, b, I, A, and ℓ.

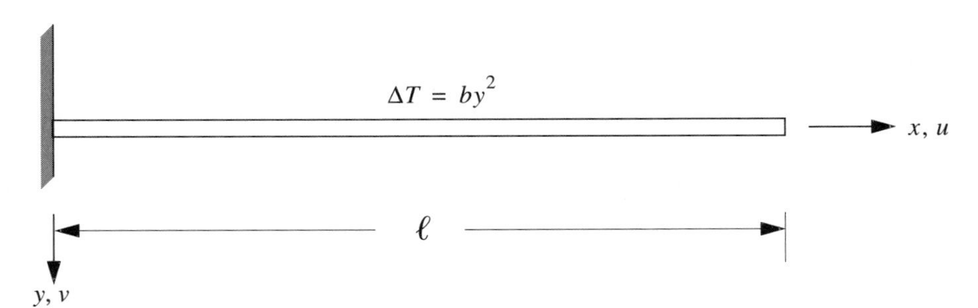

Figure 4.42 Problem 4.27.

4.28 The structure shown in Fig. 4.43 consists of a rod AB pinned at A and B, and a beam CB with the end C clamped against both rotation and translation. The rod has a cross-sectional area A_r and an elastic modulus E_r. The beam has a cross section with area A_b and second moment of area I_b, and it has an elastic modulus E_b. The structure is carrying a vertical load P applied at B. By complementary virtual work, determine

(a) The axial load N in the rod AB. Express the results in terms of L, E_b, I_b, A_b, E_r, A_r, and P.

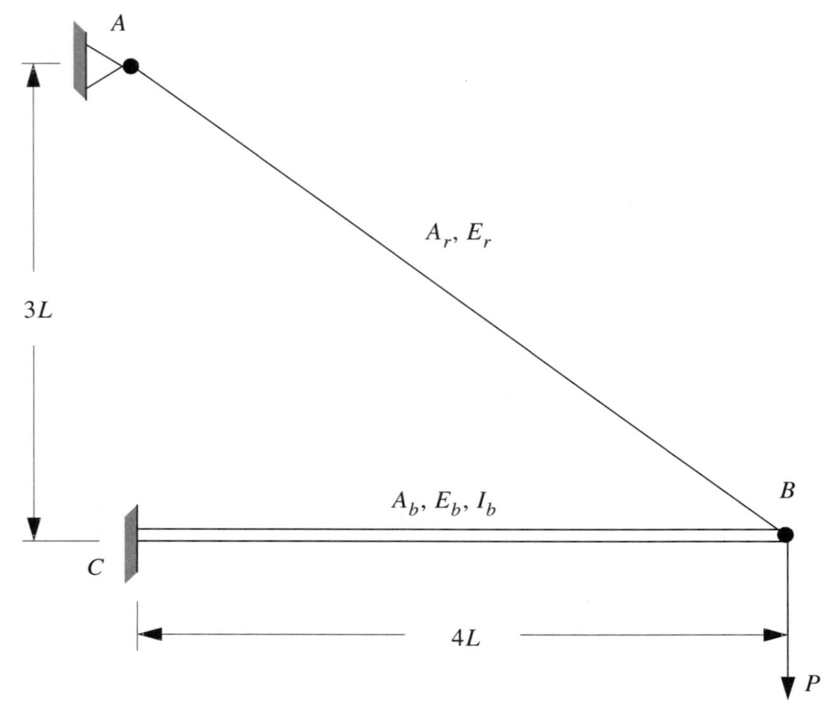

Figure 4.43 Problem 4.28.

(b) The horizontal displacement of B. Express the results in terms of L, E_b, A_b, P, and N.

(c) The vertical displacement of B. Express the results in terms of L, E_b, I_b, P, and N.

(d) The rotational displacement of B. Express the results in terms of L, E_b, I_b, P, and N.

4.29 The semicircular three-hinged arch in Fig. 4.44 is of radius R with a constant EI and is loaded by its dead weight of intensity w lb/ft of arch. Considering only flexural deformations, by using complementary virtual work in addition to using only section $\widehat{BC}$ of the arch (because of the symmetry) and using the convention for bending that compression on the outside of the arch is produced by a positive moment, determine the following:

(a) From statics, the vertical and horizontal reactions at C.

(b) The moment at any section in terms of R, w, and functions of θ.

(c) The vertical deflection at the crown B in terms of R, w, and EI.

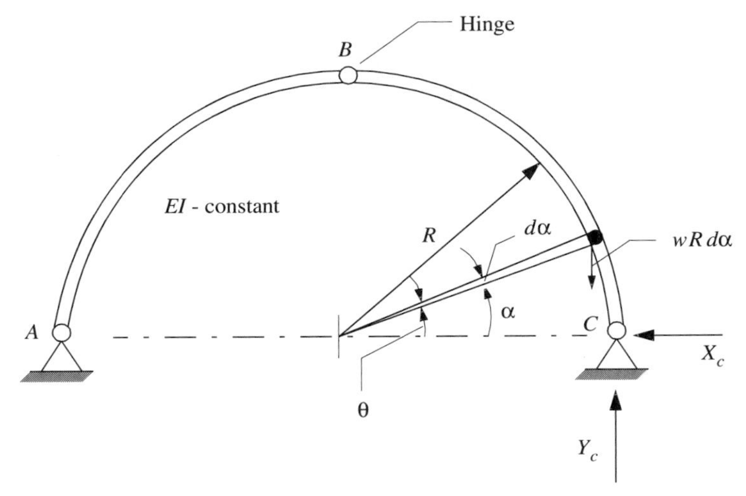

Figure 4.44 Problem 4.29.

4.30 The beam shown in Fig. 4.45 is prevented from rotating at the left end but is allowed to displace vertically at that end. The vertical displacement at the left end is resisted by a spring of modulus k. The right end of the beam is simply supported. The beam is of length ℓ and height h, has a flexural rigidity EI, and has a coefficient of thermal expansion α. The beam also has a temperature distribution through its thickness of $\Delta T = T_0(y/h)^3$. (Note: $\int y\ dA = \int y^3\ dA = 0$, $\int y^2\ dA = I$, $\int y^4\ dA = 3h^2I/20$.) By complementary virtual work, determine

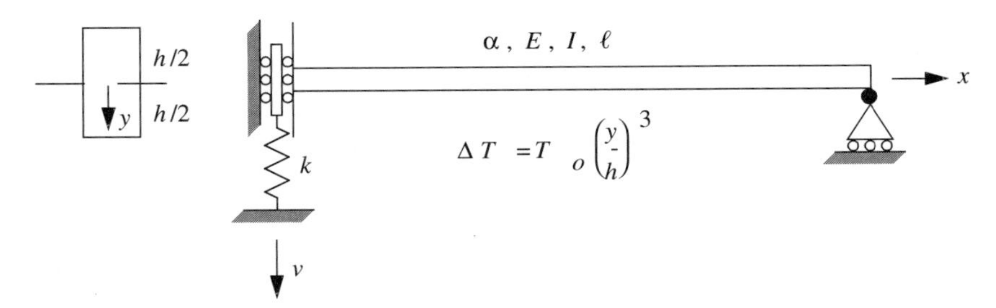

Figure 4.45 Problem 4.30.

(a) The tip deflection $v(0)$ at the left end in terms of α, EI, T_0, ℓ, h, and k.

(b) The rotation at the right end in terms of α, EI, T_0, ℓ, h, k, and $v(0)$.

4.31 At point B, the frame shown in Fig. 4.46 is prevented from vertical motion and rotation. However, point B can move horizontally, and this motion is resisted by a spring of modulus k. The horizontal member is supported at C by a simple support. Both members have flexural rigidity EI. The vertical member has a length ℓ, and the horizontal member has a length h. A horizontal load P is applied at the rigid connection between the vertical and horizontal members. By complementary virtual work, determine

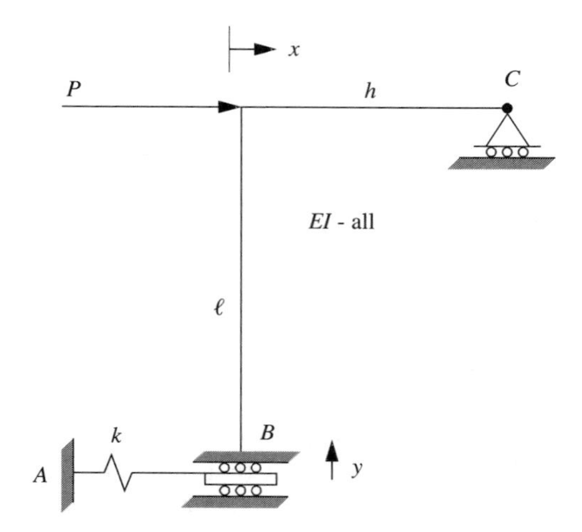

Figure 4.46 Problem 4.31.

(a) the vertical reaction R_c at C, and

(b) the horizontal displacement at C.

4.32 The fixed, semicircular end arch of radius R has a flexural rigidity EI.

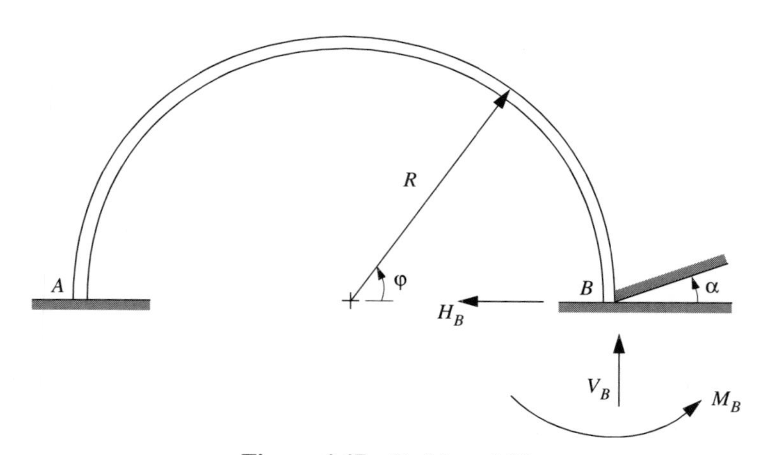

Figure 4.47 Problem 4.32.

The fixed support at the right end undergoes a counterclockwise rotation α as shown in Fig. 4.47. By complementary virtual work, determine the three reactions H_B, V_B, and M_B at the right support.

4.33 The circular fuselage frame shown in Fig. 4.48 has a radius r and a flexural rigidity EI. The straight floor beam has a length $r\sqrt{2}$ and a flexural rigidity EI. The floor is loaded by a clockwise couple T applied at the lowest point of the frame, and the couple is equilibrated by a constant shear flow q around the periphery of the frame. By complementary virtual work, determine the moment distribution in each section.

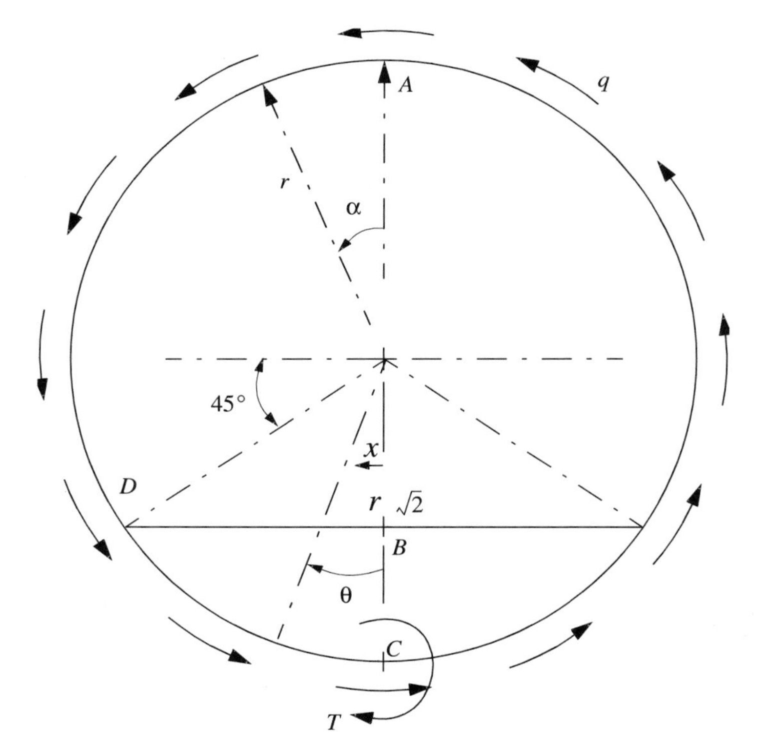

Figure 4.48 Problem 4.33.

4.34 Figure 4.49 represents a simply supported symmetric, parabolic arch expressed as $x = ky^2$, with origin at A for the left half. The right half is similar. At the base, the supports are a distance L apart. The distance from base to crown is L. The EI of the arch is constant, and the arch carries a vertical load P at the rigid connection of the crown. By complementary virtual work, determine

(a) the horizontal reaction at A, and

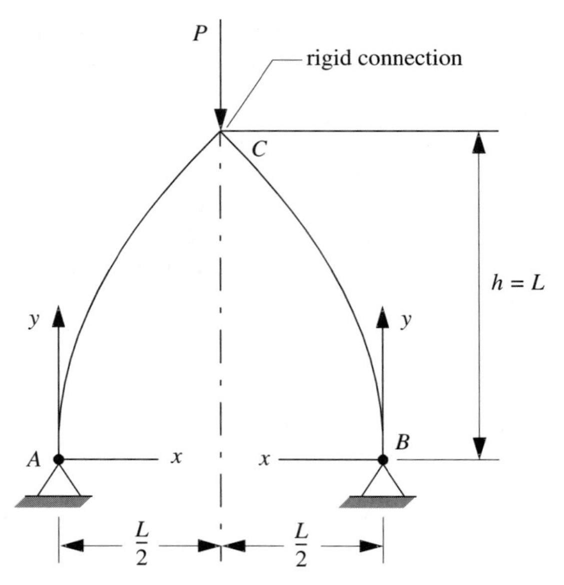

Figure 4.49 Problem 4.34.

(b) the rotation at A.

Some useful relationships are

$$C = L^2 + y^2$$

$$\int y\sqrt{C}\,dy = \frac{1}{3}\sqrt{C^3}$$

$$\int y^2\sqrt{C}\,dy = \frac{y}{4}\sqrt{C^3} - \frac{L^2 y}{8}\sqrt{C} - \frac{L^4}{8}\ln(y + \sqrt{C})$$

$$\int y^3\sqrt{C}\,dy = \frac{1}{5}\sqrt{C^5} - \frac{L^2}{3}\sqrt{C^3}$$

CHAPTER 5

SOME ENERGY METHODS

5.1 CONSERVATIVE FORCES AND POTENTIAL FUNCTIONS

In Chapter 6, we will discuss generalized nonconservative structural loading; in this chapter, however, we will concentrate on conservative loading. Such forces are characterized by the properties that they are functions only of initial and final positions. Consider the work of such a force:

$$W_e = \int_A^B \vec{F} \cdot d\vec{r}$$

The work around a closed path is

$$W_e = \oint \vec{F} \cdot d\vec{r} = 0$$

By Eq. (2.13), the above expression represents an exact differential that we can define as

$$\vec{F} \cdot d\vec{r} = -dU_e \tag{5.1}$$

where $-dU_e$ is the potential energy of the external loading.

Then, for conservative forces, we can write

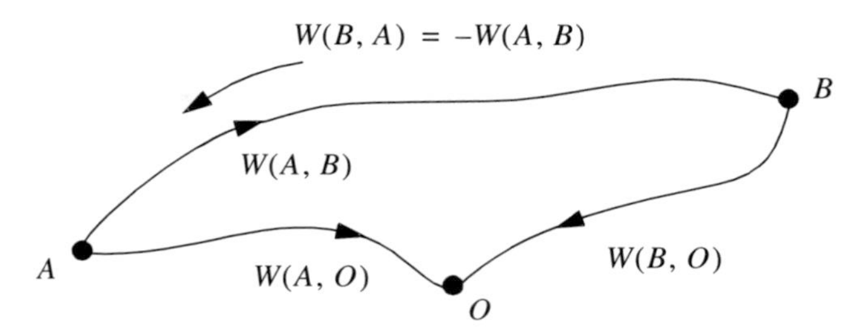

Figure 5.1 Change in potential.

$$W_e = \int_A^B \vec{F} \cdot d\vec{r} = -\int_A^B dU_e = (U_e)_A - (U_e)_B \tag{5.2}$$

This equation states that the *decrease* in potential energy U_e in moving a particle from A to B is equal to the work done *by* the conservative force field. The minus sign indicates that the force loses its potential to perform addition work as positive work is expended.

The above expressions are consistent with the formal definition of potential energy stating that potential energy is the work done in moving a particle from a point A to an arbitrarily chosen datum O, or

$$\Pi(A) = W(A, O)$$

where $\Pi(A)$ is the potential energy at configuration A and $W(A, O)$ is the work done by forces in going from A to O. Since the potential energy is relative to an arbitrary datum, its value is only known to within an arbitrary constant.

Next, consider the change in potential as a particle moves from A to B, as shown in Fig. 5.1. From Fig. 5.1, we can write for the change in potential

$$\Pi(B) - \Pi(A) = W(B, O) - W(A, O)$$

However, since we are considering conservative force, the work is independent of path, and

$$W(B, O) = W(B, A) + W(A, O)$$

The change in potential then becomes

$$\begin{aligned}\Pi(B) - \Pi(A) &= W(B, A) + W(A, O) - W(A, O) \\ &= W(B, A) \\ &= -W(A, B)\end{aligned}$$

which is consistent with Eq. (5.2).

Since conservative forces can be represented by a potential function, we have from Stokes's theorem

$$\vec{F} = -\nabla\Pi$$

or

$$F_x = -\frac{\partial\Pi}{\partial x}; \qquad F_y = -\frac{\partial\Pi}{\partial y}; \qquad F_z = -\frac{\partial\Pi}{\partial z} \tag{5.3}$$

From the preceding discussion, we are now in a position to define the potential energy of a conservative external force as

$$U_e = -W_e \tag{5.4}$$

For general body force and surface force loading, we have

$$U_e = -\int_V \int_0^{\text{final}} (X_b\, du + Y_b\, dv + Z_b\, dw)\, dV - \int_{S_\sigma} \int_0^{\text{final}} (X_s\, du + Y_s\, dv + Z_s\, dw)\, dS \tag{5.5}$$

Frequently, body forces are negligible, and the surface forces can be represented by a system of conservative concentrated forces. Then, we have

$$U_e = -(P_1\Delta_1 + \cdots + P_n\Delta_n) \tag{5.6}$$

where P_i is a concentrated force and Δ_i is the displacement at the point of application of the force and in the direction of the force.

Example 5.1 Consider the elastic potential of a linear spring as shown in Fig. 5.2. The elongation of the spring is x and is measured from the undeformed position. Consider the attachment point A that in this case represents the surrounding structure to the spring. The elastic potential of the spring is a measure of its ability to do work on its surroundings, not vice versa. Then, we are interested in the force F of the spring on the attachment point A, not the force of opposite sign that is applied to the spring by the attachment point. From Eq. (5.3) the free body, we have

$$F = -kx = -\frac{d\Pi}{dx}$$

or to within an arbitrary constant,

$$\Pi = \tfrac{1}{2}\, kx^2$$

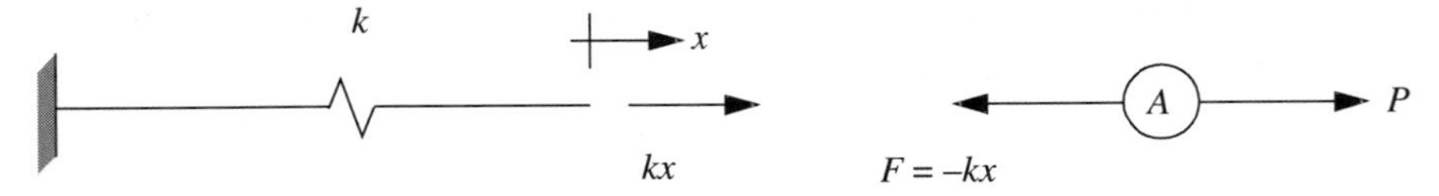

Figure 5.2 Potential in a linear spring.

Example 5.1 implies that for elastic bodies in general, perhaps an internal, elastic strain-energy expression exists. As discussed by Love in ref. [49], in 1837 Green propounded that there exists such a function that is quadratic in strain and always an exact differential. Green then used this function to derive the classical equations of elasticity for anisotropic material, resulting in twenty-one elastic constants. About eighteen years later, Thomson (Lord Kelvin) based the existence of Green's strain-energy function on the first and second laws of thermodynamics.

Consider Fig. 2.4. In the case of small displacement, the kinetic energy per unit volume can be expressed as

$$K = \frac{1}{2}\rho\left[\left(\frac{\partial u}{\partial t}\right)^2 + \left(\frac{\partial v}{\partial t}\right)^2 + \left(\frac{\partial w}{\partial t}\right)^2\right]$$

where ρ is the mass density and u, v, w are the components of displacements.[1] The increment of kinetic energy per unit volume acquired in a short time interval dt is

$$dK = \rho\left(\frac{\partial^2 u}{\partial t^2}\,du + \frac{\partial^2 v}{\partial t^2}\,dv + \frac{\partial^2 w}{\partial t^2}\,dw\right)$$

Under the assumption of small displacements, the equations of motion for the volume element are

$$\frac{\partial \sigma_{xx}}{\partial x} + \frac{\partial \tau_{xy}}{\partial y} + \frac{\partial \tau_{xz}}{\partial z} + \rho X_b = \rho\,\frac{\partial^2 u}{\partial t^2}$$

$$\frac{\partial \sigma_{yy}}{\partial y} + \frac{\partial \tau_{yx}}{\partial x} + \frac{\partial \tau_{yz}}{\partial z} + \rho Y_b = \rho\,\frac{\partial^2 v}{\partial t^2}$$

$$\frac{\partial \sigma_{zz}}{\partial z} + \frac{\partial \tau_{zx}}{\partial x} + \frac{\partial \tau_{zy}}{\partial y} + \rho Z_b = \rho\,\frac{\partial^2 w}{\partial t^2}$$

[1] If we had chosen to do the derivation in terms of large displacement, we would have used the velocity vector $\vec{v}$ in the required relationships. Also, under the large displacement assumption, we would write $d\vec{u}/dt = \partial\vec{u}/\partial t + \vec{v}\cdot \text{grad}\,\vec{u}$, where grad is in terms of spatial coordinates.

Using these equations, the increment in kinetic energy becomes

$$dK = \left(\frac{\partial \sigma_{xx}}{\partial x} + \frac{\partial \tau_{xy}}{\partial y} + \frac{\partial \tau_{xz}}{\partial z} + \rho X_b \right) du$$
$$+ \left(\frac{\partial \sigma_{yy}}{\partial y} + \frac{\partial \tau_{yx}}{\partial x} + \frac{\partial \tau_{yz}}{\partial z} + \rho Y_b \right) dv$$
$$+ \left(\frac{\partial \sigma_{zz}}{\partial z} + \frac{\partial \tau_{zx}}{\partial x} + \frac{\partial \tau_{zy}}{\partial y} + \rho Z_b \right) dw$$

The increment in external work applied to the whole body in dt is

$$\bar{d} W_e = \int_V (\rho X_b \, du + \rho Y_b \, dv + \rho Z_b \, dw) \, dV$$
$$+ \oint_{\partial V} (X_s \, du + Y_s \, dv + Z_s \, dw) \, dS$$

but from Green's theorem in the plane, we can express the surface integral as

$$\int_{\partial V} (X_s \, du + Y_s \, dv + Z_s \, dw) \, dS$$
$$= \int_V \left[\left(\frac{\partial \sigma_{xx}}{\partial x} + \frac{\partial \tau_{xy}}{\partial y} + \frac{\partial \tau_{xz}}{\partial z} \right) du + \left(\frac{\partial \sigma_{yy}}{\partial y} + \frac{\partial \tau_{yx}}{\partial x} + \frac{\partial \tau_{yz}}{\partial z} \right) dv \right.$$
$$\left. + \left(\frac{\partial \sigma_{zz}}{\partial z} + \frac{\partial \tau_{zx}}{\partial x} + \frac{\partial \tau_{zy}}{\partial y} \right) dw \right] dV$$
$$+ \oint_V \left[\sigma_{xx} \frac{\partial du}{\partial x} + \sigma_{yy} \frac{\partial dv}{\partial y} + \sigma_{zz} \frac{\partial dw}{\partial z} + \tau_{xy} \left(\frac{\partial du}{\partial y} + \frac{\partial dv}{\partial x} \right) \right.$$
$$\left. + \tau_{yz} \left(\frac{\partial dv}{\partial z} + \frac{\partial dw}{\partial y} \right) + \tau_{zx} \left(\frac{\partial dw}{\partial x} + \frac{\partial du}{\partial z} \right) \right] dV$$

The first law of thermodynamics for the whole body can now be written as

$$\int_V (dK + dU) \, dV = \bar{d} W_e + \bar{d} H$$

Collecting the above expressions, this becomes

$$\int_V dU\,dV = \int_V (\sigma_{xx}\,d\epsilon_{xx} + \sigma_{yy}\,d\epsilon_{yy} + \sigma_{zz}\,d\epsilon_{zz} + \tau_{xy}\,d\gamma_{xy} + \tau_{yz}\,d\gamma_{yz} + \tau_{zx}\,d\gamma_{zx})\,dV + \bar{d}H$$

where we have also used relationships of the form

$$\frac{\partial du}{\partial x} = d\,\frac{\partial u}{\partial x} = d\epsilon_{xx}$$

If the body is treated adiabatically,

$$\bar{d}H = 0$$

Since it is fundamental to thermodynamics that the internal energy be an exact differential, then with $\bar{d}H = 0$ the integral on the right represents an exact differential. Hence we may define the strain-energy-density function U_0 as

$$dU_0 = \sigma_{xx}\,d\epsilon_{xx} + \sigma_{yy}\,d\epsilon_{yy} + \sigma_{zz}\,d\epsilon_{zz} + \tau_{xy}\,d\gamma_{xy} + \tau_{yz}\,d\gamma_{yz} + \tau_{zx}\,d\gamma_{zx} \tag{5.7}$$

Since Eq. (5.7) represents an exact differential, we may also write

$$dU_0 = \frac{\partial U_0}{\partial \epsilon_{xx}}\,d\epsilon_{xx} + \frac{\partial U_0}{\partial \epsilon_{yy}}\,d\epsilon_{yy} + \frac{\partial U_0}{\partial \epsilon_{zz}}\,d\epsilon_{zz} + \frac{\partial U_0}{\partial \gamma_{xy}}\,d\gamma_{xy} + \frac{\partial U_0}{\partial \gamma_{yz}}\,d\gamma_{yz} + \frac{\partial U_0}{\partial \gamma_{zx}}\,d\gamma_{zx} \tag{5.8}$$

Comparing the above two expressions, we get

$$\sigma_{xx} = \frac{\partial U_0}{\partial \epsilon_{xx}}$$

$$\sigma_{yy} = \frac{\partial U_0}{\partial \epsilon_{yy}}$$

$$\sigma_{zz} = \frac{\partial U_0}{\partial \epsilon_{zz}}$$

$$\tau_{xy} = \frac{\partial U_0}{\partial \gamma_{xy}}$$

$$\tau_{yz} = \frac{\partial U_0}{\partial \gamma_{yz}}$$

$$\tau_{zx} = \frac{\partial U_0}{\partial \gamma_{zx}} \tag{5.9}$$

Reference [49] also shows that if the change in state takes place isothermally, by application of the second law of thermodynamics the function U_0 exists.

For the general case of large-strain elasticity, the function U_0 exists but is related to the stress by the relationship

$$\tilde{\mathbf{T}}_{ij} = \frac{\partial U_0}{\partial E_{ij}} \tag{5.10}$$

Where $\tilde{\mathbf{T}}_{ij}$ is the so-called Piola-Kirchhoff stress tensor of the second kind and E_{ij} is the Green's strain tensor.

For elastic media, the strain-energy-density function has been defined as

$$U_0 = \int_0^{\text{final}} \lfloor \sigma \rfloor \{d\epsilon\}$$

where

$$\lfloor \sigma \rfloor = \lfloor \sigma_{xx} \cdots \tau_{zx} \rfloor$$

is the row matrix of stress components, and

$$\{d\epsilon\} = \left\{ \begin{matrix} d\epsilon_{xx} \\ \vdots \\ d\gamma_{zx} \end{matrix} \right\}$$

is the column vector of differential total strains including temperature effects. Argyris and Kelsey in ref. [50] choose to define the strain-energy-density function strictly as the work of the elastic deformation exclusive of any thermal effects. Then, the $\{d\epsilon\}$ in the above expression is replaced by $\{d\epsilon^m\}$ for the elastic or mechanical strains. The expressions for strain energy are the same to within constants involving terms of the form $\alpha^2 \Delta T^2$. When U_0 is expressed only in terms of stress, the constant containing $\alpha^2 \Delta T^2$ terms does not explicitly contain any stress term; hence it may be considered irrelevant and dropped.

When considering thermoelastic stress, the assumed stress-strain relationship is in the form

$$\{\sigma\} = \{f(\{\epsilon\}, \Delta T)\}$$

In this form, $\{\epsilon\}$ and ΔT are treated as independent variables. However, the strain energy becomes associated with the Helmholtz free energy [51].

We now describe some examples of the strain-energy-density function. First, consider a two-dimensional Hookean material described by the following strain-stress relationships:

$$\epsilon_{xx} = \frac{\sigma_{xx}}{E} - \nu \frac{\sigma_{yy}}{E}$$

$$\epsilon_{yy} = \frac{\sigma_{yy}}{E} - \nu \frac{\sigma_{xx}}{E}$$

$$\gamma_{xy} = \frac{\tau_{xy}}{G}$$

where ν is Poisson's ratio. Then, we can write

$$U_0 = \int_0^{\text{final}} \lfloor \sigma \rfloor \{d\epsilon\}$$

$$= \int_0^{\text{final}} \left[\sigma_{xx} \left(\frac{d\sigma_{xx}}{E} - \nu \frac{d\sigma_{yy}}{E} \right) + \sigma_{yy} \left(\frac{d\sigma_{yy}}{E} - \nu \frac{d\sigma_{xx}}{E} \right) + \tau_{xy} \frac{d\tau_{yx}}{G} \right]$$

which, after integration by parts of the second term and recombining terms, yields

$$U_0 = \tfrac{1}{2} (\sigma_{xx}\epsilon_{xx} + \sigma_{yy}\epsilon_{yy} + \tau_{xy}\gamma_{xy}) \tag{5.11}$$

For a three-dimensional isotropic material, U_0 takes the form

$$U_0 = \frac{\lambda}{2} e^2 + G\left(\epsilon_{xx}^2 + \epsilon_{yy}^2 + \epsilon_{zz}^2 + \frac{1}{2} \gamma_{xy}^2 + \frac{1}{2} \gamma_{yz}^2 + \frac{1}{2} \gamma_{zx}^2 \right) - \frac{E\alpha\Delta T}{1 - 2\nu} e \tag{5.12}$$

or

$$U_0 = \frac{1}{2E}(\sigma_{xx}^2 + \sigma_{yy}^2 + \sigma_{zz}^2) - \frac{\nu}{E}(\sigma_{xx}\sigma_{yy} + \sigma_{yy}\sigma_{zz} + \sigma_{zz}\sigma_{xx}) + \frac{1}{2G}(\tau_{xy}^2 + \tau_{yz}^2 + \tau_{zx}^2) \tag{5.13}$$

where

$$e = \epsilon_{xx} + \epsilon_{yy} + \epsilon_{zz}$$

and

$$\lambda = \frac{\nu E}{(1+\nu)(1-2\nu)}$$

All of these results can be generalized into a matrix format by again considering Eq. (3.7). First, by excluding initial stress we can write

$$U_0 = \int_0^{\text{final}} \lfloor \sigma \rfloor \{d\epsilon\}$$

$$= \int_0^{\text{final}} (\lfloor \epsilon \rfloor [D] - \lfloor \epsilon_0 \rfloor [D])\{d\epsilon\}$$

For elastic materials, $[D]$ is required to be symmetric; hence, $[D] = [D]^T$, where $[D]^T$ represents the transpose. Consider the first integral

$$\int_0^{\text{final}} \lfloor \epsilon \rfloor [D]\{d\epsilon\}$$

which we can symbolically integrate by parts by defining

$$\lfloor \alpha \rfloor = \lfloor \epsilon \rfloor [D]$$

and

$$\{d\beta\} = \{d\epsilon\}$$

Now, $\lfloor \alpha \rfloor$ is a function of all the strain terms, and its total derivative $\lfloor d\alpha \rfloor$ upon term-by-term integration is

$$\lfloor d\alpha \rfloor = \left[\frac{\partial \alpha_1}{\partial \epsilon_{xx}} d\epsilon_{xx} + \frac{\partial \alpha_1}{\partial \epsilon_{yy}} d\epsilon_{yy} + \cdots, \ \frac{\partial \alpha_2}{\partial \epsilon_{xx}} d\epsilon_{xx} + \cdots \right]$$

The above expression can be collected in terms of $[D]$ and $\{d\epsilon\}$ as

$$\{d\alpha\} = [D]\{d\epsilon\}$$

and also as

$$\{\beta\} = \{\epsilon\}$$

The integration-by-parts formula becomes

$$\int \lfloor \alpha \rfloor \{d\beta\} = \lfloor \alpha \rfloor \{\beta\} - \int \lfloor \beta \rfloor \{d\alpha\}$$

or

$$2\int_0^{\text{final}} \lfloor \epsilon \rfloor [D]\{d\epsilon\} = \lfloor \epsilon \rfloor [D]\{\epsilon\}$$

Then, the strain-energy density function becomes

$$U_0 = \tfrac{1}{2}\lfloor \epsilon \rfloor [D]\{\epsilon\} - \lfloor \epsilon \rfloor [D]\{\epsilon_0\} \tag{5.14}$$

Substituting for $\{\epsilon\}$, we get

$$\begin{aligned} U_0 &= \tfrac{1}{2}\lfloor \sigma \rfloor [D]\{\sigma\} - \tfrac{1}{2}\lfloor \epsilon_0 \rfloor [D]\{\epsilon_0\} \\ &= \tfrac{1}{2}\lfloor \sigma \rfloor [D]\{\sigma\} \end{aligned} \tag{5.15}$$

where we have ignored the last term because it does not contain any explicit stress term and may be considered an arbitrary constant.

If initial stress is also specified, then the initial state is considered the reference state, and from Eq. (3.7) we have

$$\{\epsilon\} = [D]^{-1}\{\sigma - \sigma_0\} + \{\epsilon_0\}$$

and Eq. (5.14) becomes

$$U_0 = \tfrac{1}{2}\lfloor \epsilon \rfloor [D]\{\epsilon\} - \lfloor \epsilon \rfloor [D]\{\epsilon_0\} + \lfloor \epsilon \rfloor \{\sigma_0\} \tag{5.16}$$

Example 5.2 For strain energy for a nonlinear material, consider a uniaxial stress-strain law:

$$\sigma = B\epsilon^n; \qquad n \le 1$$

Then, we have

$$U_0 = \int \sigma \, d\epsilon = \int B\epsilon^n \, d\epsilon = \frac{B\epsilon^{n+1}}{(n+1)} \tag{5.17}$$

5.2 STATIONARY POTENTIAL ENERGY

Eq. (5.5) gives the potential energy of the external loading. For a set of virtual displacements, we can say

$$\delta W_e = -\delta U_e$$

The principle of virtual work states that

$$\delta W_e = \delta U$$

or for conservative external loading,

$$-\delta U_e = \delta U$$

The internal virtual work may be expressed as

$$\delta U = \int_V \lfloor \sigma \rfloor \{\delta \epsilon\} \, dV$$

where

$$\lfloor \sigma \rfloor = \lfloor \sigma_{xx} \cdots \tau_{zx} \rfloor$$

is the row matrix of stress components, and

$$\{\delta \epsilon\} = \begin{Bmatrix} \delta \epsilon_{xx} \\ \vdots \\ \delta \gamma_{zx} \end{Bmatrix}$$

However, if the strain-energy function exists, we can write

$$\delta U = \int_V \lfloor \partial U_0 \rfloor \{\delta \epsilon\} \, dV$$

where

$$\lfloor \partial U_0 \rfloor = \left[\frac{\partial U_0}{\partial \epsilon_{xx}} \cdots \frac{\partial U_0}{\partial \gamma_{zx}} \right]$$

but if we treat δ as a differential operator, the above expression is nothing more than the total first variation of U_0, and we can write

$$\begin{aligned} \delta U &= \int_V \delta U_0 \, dV \\ &= \delta \int_V U_0 \, dV \end{aligned}$$

where, now, δU can be considered the variation in the total strain energy of the body. The virtual work expression then becomes

$$\delta(U + U_e) = \delta\Pi = 0 \tag{5.18}$$

This equation expresses the fact that for an elastic structure to be in equilibrium, its variation of total potential Π must be stationary.

In general, $\Pi = f(\epsilon_{xx}, \ldots, \gamma_{zx})$, which we can also express as a function of the displacement components $\Pi = g(u, v, w)$. Then, Eq. (5.18) becomes

$$\delta\Pi = \frac{\partial\Pi}{\partial u}\,\delta u + \frac{\partial\Pi}{\partial v}\,\delta v + \frac{\partial\Pi}{\partial w}\,\delta w = 0 \tag{5.19}$$

But δu, δv, δw are independent arbitrary functions. Thus, for $\delta\Pi = 0$ we must have

$$\begin{aligned} \frac{\partial\Pi}{\partial u} &= 0 \\ \frac{\partial\Pi}{\partial v} &= 0 \\ \frac{\partial\Pi}{\partial w} &= 0 \end{aligned} \tag{5.20}$$

Next, consider the total derivative $d\Pi$:

$$d\Pi = \frac{\partial\Pi}{\partial u}\,du + \frac{\partial\Pi}{\partial v}\,dv + \frac{\partial\Pi}{\partial w}\,dw = 0 \tag{5.21}$$

However, from Eq. (5.20) we have

$$d\Pi = 0 \tag{5.22}$$

The following is the statement of stationary potential energy:

Of all possible displacements which satisfy the boundary conditions, those corresponding to equilibrium configurations makes the total potential energy stationary.

Stationary implies a local minimum, maximum, or inflection.

Example 5.3 Consider the truss shown in Fig. 5.3. The two members have equal area A and are of equal length ℓ. The load P is applied at the common connection pin and is in the direction u. Consider the stress-strain law $\sigma = B\epsilon^n$; $n \leq 1$. The strain energy is given by Eq. (5.17), and the strain in each member is given by $\epsilon = e/\ell$, where e is the member elongation. Then, the total strain energy for the truss is

$$U = 2\int U_0 \, dV = \frac{2BA}{\ell^n(n+1)} e^{n+1}; \qquad V = A\ell$$

The elongation e of each member is

$$e = u \cos \theta$$

and the total potential energy is then

$$\Pi = \frac{2BA}{\ell^n(n+1)} (u \cos \theta)^{n+1} - Pu$$

For equilibrium, we require $d\Pi/du = 0$, or

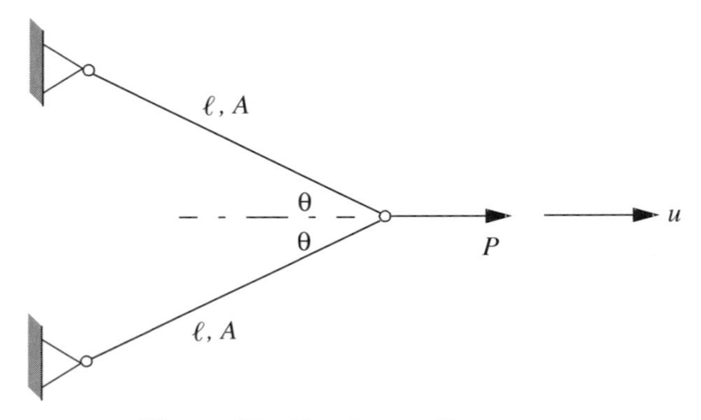

Figure 5.3 Two-bar nonlinear truss.

$$P = \frac{2BA}{\ell^n}\,(u\cos\theta)^n\cos\theta$$

as the resulting equilibrium relationship between P and u.

5.3 CASTIGLIANO'S FIRST THEOREM

Equation (5.18), which expresses the fact that for an elastic structure to be in equilibrium, the variation in its total potential energy is stationary when it is used in conjunction with Eq. (5.6), yields one of the most useful theorems of structural analysis. Consider the strain energy U to be a function of the independent displacement terms Δ_i; $i = 1, \ldots, n$. Then, we write

$$U = U(\Delta_1, \ldots, \Delta_i, \ldots, \Delta_n)$$

and Eq. (5.18) becomes

$$\delta\Pi(\Delta_1, \ldots, \Delta_i, \ldots, \Delta_n) = \delta U(\Delta_1, \ldots, \Delta_i, \ldots, \Delta_n) - \sum_{i=1}^{n} P_i\,\delta\Delta_i = 0$$

Equation (5.19) then becomes

$$\delta\Pi = \frac{\partial U}{\partial\Delta_1}\,\delta\Delta_1 + \cdots + \frac{\partial U}{\partial\Delta_i}\,\delta\Delta_i + \cdots + \frac{\partial U}{\partial\Delta_n}\,\delta\Delta_n$$
$$- P_i\,\delta\Delta_1 - \cdots - P_i\,\delta\Delta_i \cdots - P_n\,\delta\Delta_n = 0$$

Collecting coefficients for each $\delta\Delta_i$ and remembering that each $\delta\Delta_i$ is considered an arbitrary virtual displacement, we get

$$\delta\Pi = \cdots + \left(P_i - \frac{\partial U}{\partial\Delta_i}\right)\delta\Delta_i + \cdots = 0$$

or

$$P_i = \frac{\partial U}{\partial\Delta_i} \tag{5.23}$$

which is Castigliano's first theorem stating the following:

If the strain energy of a structural system is expressed in terms of n independent displacements Δ_i corresponding to a system of prescribed forces P_i, the first partial derivative of the strain energy with respect to any of these displacements Δ_i at the point i is equal to the force P_i in the direction Δ_i.

The theorem is in effect a restatement of the principle of virtual work in terms of strain energy.

Example 5.4 Consider again the truss shown in Fig. 5.3. The strain energy was shown to be

$$U = 2 \int U_0 \, dV = \frac{2BA}{\ell^n(n+1)} \, e^{n+1} = \frac{2BA}{\ell^n(n+1)} \, (u \cos \theta)^{n+1}$$

Thus:

$$P = \frac{\partial U}{\partial u} = \frac{2BA}{\ell^n} \, (u \cos \theta)^n \cos \theta$$

as before.

Example 5.5 Consider the beam shown Fig. 3.52 with the following approximate assumed solution:

$$v = a\left(1 - \sin \frac{\pi x}{2l}\right)$$

The strain energy in the beam is

$$U = \int U_0 \, dV = \int_V \int_0^{\text{final}} \sigma \, d\epsilon \, dV$$

or

$$U = \int_V \int_0^{\text{final}} E(-yv'')(-y \, dv'') \, dA \; dx = \frac{1}{2} \int EI(v'')^2 \, dx$$

Using the assumed displacement field, U becomes

$$U = \frac{EIa^2\pi^4}{64\ell^3}$$

Since a is the generalized coordinate representing the tip deflection, we have

$$P = \frac{\partial U}{\partial a} = \frac{2EIa\pi^4}{64\ell^3}$$

Hence:

$$a = \frac{32P\ell^3}{\pi^4 EI}$$

and we then have

$$v = \frac{32P\ell^3}{\pi^4 EI}\left(1 - \sin\frac{\pi x}{2l}\right)$$

Example 5.6 Now consider a uniform rod of length ℓ and area A fixed at $x = 0$, free at $x = \ell$, and under the influence of a constant temprature rise ΔT. The strain energy in the rod is given as

$$U = \int_V \int_0^{\text{final}} \sigma \, d\epsilon \, dV = \int_0^{\ell} \int_0^{\text{final}} EA(\epsilon - \alpha \Delta T)(d\epsilon)\, dx$$

or on expanding,

$$U = \int_0^{\ell} \left(\frac{EA\epsilon^2}{2} - \alpha EA\Delta T\epsilon\right) dx$$

or, finally, upon using the fact that the strain in the rod is $\epsilon = u_\ell/\ell$, we get

$$U = \frac{EAu_\ell^2}{2\ell} - \alpha EA\Delta T u_\ell$$

Castigliano's first theorem yields

$$P = 0 = \frac{\partial U}{\partial u_\ell} = \frac{2EAu_\ell}{2\ell} - \alpha EA\Delta T$$

or

$$u_\ell = \alpha \ell \Delta T$$

If the end at $x = \ell$ is clamped, we have for the reaction R

$$R = \frac{\partial U}{\partial u_\ell} = \frac{2EAu_\ell}{2\ell} - \alpha EA\Delta T$$

but $u_\ell = 0$ at $x = \ell$; hence:

$$R = -\alpha EA\Delta T$$

5.4 COMPLEMENTARY ENERGY

Complementary energy is valid for finite strain and material nonlinearity. However, for finite strain the stresses couple with the displacements, and this coupling makes complementary energy principles for finite strain more difficult to use. For small-strain problems, however, the method is extremely powerful.

Consider the terms in Eq. (5.9). If the matrix (called the Hessian matrix) from all the terms formed the second derivatives of U_0, such as $\partial^2 U_0/\partial\epsilon_{xx}^2, \ldots$ is nonzero, then the terms in Eq. (5.9) can be solved for the strains in terms of stress to obtain

$$\begin{aligned}
\epsilon_{xx} &= \frac{\partial U_0^*}{\partial \sigma_{xx}} \\
\epsilon_{yy} &= \frac{\partial U_0^*}{\partial \sigma_{yy}} \\
\epsilon_{zz} &= \frac{\partial U_0^*}{\partial \sigma_{zz}} \\
\gamma_{xy} &= \frac{\partial U_0^*}{\partial \tau_{xy}} \\
\gamma_{yz} &= \frac{\partial U_0^*}{\partial \tau_{yz}} \\
\gamma_{zx} &= \frac{\partial U_0^*}{\partial \tau_{zx}}
\end{aligned} \tag{5.24}$$

where U_0^* is called the complementary strain-energy-density function.

We will see in Chapter 6 that the strain energy for a stable equilibrium position requires that the Hessian matrix be a positive definite matrix. Indeed, from Eq. (3.34) we can define k_{uv} the force at coordinate u per unit displacement at coordinate v only. Mathematically, this is the same as the partial derivative of a force P_u at u with respect to displacement Δ_v at v, or

$$k_{uv} = \frac{\partial P_u}{\partial \Delta_v}$$

However, from Eq. (5.23) we have

$$k_{uv} = \frac{\partial}{\partial \Delta_v} \frac{\partial U}{\partial \Delta_u} = \frac{\partial^2 U}{\partial \Delta_u \partial \Delta_v}$$

and the nonvanishing Hessian matrix implies that the structure has a nonsingular stiffness matrix.

The terms in Eq. (5.24) represent a point transformation, and because of the nonvanishing Hessian matrix, they can transform by the Legendre transformation given by Eq. (2.20). Thus we have

$$U_0^* \triangleq \sum_{k=1}^{6} \sigma_k \epsilon_k - U_0 \tag{5.25}$$

where $\epsilon_1 = \epsilon_{xx}, \ldots, \epsilon_6 = \gamma_{zx}$, with similar notation for σ_k. Taking the derivative of Eq. (5.25) and using Eq. (5.7), we get

$$\begin{aligned} dU_0^* &= \epsilon_{xx}\, d\sigma_{xx} + \epsilon_{yy}\, d\sigma_{yy} + \epsilon_{zz}\, d\sigma_{zz} \\ &\quad + \gamma_{xy}\, d\tau_{xy} + \gamma_{yz}\, d\tau_{yz} + \gamma_{zx}\, d\tau_{zx} \end{aligned} \tag{5.26}$$

For a three-dimensional isotropic material, the complementary strain-energy-density function takes the form

$$\begin{aligned} U_0^* &= \frac{1}{2E}\,(\sigma_{xx}^2 + \sigma_{yy}^2 + \sigma_{zz}^2) - \frac{\nu}{E}\,(\sigma_{xx}\sigma_{yy} + \sigma_{yy}\sigma_{zz} + \sigma_{zz}\sigma_{xx}) \\ &\quad + \frac{1}{2G}\,(\tau_{xy}^2 + \tau_{yz}^2 + \tau_{zx}^2) + \alpha\Delta T(\sigma_{xx} + \sigma_{yy} + \sigma_{zz}) \end{aligned} \tag{5.27}$$

Some caution should be used when applying Eq. (5.26). The strain-energy-density function U_0 represents the area under a material stress-strain curve. Equation (5.25) implies that the complementary strain-energy density is the area above the material stress-strain curve. However, caution must be used in computing this area even for linear-elastic materials. Figure 5.4 shows four different stress-strain curves for a linear-elastic material. Curve (a) considers a straight tensile loading; curve (b), a temperature increase followed by a tensile loading; curve (c), a tensile loading followed by a temperature increase; and curve (d), a prestress loading followed by a tensile loading. Only for curve (a) does $U_0 = U_0^*$; in the other three curves, $U_0 \neq U_0^*$.

Consider a uniaxial stress-strain law:

$$\epsilon = \frac{\sigma}{E} + \epsilon_0; \qquad d\epsilon = \frac{d\sigma}{E}$$

and $\epsilon_0 = \alpha\Delta T$. We then have

$$U_0 = \int \sigma\, d\epsilon = \frac{\sigma^2}{2E} = \frac{E\epsilon^2}{2} - E\epsilon_0\epsilon$$

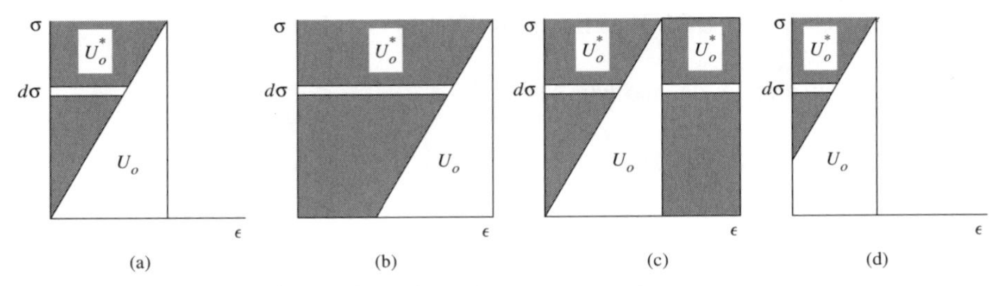

Figure 5.4 Complementary strain energy.

and

$$U_0^* = \int \epsilon \, d\sigma = \frac{\sigma^2}{2E} + \epsilon_0 \sigma = \frac{E\epsilon^2}{2}$$

For conservative systems, we may define the complementary potential energy of the external forces as

$$U_e^* = -W_e^* \tag{5.28}$$

For general body force and surface force loading, we have

$$\begin{aligned} U_e^* = &-\int_V \int_0^{\text{final}} (u \; dX_b + v \, dY_b + w \; dZ_b) \, dV \\ &- \oint_{\partial V} \int_0^{\text{final}} (u \; dX_s + v \, dY_s + w \; dZ_s) \, dS \end{aligned} \tag{5.29}$$

Notice that in addition to permitting variations in forces on the geometrical boundary where forces are arbitrary. Eq. (5.29) considers variations in the body forces and mechanical boundary conditions.

Frequently, body forces are negligible and surface forces can be represented by a system of conservative, concentrated forces. We then have

$$U_e^* = -(\Delta_1 P_1 + \cdots + \Delta_n P_n) \tag{5.30}$$

where the Δ_i are the displacements and the P_i are the concentrated forces at the point of application of the displacement and in the direction of the displacement. The displacements Δ_i are regarded as independent of the forces P_i. The P_i consist of two types of point loads: loads on the boundary S_σ, which we are for the time being allowed to vary, and reactive loads on the boundary S_u, which may always assume an arbitrary variation.

5.5 STATIONARY COMPLEMENTARY POTENTIAL ENERGY

For a set of virtual loads, we can say

$$\delta W_e^* = -\delta U_e^*$$

The principle of complementary virtual work states that

$$\delta W_e^* = \delta U^*$$

or for conservative external loading,

$$-\delta U_e^* = \delta U^*$$

The internal complementary virtual work may be expressed as

$$\delta U^* = \int_V \lfloor \epsilon \rfloor \{\delta \sigma\}\, dV$$

where

$$\lfloor \epsilon \rfloor = \lfloor \epsilon_{xx} \ldots \gamma_{zx} \rfloor$$

is the row matrix of strain components, and

$$\{\delta \sigma\} = \begin{Bmatrix} \delta \sigma_{xx} \\ \vdots \\ \delta \tau_{zx} \end{Bmatrix}$$

Because of Eq. (5.24), we can write

$$\delta U^* = \int_V \lfloor \partial U_0^* \rfloor \{\delta \sigma\}\, dV$$

where

$$\lfloor \partial U_0^* \rfloor = \left[\frac{\partial U_0^*}{\partial \sigma_{xx}} \cdots \frac{\partial U_0^*}{\partial \tau_{zx}} \right]$$

However, if we treat δ as a differential operator, the above expression is nothing more than the total first variation of U_0^*, and we can write

$$\delta U^* = \int_V \delta U_0^* \, dV$$

$$= \delta \int_V U_0^* \, dV$$

where now δU^* can be considered the variation in the total complementary strain energy of the body. The complementary virtual work expression then becomes

$$\delta(U^* + U_e^*) = \delta\Pi^* = 0 \tag{5.31}$$

This equation expresses the fact that if an elastic structure is to satisfy compatibility, its variation of total complementary potential Π^* must be stationary.

The displacements in a deformable system are compatible and consistent with the constraints if the first variation in the total complementary potential energy is zero for every virtual force system satisfying equilibrium.

Stationary implies a local minimum, maximum, or inflection.

The complementary potential energy can be put into the following form:

$$\Pi^* = \int_V U_0^* \, dV - \int_V \int_0^{\text{final}} (u \; dX_b + v \, dY_b + w \; dZ_b) \, dV - \oint_{\partial V} \int_0^{\text{final}} (u \; dX_s + v \, dY_s + w \; dZ_s) \, dS \tag{5.32}$$

Equations (2.21)–(2.23) must be satisfied throughout the deformed volume of the body, whereas Eqs. (2.24)–(2.26) must be satisfied on the boundary. A unique solution of these equations must satisfy the conditions of compatibility. Equations (4.1)–(4.3), however, have only the restriction that they are self-equilibrating; they need not satisfy compatibility.

If we want to compare stress fields that satisfy all the equilibrium but not necessarily the conditions of compatibility, the above surface integral can be broken into two parts.

$$\partial V = S_\sigma + S_u$$

On S_u, since the displacements are considered prescribed, the reactive forces dX_s, dY_s, and dZ_s can be considered independent of the displacements and integrated out. Then, Eq. (5.32) becomes

$$\Pi^* = \int_V U_0^* \, dV - \int_{S_u} (uX_s + vY_s + wZ_s) \, dS \tag{5.33}$$

where also the integral involving the body forces has been dropped because it, too, is considered prescribed. Thus any variation of stress that satisfies the equilibrium equations given by Eq. (4.1)–(4.3), with the δX_b, δY_b, $\delta Z_b = 0$ and the boundary conditions on S_σ, will make $\delta\Pi^* = 0$. Hence:

Among all states of stress that satisfy the differential equation of equilibrium and the boundary conditions where stresses are prescribed, that which represents the actual equilibrium state provides a stationary value of the complementary energy as given by Eq. (5.33).

5.6 ENGESSER-CROTTI THEOREM

Consider a constrained nonlinear-elastic body undergoing point displacements Δ_i with the applied force system P_i moving through the displacements and with the Δ_i independent of the P_i. We may then consider the complementary strain energy U^* to be a function of the independent force terms P_i; $i = 1, \ldots, n$, and we write

$$U^* = U^*(P_1, \ldots, P_i, \ldots, P_n)$$

and Eq. (5.32) takes the following form:

$$\Pi^*(P_1, \ldots, P_i, \ldots, P_n) = U^*(P_1, \ldots, P_i, \ldots, P_n) - \sum_{i=1}^{n} P_i \Delta_i$$

Because we are assuming that the loads and displacements are independent, we can write

$$\begin{aligned} \delta\Pi^* &= \frac{\partial U^*}{\partial P_1} \delta P_1 + \cdots + \frac{\partial U^*}{\partial P_i} \delta P_i + \cdots + \frac{\partial U^*}{\partial P_n} \delta P_n \\ &\quad - \Delta_i \delta P_1 - \cdots - \Delta_i \delta P_i - \cdots - \Delta_n \delta P_n \\ &= 0 \end{aligned}$$

Collecting coefficients for each δP_i and remembering that each δP_i is considered an arbitrary virtual force, we get

$$\delta \Pi^* = \cdots + \left(\Delta_i - \frac{\partial U^*}{\partial P_i} \right) \delta P_i + \cdots = 0$$

or

$$\Delta_i = \frac{\partial U^*}{\partial P_i} \tag{5.34}$$

which is the Engesser-Crotti theorem stating the following:

If the complementary strain energy of a structural system is expressed in terms of the n independent forces corresponding to a system of n prescribed displacements Δ_i, the first partial derivative of the complementary strain energy with respect to any force P_i at the point i is equal to the displacement Δ_i in the direction of the force P_i.

If we wish to consider a reaction force R_i on the geometric boundary S_u where the displacements are prescribed, then U^* becomes

$$U^* = U^*(P_1, \ldots, P_i, \ldots, P_n, R_i)$$

If $\hat{\Delta}_i$ is the prescribed displacement, Eq. (5.34) becomes

$$\hat{\Delta}_i = \frac{\partial U^*}{\partial R_i}$$

If $\hat{\Delta}_i = 0$, we have

$$0 = \frac{\partial U^*}{\partial R_i}$$

If the elastic material is linear, and if *there are no initial strains or initial stresses present*, then Eq. (5.34) becomes

$$\Delta_i = \frac{\partial U}{\partial P_i}$$

which is Castigliano's second theorem.

Equation (5.34) is valid for nonlinear materials, but it requires small-strain theory because of the assumption that U^* can be written as a function of the P_i independent of the displacements Δ_i. This restriction can, however, be removed. For finite strain elasticity, we could start with Eq. (5.10). However, such a discussion at the large-strain level is beyond the present scope; instead, we will start with Eq. (5.23), which is valid for nonlinear material and large-displacement, large-rotation, small-strain, geometrically nonlinear conditions.

The strain energy U, at least up to the point of structural instability, can be shown to have a nonsingular Hessian matrix. We then can define the Legendre transformation as

$$U^*(P_1, \ldots, P_i, \ldots, P_n) = \sum_{k=1}^{n} P_k \Delta_k - U(\Delta_1, \ldots, \Delta_i, \ldots, \Delta_n) \tag{5.35}$$

Then, by taking the partial derivative of U^* with respect to P_i, we get the following:

$$\begin{aligned} \frac{\partial U^*}{\partial P_i} &= \frac{\partial}{\partial P_i} \left(\sum_{k=1}^{n} P_k \Delta_k \right) - \frac{\partial U}{\partial P_i} \\ &= \Delta_i + \sum_{k=1}^{n} P_k \frac{\partial \Delta_k}{\partial P_i} - \sum_{k=1}^{n} \frac{\partial U}{\partial \Delta_k} \frac{\partial \Delta_k}{\partial P_i} \\ &= \Delta_i + \sum_{k=1}^{n} \left[\frac{\partial \Delta_k}{\partial P_i} \left(P_k - \frac{\partial U}{\partial \Delta_k} \right) \right] \end{aligned}$$

However, from Eq. (5.23),

$$P_k - \frac{\partial U}{\partial \Delta_k} = 0$$

and we again have

$$\Delta_i = \frac{\partial U^*}{\partial P_i}$$

or

$$\Delta_i = \frac{\partial}{\partial P_i} \left(\sum_{k=1}^{n} P_k \Delta_k - U \right) \tag{5.36}$$

Expanding out the right-hand side of Eq. (5.36), we get the following for a very general form of the Engesser-Crotti theorem:

$$\frac{\partial U}{\partial P_i} - \sum_{k=1}^{n} P_k \frac{\partial \Delta_k}{\partial P_i} = 0 \tag{5.37}$$

This equation shows very clearly the removal of the requirement that the Δ_i and P_i be independent.

Example 5.7 Consider the beam shown in Fig. (3.22). It is symmetric and clamped at $x = 0$, simply supported at end $x = l$, and loaded in the xy plane by a uniform load $p(x)$. The beam is also subject to a temperature distribution $\Delta T = T_0[1 + y/(2h)]$, where T_0 is a known constant and $2h$ is the depth of the beam's cross section. Let us determine the vertical reaction of the right support.

In general, the strain for such a beam is given as

$$\epsilon = \frac{\sigma}{E} + \alpha\Delta T = u' - yv''$$

and the stress can be written as

$$\sigma = Eu_0' - Eyv'' - \alpha E\Delta T$$

Using Eq. (3.63), we can write

$$u_0' = \frac{N + N_T}{AE}$$

$$v'' = -\frac{M + M_T}{EI}$$

where we have dropped the subscript z on M and I. The stress then becomes

$$\sigma = -\alpha E\Delta T + \frac{N + N_T}{A} + \frac{M + M_T}{I}\,y$$

The strain is

$$\epsilon = \frac{N + N_T}{AE} + \frac{M + M_T}{EI}\,y$$

and $d\sigma$ is

$$d\sigma = \frac{dN}{A} + \frac{dM}{I}\,y$$

where we have used the fact that N_T, M_T, and $\alpha E\Delta T$ are prescribed functions. The complementary strain energy becomes

$$\begin{aligned} U^* &= \int_V \int_0^{\text{final}} \epsilon\, d\sigma\, dV \\ &= \int_A \int_x \int_{\text{load}} \left(\frac{N + N_T}{AE} + \frac{M + M_T}{EI}\,y\right)\left(\frac{dN}{A} + \frac{dM}{I}\,y\right) dA\ dx \end{aligned}$$

or

$$U^* = \int \frac{N^2}{2AE}\,dx + \int \frac{N_T N}{AE}\,dx + \int \frac{M^2}{2EI}\,dx + \int \frac{M_T M}{EI}\,dx \tag{5.38}$$

Then, Eq. (5.34) for a general point load Q becomes

$$\Delta_Q = \int \frac{N}{AE}\frac{\partial N}{\partial Q}\,dx + \int \frac{N_T}{AE}\frac{\partial N}{\partial Q}\,dx + \int \frac{M}{EI}\frac{\partial M}{\partial Q}\,dx + \int \frac{M_T}{EI}\frac{\partial M}{\partial Q}\,dx \tag{5.39}$$

Equation (5.39) should now be compared with Eq. (4.14) reduced to the corresponding terms. Specific note should be taken of the terms $(\overline{N}, \overline{M})$ and $(\partial N/\partial Q, \partial M/\partial Q)$. In general, then, to determine Δ_Q we need only to look at the right-hand side of the equations for δU^* in Chapter 4 and replace the bared terms with the appropriate partial derivative.

For the beam in question, we have

$$N = 0$$

$$M = R(\ell - x) - \frac{p(\ell - x)^2}{2}$$

$$N_T = AE\alpha T_0$$

$$M_T = \frac{EI\alpha T_0}{2h}$$

$$\frac{\partial N}{\partial R} = 0$$

$$\frac{\partial M}{\partial R} = (\ell - x)$$

Then,

$$\Delta_R = \frac{\partial U^*}{\partial R} = 0$$

since vertical movement is constrained at the right end. Hence we have

$$\int_0^\ell \left(\frac{R(\ell - x)}{EI} - \frac{p(\ell - x)^2}{2EI} + \frac{\alpha T_0}{2h}\right)(\ell - x)\,dx = 0$$

or

$$R = \frac{3p\ell}{8} - \frac{3\alpha EIT_0}{4h\ell}$$

for the value of the right-hand vertical reaction.

5.7 VARIATIONAL STATEMENTS

Consider the following differential equation for the beam shown in Fig. 3.17:

$$\frac{d^2}{dx^2}(EIv'') - p = 0 \tag{5.40}$$

with the boundary conditions:

$$\frac{d}{dx}(EIv'') + V = 0$$

$$EIv'' + M = 0 \tag{5.41}$$

If we were to approximate the solution of this set of equations, then the approximating function would need to be continuous through the first three derivatives. We can often find integral statements that are equivalent to the differential equations. These have the advantage of requiring a lower order of continuity in the choice of the approximating function.

One such way of obtaining an integral form is to first consider a function $\delta\Pi$ formed from Eq. (5.40):

$$\delta\Pi = \int_A^B \left[\frac{d^2}{dx^2}(EIv'') - p\right] \delta v \, dx = 0 \tag{5.42}$$

where δv is a variation in the dependent variable. The expression is, of course, equal to zero because the expression in the square brackets is equal to zero.

Then, integrate by parts twice: first by letting

$$\alpha = \delta v = \overline{v}; \qquad d\beta = d\left[\frac{d}{dx}(EIv'')\right]$$

$$d\alpha = \frac{d\overline{v}}{dx}\,dx = \overline{v}'\,dx; \qquad \beta = \frac{d}{dx}(EIv'')$$

and again by letting

$$\alpha = \overline{v}'; \qquad d\beta = d(EIv'')$$
$$d\alpha = \overline{v}''\,dx; \qquad \beta = (EIv'')$$

to get

$$\delta\Pi = \frac{d}{dx}(EIv'')\overline{v}\Big|_A^B - EIv''\overline{v}'\Big|_A^B + \int_A^B EIv''\overline{v}''\,dx - \int_A^B p\overline{v}\,dx = 0 \tag{5.43}$$

Next, note the following identity:

$$\tfrac{1}{2}\,\delta(v'')^2 = v''\delta v'' = v''\overline{v}''$$

Substituting this into the first integral of the above equation, we get

$$\delta\Pi = \frac{d}{dx}(EIv'')\overline{v}\Big|_A^B - EIv''\overline{v}'\Big|_A^B + \int_A^B \frac{1}{2}\,EI\,\delta(v'')^2\,dx - \int_A^B p\,\delta v\,dx = 0 \tag{5.44}$$

Now for the integral terms of Eq. (5.44), the operator δ only operates on v or v'', and this we can factor out to obtain

$$\delta\Pi = \frac{d}{dx}(EIv'')\overline{v}\Big|_A^B - EIv''\overline{v}'\Big|_A^B + \delta\left[\int_A^B \frac{1}{2}\,EI(v'')^2\,dx - \int_A^B pv\,dx\right] = 0 \tag{5.45}$$

From Eq. (3.71), we have for our simplified case

$$\frac{d}{dx}(EIv'')\,\delta v' = -V\delta v' = -\delta(Vv')$$

and

$$EIv''\;\delta v = -M\;\delta v = -\delta(Mv)$$

where the variation is only over displacement, not load. Using the above relationships in Eq. (5.45) and expanding the boundary terms, we get the following result:

$$\delta\Pi = \delta\left\{(V_A v_A - V_B v_B) + (M_B v'_B - M_A v'_A) + \int_A^B \left[\frac{EI}{2}(v'')^2 - pv\right] dx\right\} = 0$$

(5.46)

We identify Π as the total potential energy of the beam and the equation for $\delta\Pi$ as the variational statement. In the integrated form of this example, the approximating function now needs to be only continuous through the first derivative. If we were seeking exact solutions in the context of continuum mechanics, we would seek solutions that satisfy the differential equation everywhere throughout the domain and the boundary conditions exactly.

For the finite element method, however, the domain is discretized into many small elements and the variation of the solution over the boundary is small. The boundary conditions are then approximated by a set of generalized coordinates at the boundary nodes and by some smooth function interpolated between the boundaries. Thus we allow local approximate solutions to build up a solution throughout the entire domain.

As shown in earlier examples for the rod element, such as Eq. (3.38), the solution for the above beam is approximated as

$$v = \sum_{i=i}^{n} a_i \phi_i(x)$$

where the a_i are considered generalized coordinates and the $\phi_i(x)$ are assumed shape functions. Because Eq. (5.46) represents a restatement of the principle of stationary potential energy, we can use it to generate a system of equations to solve for the n generalized coordinates by substituting the approximating function into Eq. (5.46) and using the following relationships:

$$\frac{\partial \Pi}{\partial a_1} = 0$$

$$\vdots$$

$$\frac{\partial \Pi}{\partial a_i} = 0$$

$$\vdots$$

$$\frac{\partial \Pi}{\partial a_n} = 0 \qquad (5.47)$$

The above variational statement technique is often call the Ritz variational method.

A major difficulty in obtaining variational statements was that when we went from Eq. (5.43) to Eq. (5.46), we had to recognize some identities that for more

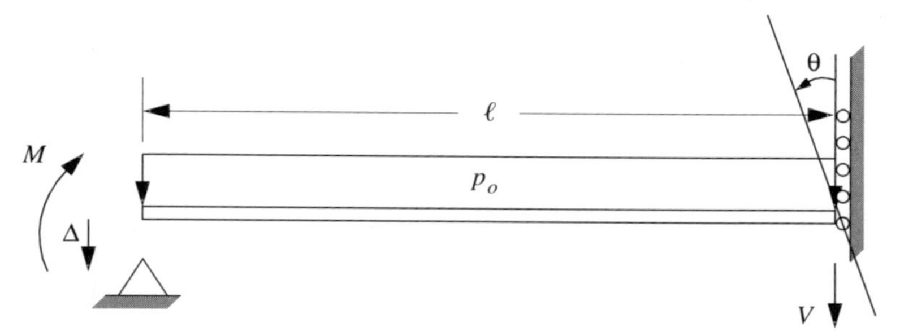

Figure 5.5 Beam with general boundary conditions.

difficult problems would not be so easy (if possible at all) to obtain. From this, we introduce the Galerkin method.

5.8 THE GALERKIN METHOD

Consider the beam shown in Fig. 5.5. The differential equation is

$$\frac{d^2}{dx^2}\, EIv''(x) - p(x) = 0 \tag{5.48}$$

and has the following boundary conditions:

$$v(0) = \Delta; \qquad v'(\ell) = \theta \tag{5.49}$$

$$EIv''(0) = -M; \qquad \frac{d}{dx}\, EIv''(\ell) = -V \tag{5.50}$$

Let

$$v = \phi_0(x) + \phi(x)$$

be a trial function that satisfies the differential equation and the kinematic boundary conditions $v(0) = \Delta$ and $v'(\ell) = \theta$, where $\phi_0(x)$ has been added to expressly satisfy the kinematic boundary conditions in Eq. (5.49). Also require that v'' be square-integrable or

$$\int_0^{\ell} (v'')^2\, dx < \infty$$

to ensure that the solutions are bounded.

Next, let

$$W = \phi(x)$$

be a weighting function that satisfies the requirements that it or its appropriate derivatives are zero at the location of kinematic boundary conditions. In the case of Fig. 5.5, we would have $W(0) = 0$ and $W'(\ell) = 0$. Therefore, the $\phi(x)$ are chosen so that they are identically zero at the kinematic boundary conditions. In general,

$$v(x) = \phi_0(x) + W(x)$$

form the functional

$$\int_0^\ell \left[\frac{d^2}{dx^2} (EIv'') - p \right] W \, dx \tag{5.51}$$

and integrate by parts to obtain

$$\frac{d}{dx} (EIv'')W|_0^\ell - EIv''W'|_0^\ell + \int_0^\ell EIv''W'' \, dx - \int_0^\ell pW \, dx = 0$$

Applying the kinematic conditions and moving all load terms to the right-hand side, we get

$$\int_0^\ell EIv''W'' \, dx = \int_0^\ell pW \, dx + M(0)W(0) + V(\ell)W'(\ell) \tag{5.52}$$

where the left-hand side represents the internal loading and the right-hand side represents the external loading. Reference [52] demonstrates that integration by parts of Eq. (5.52) and applying the square integrable property will yield the original differential equation and boundary conditions. Equation (5.52) is often called the weak form of the differential equation.

Comparing W with δv of the previous section, we see that we can consider W as a generalized virtual displacement. Hence we have generalized mathematical statement of the principle of virtual work.

Next, let v be an approximate solution to the differential Eq. (5.48), expressed as

$$\begin{aligned} v(a; x) &= \phi_0(x) + \sum_{i=1}^{n} a_i \phi_i(x) \\ &= \phi_0(x) + W(a; x) \end{aligned} \tag{5.53}$$

where the notation $v(a; x)$ is used to imply the v is a function of both x and the generalized coordinates a_i. Substituting Eq. (5.53) into Eq. (5.48) generally will not satisfy the differential equation, and we can write

$$R(a;x) = \frac{d^2}{dx^2} EIv'' - p(x) \neq 0 \tag{5.54}$$

$R(a;x)$ is called the residual. The Galerkin or Bubnov-Galerkin method [52] (to distinguish it from the broader Petrov-Galerkin method) seeks to minimize the error $R(a;x)$ for each generalized coordinate a_i over the entire domain with respect to each weighting function. For the finite element, the domain is the element.

Thus, for the Ritz method we try to minimize the potential energy; for the Galerkin method, we try to minimize the error. Assuming Eq. (5.53), the Galerkin counterpart to Eq. (5.47) is

$$\begin{aligned}
&\int_0^\ell R(a;x)\phi_1(x)\,dx = 0 \\
&\qquad\vdots \\
&\int_0^\ell R(a;x)\phi_i(x)\,dx = 0 \\
&\qquad\vdots \\
&\int_0^\ell R(a;x)\phi_n(x)\,dx = 0
\end{aligned} \tag{5.55}$$

As an example, consider the function

$$v = a_1 + a_2 x + a_3 x^2 + a_4 x^3 + a_5 x^4 \tag{5.56}$$

as an approximating function to the beam in Fig. 5.5. Since we are going to use Eq. (5.55), we need to satisfy both Eqs. (5.49) and (5.50). This can be done by substituting Eq. (5.56) into both the kinematic and natural boundary conditions to get

$$a_1 = \Delta; \qquad a_2 = \theta + \frac{M\ell}{EI} + \frac{V\ell^2}{2EI} + 8a_5\ell^3$$

$$a_3 = -\frac{M}{2EI}; \qquad a_4 = -\frac{V}{6EI} - 4a_5\ell$$

Then, Eq. (5.56) becomes

$$v = \phi_0 + a_1\phi_1 \tag{5.57}$$

where

$$\phi_0 = \Delta + \left(\theta + \frac{M\ell}{EI} + \frac{V\ell^2}{2EI}\right) x - \frac{M}{2EI}\, x^2 - \frac{V}{6EI}\, x^3 \tag{5.58}$$

$$\phi_1 = 8\ell^3 x - 4\ell x^3 + x^4 \tag{5.59}$$

and a_5 has been renamed to a_1 for convenience. The expression for $R(a;x)$ becomes

$$\begin{aligned} R(a;x) &= EIv^{iv} - p \\ &= EI(\phi_0^{iv} + a_1\phi_1^{iv}) - p_0 \\ &= 24EIa_1 - p_0 \end{aligned} \tag{5.60}$$

Then Eq. (5.55) becomes

$$(24EIa_1 - p_0) \int_0^\ell \phi_1 \, dx = 0$$

that when integrated out yields

$$a_1 = \frac{p_0}{24EI}$$

Next, let us use the concept introduced with Eq. (5.52) by integrating by parts each of the n terms of Eq. (5.55). Remembering that in Eq. (5.52) ϕ_0 need only satisfy exactly the kinematic boundary conditions, we get the following for ith Eq. (5.55) after two integrations by parts:

$$\begin{aligned} \int_0^\ell EIv''\phi_i'' \, dx = \int_0^\ell & p\phi_i \, dx \\ &+ EIv''(\ell)\phi_i'(\ell) - EIv''(0)\phi_i'(0) \\ &- \frac{d}{dx}\, EIv''(\ell)\phi_i(\ell) + \frac{d}{dx}\, EIv''(0)\phi_i(0) \end{aligned} \tag{5.61}$$

or on substituting Eq. (5.53) into the left-hand side:

$$\sum_{j=1}^{n} a_j \int_0^\ell EI\phi_j''\phi_i''\,dx = \int_0^\ell p\phi_i\,dx - \int_0^\ell EI\phi_0''\phi_i''\,dx + EIv''(\ell)\phi_i'(\ell) - EIv''(0)\phi_i'(0) - \frac{d}{dx}EIv''(\ell)\phi_i(\ell) + \frac{d}{dx}EIv''(0)\phi_i(0) \quad (5.62)$$

There are such n terms for a_i—one for each ϕ_i representing n algebraic equations in n unknowns. Call the left-hand side the stiffness k_{ij} and the right-hand side the load F_i. Then, typical terms are

$$k_{ij} = \int_0^\ell EI\phi_j''\phi_i''\,dx \quad (5.63)$$

$$F_i = \int_0^\ell p\phi_i\,dx - \int_0^\ell EI\phi_0''\phi_i''\,dx + EIv''(\ell)\phi_i'(\ell) - EIv''(0)\phi_i'(0) - \frac{d}{dx}EIv''(\ell)\phi_i(\ell) + \frac{d}{dx}EIv''(0)\phi_i(0) \quad (5.64)$$

The process of determining ϕ_0 to determine the kinematic boundary conditions and ϕ_i to be zero at the kinematic boundaries is at best a cumbersome operation. However, once our equations have been reduced to a set of linear algebraic equations, these constraint requirements can be applied during the solution process rather than before the process. Thus we will drop ϕ_0 from the equation for F_i and apply the kinematic conditions as a constraint on the solution. Note that the natural boundary conditions are now explicitly represented in F_i; hence explicitly known values can be substituted where they are specified. The constraining process will remove them where they are unspecified.

To demonstrate, consider again the beam shown in Fig. 5.5 but with the following assumed displacement:

$$v = a_1 + a_2x + a_3x^2 + a_4x^3 \quad (5.65)$$

$$v' = a_2 + 2a_3x + 3a_4x^2 \quad (5.66)$$

Define the ϕ_i as

$$\phi_1 = 1 \qquad \phi_2 = x \qquad \phi_3 = x^2 \qquad \phi_4 = x^3 \quad (5.67)$$

Using the above expressions for ϕ_i in Eqs. (5.63) and (5.64), we get

$$\begin{bmatrix} 0 & 0 & 0 & 0 \\ 0 & 0 & 0 & 0 \\ 0 & 0 & 4EI\ell & 6EI\ell^2 \\ 0 & 0 & 6EI\ell^2 & 12EI\ell^3 \end{bmatrix} \begin{Bmatrix} a_1 \\ a_2 \\ a_3 \\ a_4 \end{Bmatrix} = \begin{Bmatrix} F_1 \\ F_2 \\ F_3 \\ F_4 \end{Bmatrix} \tag{5.68}$$

where

$$\begin{Bmatrix} F_1 \\ F_2 \\ F_3 \\ F_4 \end{Bmatrix} = \begin{Bmatrix} p_0\ell \\ p_0\ell^2/2 \\ p_0\ell^3/3 \\ p_0\ell^4/4 \end{Bmatrix} + \begin{Bmatrix} EIv''(\ell)\phi_1'(\ell) - EIv''(0)\phi_1'(0) \\ EIv''(\ell)\phi_2'(\ell) - EIv''(0)\phi_2'(0) \\ EIv''(\ell)\phi_3'(\ell) - EIv''(0)\phi_3'(0) \\ EIv''(\ell)\phi_4'(\ell) - EIv''(0)\phi_4'(0) \end{Bmatrix}$$

$$- \begin{Bmatrix} EIv'''(\ell)\phi_1(\ell) - EIv'''(0)\phi_1(0) \\ EIv'''(\ell)\phi_2(\ell) - EIv'''(0)\phi_2(0) \\ EIv'''(\ell)\phi_3(\ell) - EIv'''(0)\phi_3(0) \\ EIv'''(\ell)\phi_4(\ell) - EIv'''(0)\phi_4(0) \end{Bmatrix} \tag{5.69}$$

Applying the kinematic boundary conditions $v(0) = \Delta$, we get

$$a_1 = \Delta \tag{5.70}$$

and for $v'(\ell) = \theta$, we get

$$a_2 = -2\ell a_3 - 3\ell^2 a_4 + \theta \tag{5.71}$$

Equation (5.70) is called a single-point constraint and Eq. (5.71) is called a multipoint constraint. Equation (5.71) can be represented in matrix form as

$$a_2 = \lfloor 0 \quad -2\ell \quad -3\ell^2 \rfloor \begin{Bmatrix} a_1 \\ a_3 \\ a_4 \end{Bmatrix} + \theta \tag{5.72}$$

Both Eqs. (5.70) and (5.72) can be generalized for the case of several single-point and multipoint constraint conditions. Let s be the number of single-point constraint relations; then, we can write the following for the collection of single-point constraints:

$$\{a_s\} = \{\Delta_s\} \tag{5.73}$$

Let m represent the number of multipoint equations and n the remaining independent generalized coordinates; then, we can write the following for the multipoint constraints

$$\{a_m\} = [G_{mn}]\{a_n\} + \{\theta_m\} \tag{5.74}$$

By column and row operations, we may partition Eq. (5.68) into an m set and an n set to obtain

$$\begin{bmatrix} K_{nn} & K_{nm} \\ K_{mn} & K_{mm} \end{bmatrix} \begin{Bmatrix} a_n \\ a_m \end{Bmatrix} = \begin{Bmatrix} F_n \\ F_m \end{Bmatrix} \tag{5.75}$$

Next, we apply the constraints of Eq. (5.74). It is, however, important to realize that when we apply the constraint to the a_m generalized coordinates, forces of constraint must be generated on the right-hand-side load vector because it is these forces that actually apply the constraint. Also, Eq. (5.74) represents a coordinate transform between the m and n generalized coordinates. Following Eq. (3.48), the conjugate relationship is then

$$\{\lambda_n\} = -[G_{mn}]^{\mathrm{T}}\{\lambda_m\} \tag{5.76}$$

where the minus sign comes from the fact that we are imposing constraints and that the λ_n forces are reactive forces to the λ_m constraint forces. Then, Eq. (5.75) becomes

$$\begin{bmatrix} K_{nn} & K_{nm} \\ K_{mn} & K_{mm} \end{bmatrix} \begin{Bmatrix} a_n \\ G_{mn}a_n + \theta_m \end{Bmatrix} = \begin{Bmatrix} F_n - G_{mn}^{\mathrm{T}}\lambda_m \\ F_n + \lambda_m \end{Bmatrix} \tag{5.77}$$

Solving the lower set of equations, we get for λ_m

$$\{\lambda_m\} = [K_{mn} + K_{mm}G_{mn}]\{a_n\} + [K_{mm}]\{\theta_m\} - \{P_m\}$$

and substituting for λ_m into the upper partition, we get

$$[\hat{K}_{nn}]\{a_n\} = \{\hat{F}_n\}$$

where

$$[\hat{K}_{nn}] = [K_{nn} + K_{nm}G_{mn} + G_{mn}^{\mathrm{T}}K_{mn} + G_{mn}^{\mathrm{T}}K_{mm}G_{mn}] \tag{5.78}$$

and

$$\{\hat{F}_n\} = \{F_n + G_{mn}^{\mathrm{T}}P_m - K_{nm}\theta_m - G_{mn}^{\mathrm{T}}K_{mm}\theta_m\} \tag{5.79}$$

Applying these results to Eq. (5.68), we get

$$\begin{bmatrix} 0 & 0 & 0 \\ 0 & 4EI\ell & 6EI\ell^2 \\ 0 & 6EI\ell^2 & 12EI\ell^3 \end{bmatrix} \begin{Bmatrix} a_1 \\ a_3 \\ a_4 \end{Bmatrix} = \begin{Bmatrix} F_1 \\ F_3 - 2\ell F_2 \\ F_4 - 3\ell^2 F_2 \end{Bmatrix} \tag{5.80}$$

Next, we can apply the single-point constraints by partitioning the $\hat{K}_{nn}$ matrix into the s and f sets or the free set to obtain

$$\begin{bmatrix} K_{ff} & K_{fs} \\ K_{sf} & K_{ss} \end{bmatrix} \begin{Bmatrix} a_f \\ a_s \end{Bmatrix} = \begin{Bmatrix} F_f \\ F_s \end{Bmatrix} \tag{5.81}$$

Then, apply Eq. (5.73) to get

$$\begin{bmatrix} K_{ff} & K_{fs} \\ K_{sf} & K_{ss} \end{bmatrix} \begin{Bmatrix} a_f \\ \Delta_s \end{Bmatrix} = \begin{Bmatrix} F_f \\ F_s + \lambda_s \end{Bmatrix} \tag{5.82}$$

where λ_s represents the forces of constraint required to impose the single-point constraint. Since the a_s generalized coordinates are not related to the a_f generalized coordinates, there are no reactive forces in the F_f loads. Solving the partition equations, we get

$$\{\lambda_s\} = [K_{sf}]\{a_f\} + [K_{ss}]\{\Delta_s\} - \{F_s\}$$

and

$$[K_{ff}]\{a_f\} = \{F_f - K_{fs}\Delta_s\}$$

Then, Eq. (5.80) becomes

$$\begin{bmatrix} 4EI\ell & 6EI\ell^2 \\ 6EI\ell^2 & 12EI\ell^3 \end{bmatrix} \begin{Bmatrix} a_3 \\ a_4 \end{Bmatrix} = \begin{Bmatrix} F_3 - 2\ell F_2 \\ F_4 - 3\ell^2 F_2 \end{Bmatrix} \tag{5.83}$$

or

$$\begin{Bmatrix} a_3 \\ a_4 \end{Bmatrix} = \frac{1}{12EI\ell^4} \begin{bmatrix} 4\ell & 6\ell^2 \\ 6\ell^2 & 12\ell^3 \end{bmatrix} \begin{Bmatrix} F_3 - 2\ell F_2 \\ F_4 - 3\ell^2 F_2 \end{Bmatrix} \tag{5.84}$$

It should be observed how the above operations remove from the reduced equations the unimposed natural boundary conditions. For example, considering the $EIv''(\ell)\phi_4'(\ell)$ term of the load vector, we get

$$EIv''(\ell)\phi_4'(\ell) - 3\ell^2 EIv''(\ell)\phi_2'(\ell)$$

but from Eq. (5.67), we have $\phi_4'(\ell) = 3\ell^2$ and $\phi_2'(\ell) = 1$ so that the unspecified natural boundary condition cancels out.

Substituting the specified boundaries yields

$$a_3 = -\frac{p_0\ell}{24EI} - \frac{M}{2EI} \tag{5.85}$$

$$a_4 = -\frac{p_0\ell}{12EI} - \frac{V}{6EI} \tag{5.86}$$

and Eqs. (5.65) and (5.66) become

$$v = \Delta + a_3(x - 2\ell)x + a_4(x^2 - 3\ell^2)x \tag{5.87}$$

$$v' = 2a_3(x - \ell) + 3a_4(x^2 - \ell^2) \tag{5.88}$$

Equations (5.63) and (5.64) are also good in the case of finite elements. Using the sign convention of Fig. 3.44, consider a finite element beam with a uniform load distribution. The x and nondimensional coordinate system s are measured from the left end of the beam element. The x and s systems are related by

$$x = \ell s; \qquad 0 \le s \le 1$$

This coordinate system, rather than one at the center, was chosen so that the selected interpolation polynomials would be in standard form. If a nondimensional system ξ at the center is desired, then $\xi = 2s - 1$.

To apply the Galerkin method to a finite element, it is only necessary to define the domain over the element. To do this, consider the shape functions

$$N_i = a_1 + a_2 s + a_3 s^2 + a_4 s^3$$

and

$$N_i' = \frac{1}{\ell}\frac{dN_i}{ds} = \frac{a_2}{\ell} + \frac{2a_3}{\ell}s + \frac{3a_4}{\ell}s^2$$

Then, assume the displacement in the form

$$v = N_1 v_1 + N_2 v_1' + N_3 v_2 + N_4 v_2' \tag{5.89}$$

The a_i of the shape functions can be determined such that

$$
\begin{aligned}
N_1(0) &= 1; & N_1(1) &= 0 \\
N_1'(0) &= 0; & N_1'(1) &= 0 \\
N_2(0) &= 0; & N_2(1) &= 0 \\
N_2'(0) &= 1; & N_2'(1) &= 0 \\
N_3(0) &= 0; & N_3(1) &= 1 \\
N_3'(0) &= 0; & N_3'(1) &= 0 \\
N_4(0) &= 0; & N_4(1) &= 0 \\
N_4'(0) &= 0; & N_4'(1) &= 1
\end{aligned}
$$

The results are first-order Hermite interpolation polynomials that take the form

$$
\begin{aligned}
N_1 &= 1 - 3s^2 + 2s^3 \\
N_2 &= \ell s(s-1)^2 \\
N_3 &= s^2(3-2s) \\
N_4 &= \ell s^2(s-1)
\end{aligned}
\tag{5.90}
$$

A plot of these functions is shown in Fig. 5.6.

Substituting these functions into Eqs. (5.63) and (5.64), we get the following element relationships:

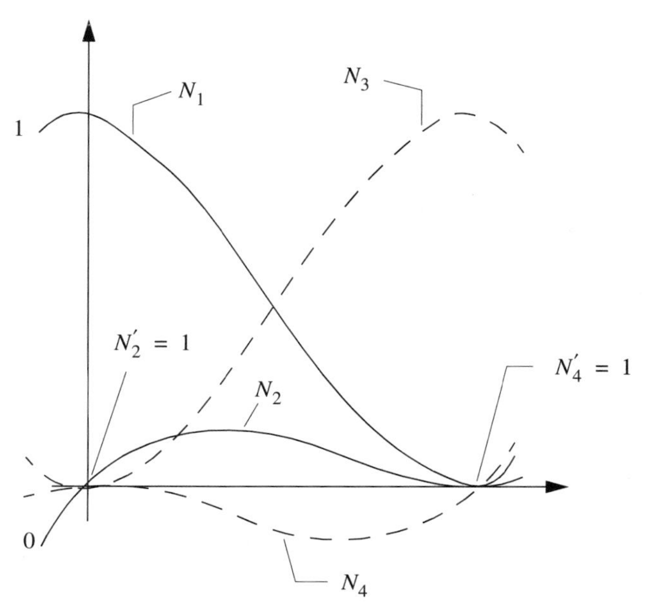

Figure 5.6 First-order Hermite polynomials.

$$\frac{EI}{\ell^3}\begin{bmatrix} 12 & 6\ell & -12 & 6\ell \\ 6\ell & 4\ell^2 & -6\ell & 2\ell^2 \\ -12 & -6\ell & 12 & -6\ell \\ 6\ell & 2\ell^2 & -6\ell^2 & 4\ell^2 \end{bmatrix}\begin{Bmatrix} v_1 \\ v_1' \\ v_2 \\ v_2' \end{Bmatrix} = \begin{Bmatrix} p_0\ell/2 + V_1 \\ p_0\ell^2/12 + M_1 \\ p_0\ell/2 + V_2 \\ -p_0\ell^2/12 + M_2 \end{Bmatrix} \tag{5.91}$$

5.9 DERIVED VARIATIONAL PRINCIPLES

Derived stationary principles incorporate all of the governing equations of elasticity into a single functional that is made stationary. The procedure yields differential equations and boundary conditions but does not explicitly yield solutions to a structural problem. They are, however, very important in obtaining numerical approximations to differential equations and form the basis of a variety of finite element mixed and hybrid models.

Reference [53] defines a function Π_I that incorporates the strain-energy-density function (in finite elasticity, a free-energy function), the strain-displacement relationships, the kinematic boundary conditions, and the potential of both the body forces and the surface forces. The function, then, is dependent on eighteen independent variables:

$$\Pi_I(\epsilon_x, \epsilon_y, \ldots, \gamma_{xy}, u, v, w, \lambda_x, \lambda_y, \ldots, \lambda_{xy}, \lambda_u, \lambda_v, \lambda_w)$$

where the λs are Lagrange multipliers.

The function takes the following form:

$$\begin{aligned} \Pi_I = & \int_V [U_0(\epsilon_x, \epsilon_y, \ldots, \gamma_{xy}) - (\hat{X}_b u + \hat{Y}_b v + \hat{Z}_b w)]\, dV \\ & - \int_V \left[\left(\epsilon_x - \frac{\partial u}{\partial x}\right)\lambda_x + \left(\epsilon_y - \frac{\partial v}{\partial y}\right)\lambda_y + \cdots + \left(\gamma_{xy} - \frac{\partial v}{\partial x} - \frac{\partial u}{\partial y}\right)\lambda_{xy}\right] dV \\ & - \int_{S_\sigma} [\hat{X}_s u + \hat{Y}_s v + \hat{Z}_s w]\, dS \\ & - \int_{S_u} [(u - \hat{u})\lambda_u + (v - \hat{v})\lambda_v + (w - \hat{w})\lambda_w]\, dS \end{aligned} \tag{5.92}$$

where we have used Eqs. (2.39)–(2.44) for strain-displacement constraints and the kinematic constraints

$$u = \hat{u}; \qquad v = \hat{v}; \qquad w = \hat{w} \quad \text{on} \quad S_u$$

where $\hat{u}$, $\hat{v}$, $\hat{w}$ are prescribed displacements, $\hat{X}_b$, $\hat{Y}_b$, $\hat{Z}_b$ are prescribed body forces, and $\hat{X}_s$, $\hat{Y}_s$, $\hat{Z}_s$ are prescribed surface forces on S_σ. Stationarity is obtained by taking the first variation of Π_I and requiring it to be zero.

The process of taking the variations is straightforward except for terms such as

$$\int_V \lambda_x \frac{\partial \delta u}{\partial x} \, dV$$

These types of terms must be integrated by parts to yield terms of the form

$$\int_{\partial V = S_\sigma + S_u} \lambda_x l \, \delta u \, dS - \int_V \frac{\partial \lambda_x}{\partial x} \, \delta u \, dV$$

Collecting all such terms and defining

$$\begin{aligned}
\mathcal{X}_s &= l\lambda_x + m\lambda_{xy} + n\lambda_{xz} \\
\mathcal{Y}_s &= l\lambda_{yx} + m\lambda_y + n\lambda_{yz} \\
\mathcal{Z}_s &= l\lambda_{zx} + m\lambda_{zy} + n\lambda_z
\end{aligned}$$

we get

$$\begin{aligned}
\delta\Pi_I = &\int_V \left[\left(\frac{\partial U_0}{\partial \epsilon_x} - \lambda_x \right) \delta\epsilon_x + \cdots + \left(\frac{\partial U_0}{\partial \gamma_{xy}} - \lambda_{xy} \right) \delta\gamma_{xy} \right] dV \\
&- \int_V \left[\left(\epsilon_x - \frac{\partial u}{\partial x} \right) \delta\lambda_x + \cdots + \left(\gamma_{xy} - \frac{\partial v}{\partial x} - \frac{\partial u}{\partial y} \right) \delta\lambda_{xy} \right] dV \\
&- \int_V \left[\left(\frac{\partial \lambda_x}{\partial x} + \frac{\partial \lambda_{xy}}{\partial y} + \frac{\partial \lambda_{zz}}{\partial z} + \hat{X}_b \right) \delta u + \cdots + (\cdots)\, \delta w \right] dV \\
&- \int_{S_u} [(u - \hat{u})\, \delta\lambda_u + (v - \hat{v})\, \delta\lambda_v + (w - \hat{w})\, \delta\lambda_w] \, dS \\
&+ \int_{S_\sigma} [(\mathcal{X}_s - \hat{X}_s)\, \delta u + (\mathcal{Y}_s - \hat{Y}_s)\, \delta v + (\mathcal{Z}_s - \hat{Z}_s)\, \delta w] \, dS \\
&+ \int_{S_u} [(\mathcal{X}_s - \lambda_u)\, \delta u + (\mathcal{Y}_s - \lambda_v)\, \delta v + (\mathcal{Z}_s - \lambda_w)\, \delta w] \, dS
\end{aligned} \tag{5.93}$$

The stationarity requirement gives, for example,

$$\frac{\partial U_0}{\partial \epsilon_x} = \lambda_x, \ldots, \frac{\partial U_0}{\partial \gamma_{xy}} = \lambda_{xy}$$

which, by Eq. (5.9), identifies the $\lambda_x, \ldots, \lambda_{xy}$ as the stresses in V. This implies that $X_s = X_s$ and so forth and $\lambda_u = X_s$ and so forth on S_u.

Other variational principles can be obtained from Eq. (5.92). For example, assume that the strains are not independent but are functions of the stresses. Then, by Eqs. (5.24) and (5.25), Eq. (5.92) takes the form

$$\begin{aligned}\Pi_R = &\int_V \left[\sigma_x \frac{\partial u}{\partial x} + \sigma_y \frac{\partial v}{\partial y} + \cdots + \tau_{xy}\left(\frac{\partial v}{\partial x} + \frac{\partial u}{\partial y}\right)\right] dV \\ &- \int_V [U_0^*(\sigma_x, \sigma_y, \ldots, \tau_{xy}) - (\hat{X}_b u + \hat{Y}_b v + \hat{Z}_b w)]\, dV \\ &- \int_{S_\sigma} (\hat{X}_s u + \hat{Y}_s v + \hat{Z}_s w)\, dS \\ &- \int_{S_u} [(u - \hat{u})\lambda_u + (v - \hat{v})\lambda_v + (w - \hat{w})\lambda_w]\, dS \end{aligned} \tag{5.94}$$

where

$$\Pi_R(\sigma_x, \sigma_y, \ldots, \tau_{xy}, u, v, w, \lambda_u, \lambda_v, \lambda_w)$$

is a function of twelve variables and is called the Hellinger-Reissner principle.

Consider the equivalent expression of Eq. (5.92) for the beam shown in Fig. 5.7. The total potential energy for this beam can be expressed as

$$\Pi = \frac{1}{2} \int EI(v'')^2\, dx - \int \hat{p} v\, dx + \hat{M} v'(\ell) \tag{5.95}$$

Define an auxilikary function

$$\kappa = v''$$

and then define a functional Π_I as a function of the variables

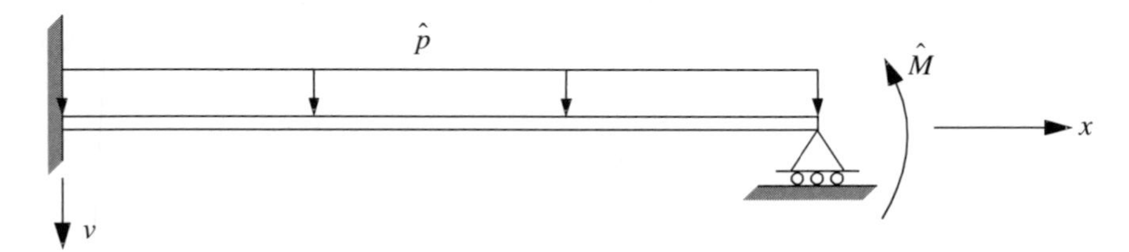

Figure 5.7 Beam for the derived variational principle.

$$\Pi_I(\kappa, v, \lambda, \mu_1, \mu_2, \mu_3)$$

to get

$$\Pi_I = \frac{1}{2}\int EI\kappa^2\,dx - \int \hat{p}v\,dx + \hat{M}v'(\ell)$$
$$+ \int (\kappa - v'')\lambda\,dx + \mu_1 v(0) + \mu_2 v'(0) + \mu_3 v(\ell) \tag{5.96}$$

In the above expressions, $\hat{p}$ and $\hat{M}$ are the prescribed loadings, and λ and μ are the Lagrange multipliers—λ imposing the auxiliary function constraint, μ imposing the kinematic boundary conditions $v(0) = v'(0) = v(\ell) = 0$.

Next, we can take the variation of Eq. (5.96) to obtain the stationary conditions. The variation is straightforward except for the integral containing the $\delta v''$ term that must be twice integrated by parts. The result is

$$\delta\Pi_I = \int [(EI\kappa + \lambda)\,\delta\kappa - (\lambda'' + \hat{p})\,\delta v + (\kappa - v'')\,\delta\lambda]\,dx$$
$$+ [\mu_1 - \lambda'(0)]\,\delta v(0) + [\mu_2 + \lambda(0)]\,\delta v'(0)$$
$$+ [\mu_3 - \lambda'(\ell)]\,\delta v(\ell) - [\lambda(\ell) - \hat{M}]\,\delta v'(\ell)$$
$$+ v(0)\,\delta\mu_1 + v'(0)\,\delta\mu_2 + v(\ell)\,\delta\mu_3 \tag{5.97}$$

The stationary conditions are now identified in the interior as

$$EI\kappa + \lambda = 0 \tag{5.98}$$
$$\lambda'' + \hat{p} = 0 \tag{5.99}$$
$$\kappa - v'' = 0 \tag{5.100}$$

and on the boundary as

$$[\lambda(\ell) - \hat{M}] = 0 \tag{5.101}$$
$$[\mu_1 - \lambda'(0)] = 0 \tag{5.102}$$
$$[\mu_2 + \lambda(0)] = 0 \tag{5.103}$$
$$[\mu_3 + \lambda'(\ell)] = 0 \tag{5.104}$$
$$v(0) = v'(0) = v(\ell) = 0 \tag{5.105}$$

If in Eq. (5.97) we require the coefficient of $\delta\kappa$ to vanish, then by direct substitution into Eq. (5.96) of the relationship

$$\kappa = -\frac{\lambda}{EI}$$

we obtain the functional

$$\Pi_{R1} = -\int v''\lambda \, dx - \int \frac{\lambda^2}{2EI} \, dx - \int \hat{p}v \, dx + \hat{M}v'(\ell) + \mu_1 v(0) + \mu_2 v'(0) + \mu_3 v(\ell) \tag{5.106}$$

The variation of this functional yields the interior stationary conditions

$$\left[\frac{\lambda}{EI} + v''\right] = 0 \tag{5.107}$$

$$[\lambda'' + \hat{p}] = 0 \tag{5.108}$$

and on the boundary

$$[-\lambda(\ell) + \hat{M}] = 0 \tag{5.109}$$

$$[-\lambda'(0) + \mu_1] = 0 \tag{5.110}$$

$$[-\lambda(0) + \mu_2] = 0 \tag{5.111}$$

$$[-\lambda'(\ell) + \mu_3] = 0 \tag{5.112}$$

$$v(0) = v'(0) = v(\ell) = 0 \tag{5.113}$$

If we now integrate by parts the first term of Eq. (5.106) and use the condition expressed by Eq. (5.101), we get the functional

$$\Pi_{R2} = \int v'\lambda' \, dx - \int \frac{\lambda^2}{2EI} \, dx - \int \hat{p}v \, dx + \lambda(0)v'(0) + \mu_1 v(0) + \mu_2 v'(0) + \mu_3 v(\ell) \tag{5.114}$$

with the interior stationary conditions

$$\left[\frac{\lambda}{EI} + v''\right] = 0 \tag{5.115}$$

$$[\lambda'' + \hat{p}] = 0 \tag{5.116}$$

and on the boundary

$$[-\lambda'(0) + \mu_1] = 0 \tag{5.117}$$

$$[-\lambda(0) + \mu_2] = 0 \tag{5.118}$$

$$[-\lambda'(\ell) + \mu_3] = 0 \tag{5.119}$$

$$v(0) = v'(0) = v(\ell) = 0 \tag{5.120}$$

If we now let $\lambda = M$ in Eqs. (5.106) and (5.114), we can define the Reissner principle for the beam in either of the two forms

$$\Pi_R = -\int v''M\,dx - \int \frac{M^2}{2EI}\,dx - \int \hat{p}v\,dx \tag{5.121}$$

and

$$\Pi_R^* = \int v'M'\,dx - \int \frac{M^2}{2EI}\,dx - \int \hat{p}v\,dx \tag{5.122}$$

The variation of either of these expressions yields the equilibrium and compatibility equations for the beam as well as the appropriate forced and natural boundary conditions.

For more complex problems, such as shell theory, equations that take the general form of the first term of Eq. (5.122) would be preferable over the form of the first term of Eq. (5.121) because the v' and M' are equal in derivatives of assumed shape functions.

Example 5.8 For a very simple example, consider a simply supported beam under a uniformly distributed load $\hat{p}$ with the following assumed functions for v and M:

$$v = c_1 x(x - \ell)$$
$$M = c_2 x(x - \ell)$$

Substituting these relations into Eq. (5.122) and using the fact that

$$\frac{\partial \Pi_R^*}{\partial c_1} = 0$$

$$\frac{\partial \Pi_R^*}{\partial c_2} = 0$$

we get

$$c_1 = \frac{\hat{p}\ell^2}{20EI}$$

and

$$c_2 = \frac{\hat{p}}{2}$$

PROBLEMS

5.1 For the n-member truss shown in Fig. 5.8, determine by using Castigliano's first theorem the equilibrium expressions relating the unknown displacements u, v to the applied load P. Each member has a length ℓ_m and an axial rigidity $A_m E_m$.

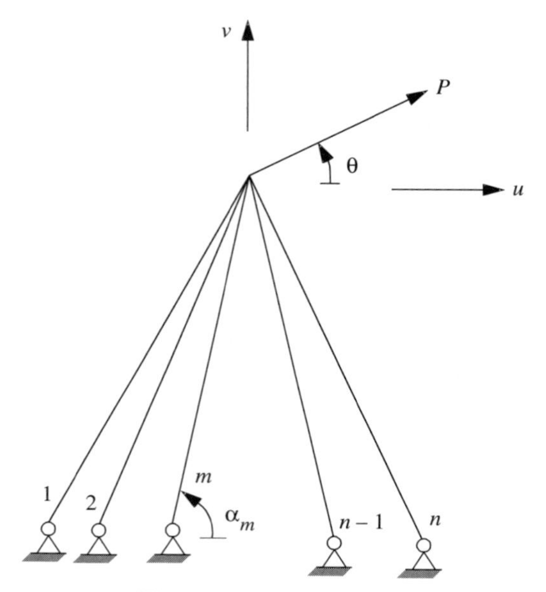

Figure 5.8 Problem 5.1.

5.2 The thin, semicircular beam of radius R shown in Fig. 5.9 has a uniform cross section with a flexural rigidity EI. The beam is fixed at A and carries a vertical load P at its free end B. Using the Engesser-Crotti theorem, determine

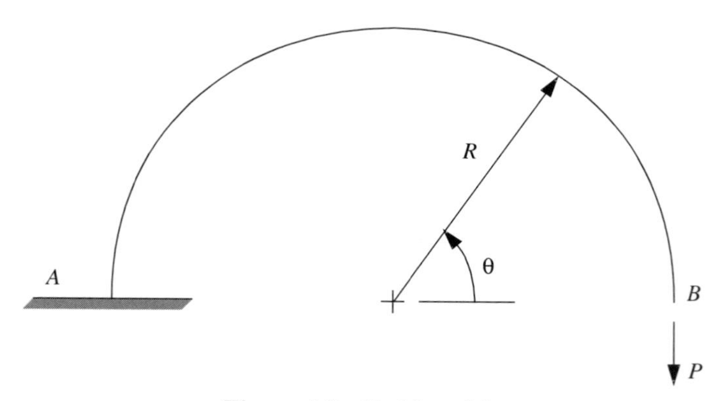

Figure 5.9 Problem 5.2.

(a) the vertical deflection at B, and

(b) the horizontal deflection at B.

5.3 The frame shown in Fig. 5.10 is built in at A, has a lateral simple support at B, and carries a vertical load P at C. The flexural rigidity of the frame is EI. Its dimensions are as shown. Using the Engesser-Crotti theorem, determine the horizontal reaction at B.

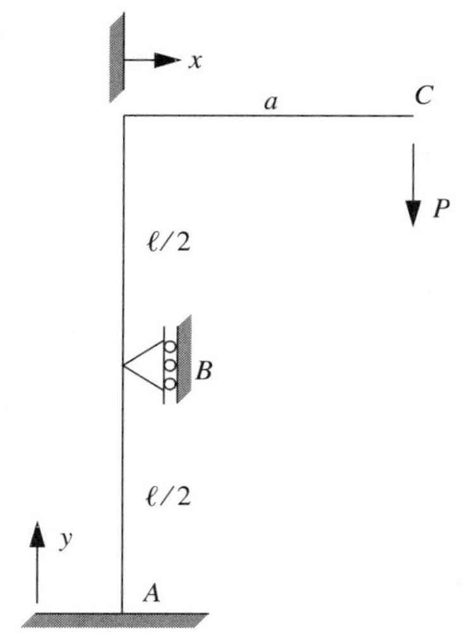

Figure 5.10 Problem 5.3.

5.4 The semicircular frame of radius R shown in Fig. 5.11 has a uniform cross section with a flexural rigidity EI. The frame is pinned as shown and carries a vertical load P at its crown. Using the Engesser-Crotti theorem, determine the horizontal support reactions.

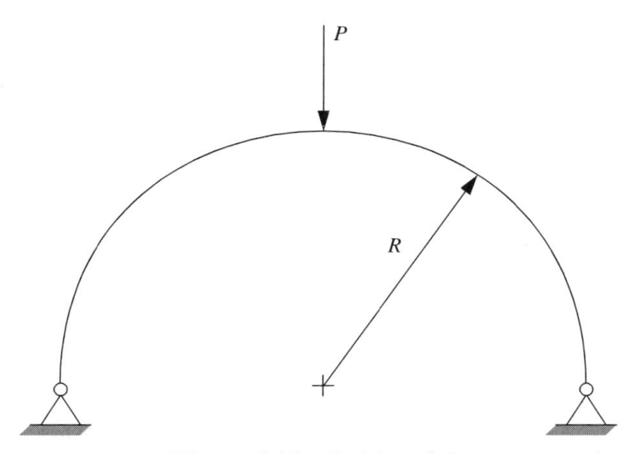

Figure 5.11 Problem 5.4.

5.5 The beam in Fig. 5.12 is clamped at both ends. It is of length ℓ with a flexural rigidity EI and is loaded with a uniformly distributed load of intensity p_0. Using the Engesser-Crotti theorem, determine the moment at the left end.

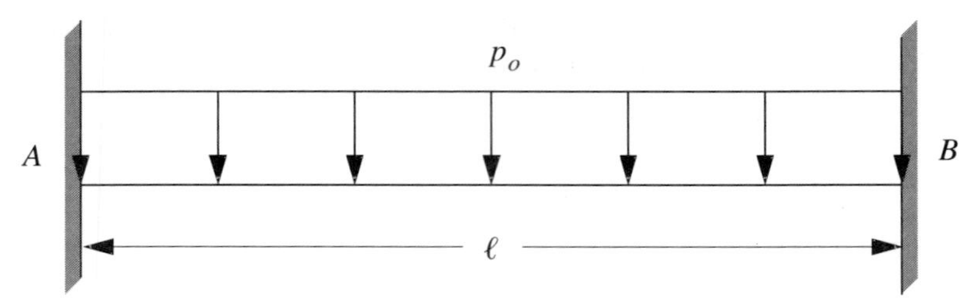

Figure 5.12 Problem 5.5.

5.6 The circular fuselage frame of radius r in Fig. 5.13 is subject to a pure torque T as shown. The torque is reacted in the fuselage by a uniform shear flow $q = T/(2\pi r^2)$. The frame has a uniform cross section with a flexural rigidity EI. Using the Engesser-Crotti theorem, determine the bending moment distribution as a function of T and ϕ.

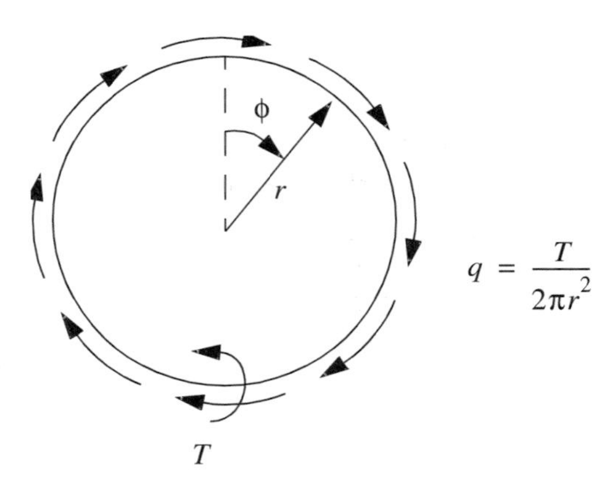

Figure 5.13 Problem 5.6.

5.7 The three-member truss in Fig. 5.14 has a vertical load P applied at pin D. All three members have an axial rigidity AE. Members AD and CD are of length ℓ and make angles ϕ with the horizontal. The vertical member BD has a coefficient of thermal expansion α and undergoes a uniform temperature increase $\Delta T = T_0$. By Castigliano's first theorem, determine the vertical displacement v of the pin D.

5.8 The semicircular frame of radius R shown in Fig. 5.15 has a uniform cross section with flexural rigidity EI. The frame is pinned at A, is on rollers at B, and carries a vertical load P at its crown. Using the Engesser-Crotti theorem, determine

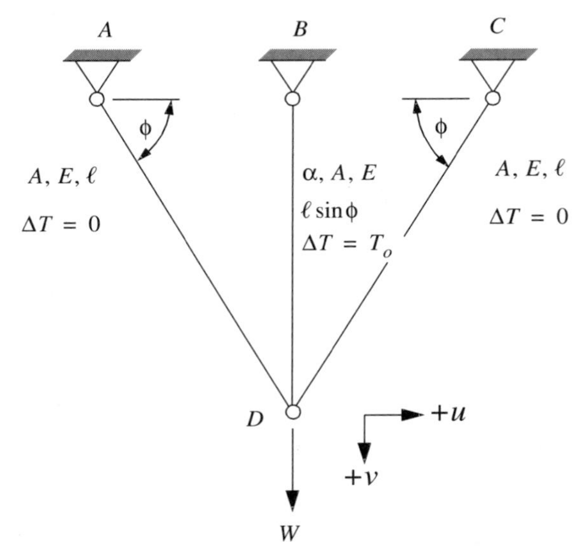

Figure 5.14 Problem 5.7.

(a) the vertical deflection under the load P, and

(b) the horizontal deflection at the roller support B.

Is a vertical plane that passes through the center of the frame a plane of symmetry?

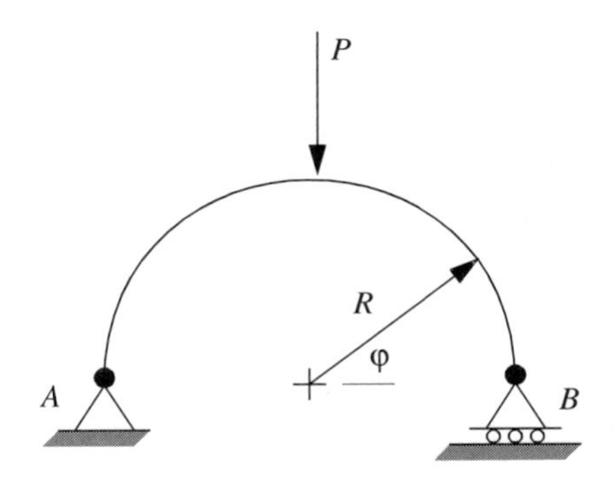

Figure 5.15 Problem 5.8.

5.9 Figure 5.16 shows a thin beam of uniform cross section. The beam consists of a semicircular portion of radius r and two straight portions of length ℓ. Its flexural rigidity is EI, and it carries at its ends two equal and opposite vertical loads P tending to separate the ends. Using the Engesser-Crotti theorem, determine

(a) the increase in distance between the ends A and D, and

(b) the relative rotation between the ends A and D.

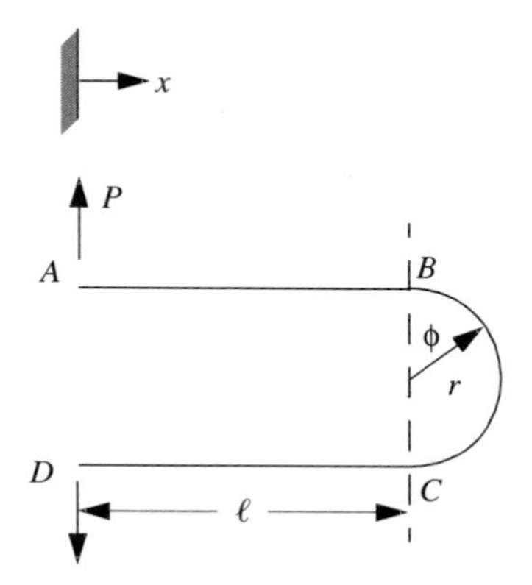

Figure 5.16 Problem 5.9.

5.10 For the frame in Fig. 5.17, the horizontal members have a flexural rigidity EI and the vertical members have a flexural rigidity EI_1. The frame has the dimensions shown and is load by two equal and opposite loads as shown. Using the Engesser-Crotti theorem, determine the bending moment at the midsection of the vertical members.

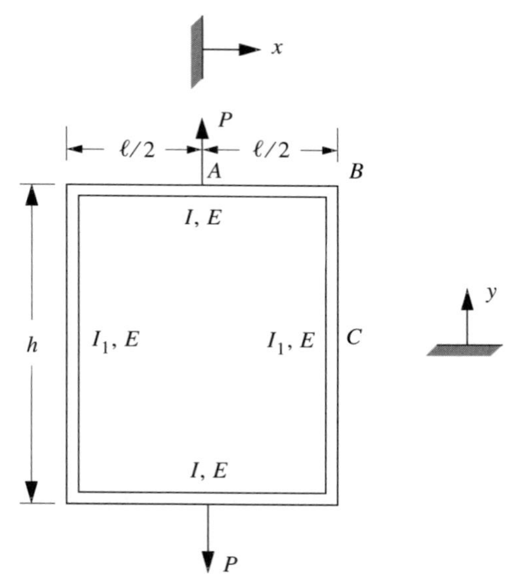

Figure 5.17 Problem 5.10.

5.11 The beam shown in Fig. 5.18 is simply supported at both ends. The left end also has a rotational spring of modulus $k = EI/\ell$ to provide some rotational constraint. The beam is of length ℓ and has a flexural rigidity EI. A uniform distributed load of magnitude p is applied along the span. Using the Engesser-Crotti theorem, determine the vertical reaction at the right end.

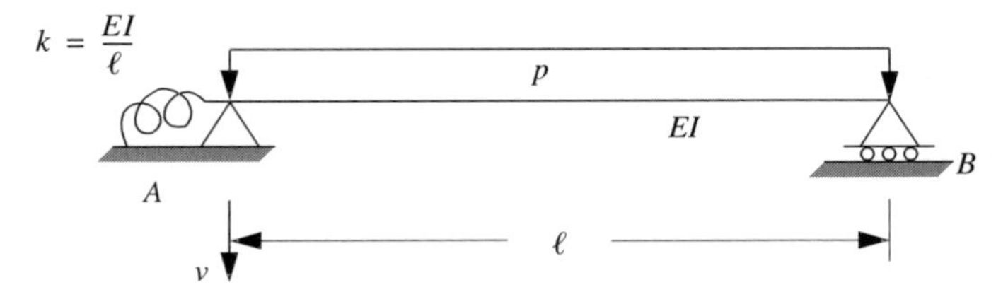

Figure 5.18 Problem 5.11.

5.12 The beam shown in Fig. 5.19 is loaded with a triangular load distribution with the load intensity p_0 at the right end. The beam has a length ℓ, a height h, and a flexural rigidity EI, and it is subject to the temperature change $\Delta T = T_0 xy/(\ell h)$. The coefficient of thermal expansion is α. Using the Engesser-Crotti theorem, determine the rotation at the right end.

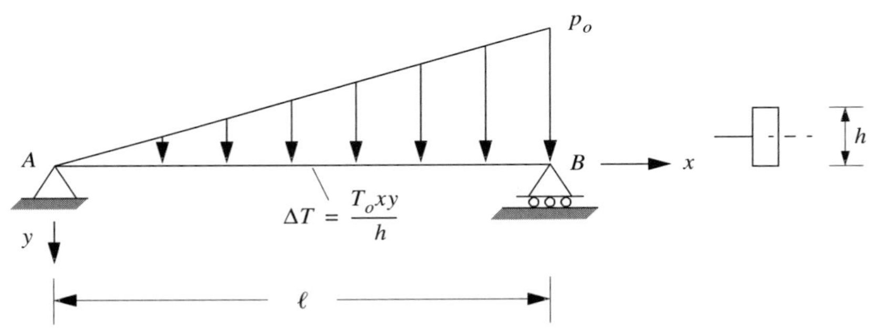

Figure 5.19 Problem 5.12.

5.13 The frame shown in Fig. 5.20 has its outer surface heated to a temperature T_1 and its inner surface heated to a temperature T_2 so that

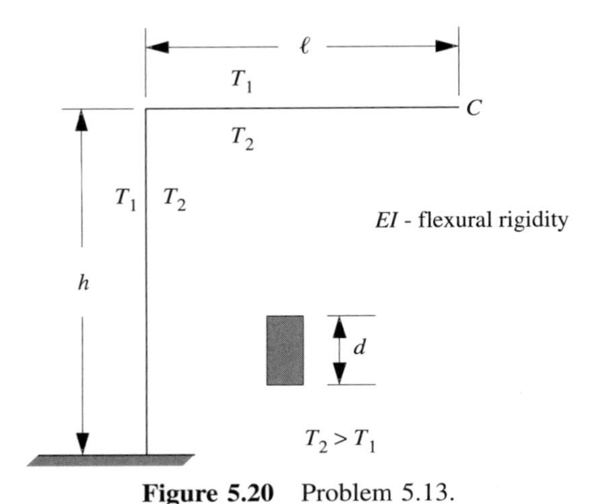

Figure 5.20 Problem 5.13.

the temperature change producing bending is given by the relationship $\Delta T = (T_2 - T_1)y/d$, where $T_2 > T_1$. The frame has a flexural rigidity EI and the dimensions shown in Fig. 5.20. Using the Engesser-Crotti theorem, determine

(a) the horizontal displacement at C;

(b) the vertical displacement at C; and

(c) the rotation at C.

5.14 The frame shown in Fig. 5.21 is fixed at A and simply supported at C. The horizontal span of length ℓ carries a uniform load of intensity p_0. The vertical member is of length h. Both members have a flexural rigidity EI. Using the Engesser-Crotti theorem, determine the vertical reaction at C.

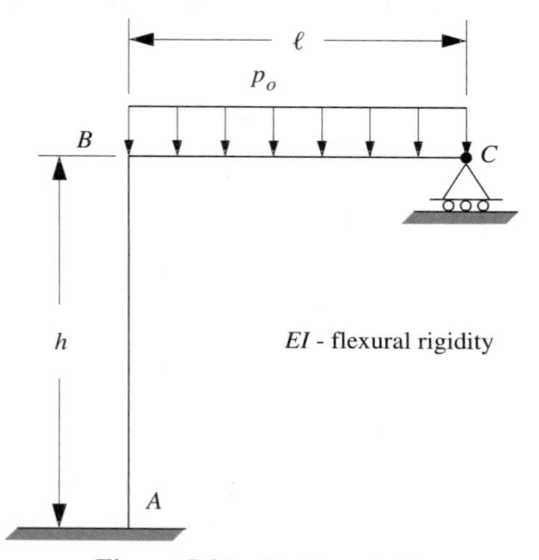

Figure 5.21 Problem 5.14.

5.15 The beam shown in Fig. 5.22 has a flexural rigidity EI, is of length 2ℓ and depth h, and is clamped at its left end. It has a coefficient of thermal expansion α. Its top surface is heated to a temperature T_1 and its bottom surface is heated to a temperature T_2 so that the temperature change producing bending is given by the relationship $\Delta T = (T_2 - T_1)y/d$,

Figure 5.22 Problem 5.15.

where $T_2 > T_1$. Using the Engesser-Crotti theorem, determine the vertical deflection of the beam at its midpoint ℓ.

5.16 The frame in Fig. 5.23 has a constant flexural rigidity EI in each member. Member AB is of length ℓ and members BC and CD are of length 2ℓ. Member BC is of depth d, has a coefficient of thermal expansion α, and is subject to the temperature change $\Delta T = T_0 y/d$. The frame is pinned at A and has a roller support at D. Using the Engesser-Crotti theorem, determine

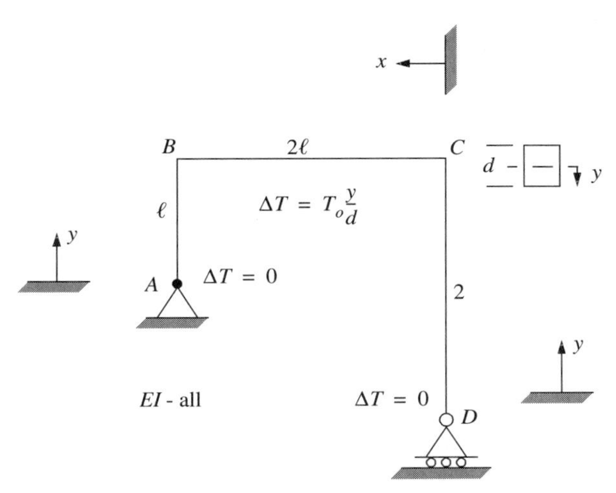

Figure 5.23 Problem 5.16.

(a) the horizontal displacement at C;

(b) the vertical displacement at C; and

(c) the rotation at C.

5.17 The circular frame in Fig. 5.24 is subject to the applied load shown. It

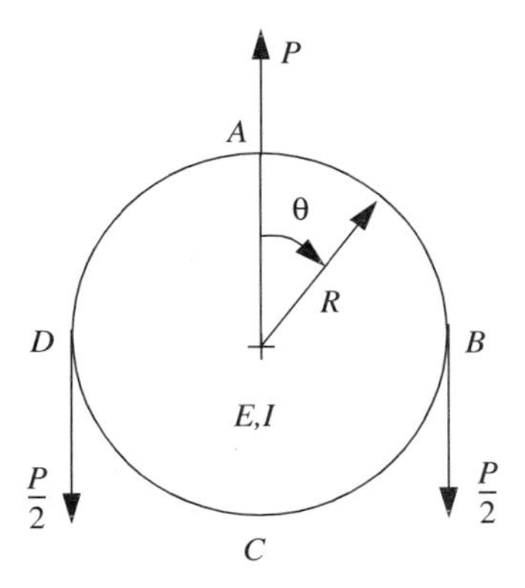

Figure 5.24 Problem 5.17.

is of radius R and has a flexural rigidity EI. Using the Engesser-Crotti theorem, determine the internal loads at A. Is a plane through DB suitable as a plane of symmetry?

5.18 Consider the problem of a uniform bar of length ℓ in Fig. 5.25, under a distributed axial load $p = p(x)$ and an applied force F at $x = \ell$. The axial displacement is governed by the equation

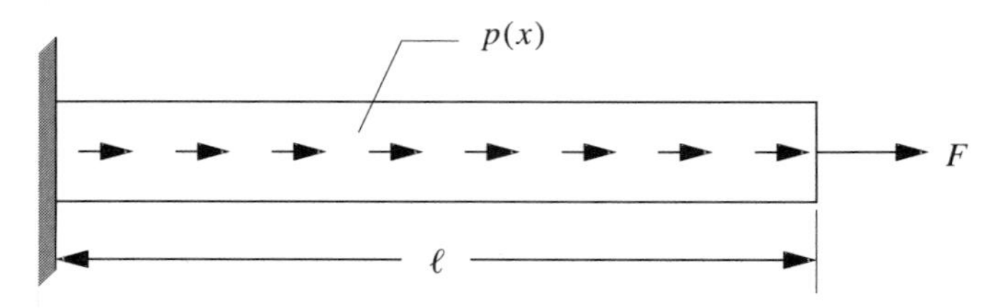

Figure 5.25 Problem 5.18.

$$AE\,\frac{d^2u}{dx^2} + p(x) = 0$$

Derive the equivalent variational statement.

5.19 Consider the problem of a uniform bar of length ℓ in Fig. 5.26, under a distributed axial load $p = p(x)$. At $x = \ell$, there is a spring of modulus k. The axial displacement is governed by the equation

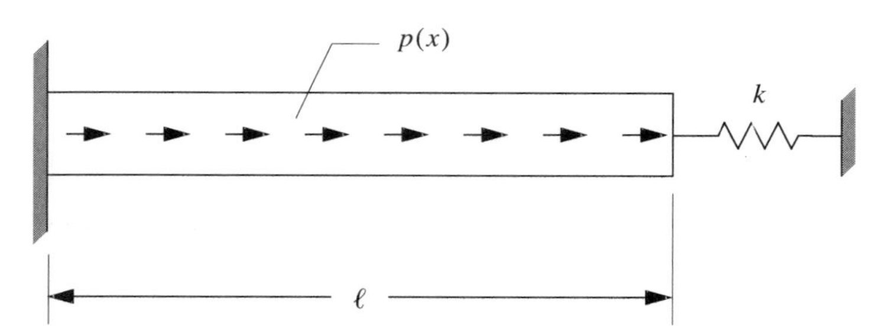

Figure 5.26 Problem 5.19.

$$AE\,\frac{d^2u}{dx^2} + p(x) = 0$$

Derive the equivalent variational statement.

5.20 The beam in Fig. 5.27 is linearly elastic, has a flexural rigidity EI, and is supported and loaded as shown. Assume the shape function

$$v(x) = a \sin\frac{\pi x}{\ell}$$

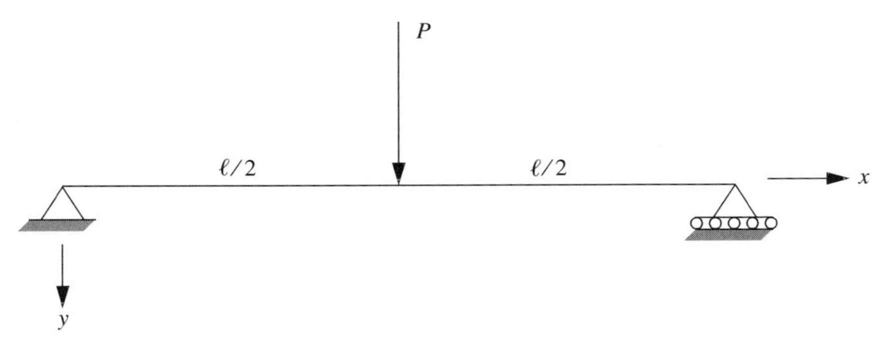

Figure 5.27 Problem 5.20.

where a is an unknown parameter to be determined. By using stationary potential energy:

(a) Determine the parameter a and compare with the strength solution.

(b) Based on the value of a, compare the moment M at $x = \ell/2$ with its strength value.

5.21 The beam in Fig. 5.28 has a flexural rigidity EI a length ℓ, and a uniform load of intensity p_0. Its left end is clamped and its right end rests on a simple support. Additionally, the right end has a rotation spring of modulus k to provide some rotational resistance. The transverse displacement is governed by the equation

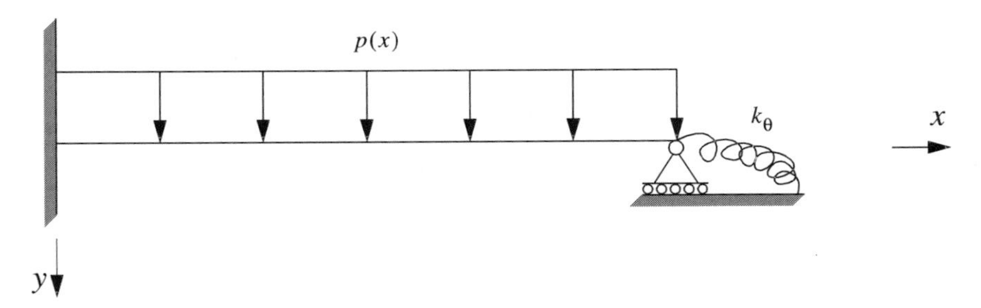

Figure 5.28 Problem 5.21.

$$EI \frac{d^4 v(x)}{dx^4} - p(x) = 0$$

Derive the equivalent variational statement.

5.22 Consider the problem of a uniform bar of length ℓ and depth h in Fig. 5.29, under a temperature change $\Delta T = T_0 x/h$. The axial displacement is governed by the equation

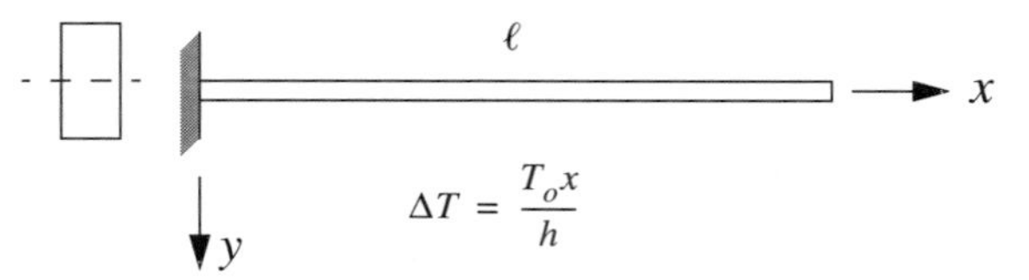

Figure 5.29 Problem 5.22.

$$AE\,\frac{d^2u}{dx^2} - \frac{AE\alpha T_0}{h} = 0$$

(a) Derive the equivalent variational statement.

(b) Compare the result with the expression obtained for strain energy using

$$U = \int_V \left(\int_0^{\epsilon} \sigma\, d\epsilon \right) dV$$

5.23 The rod in Fig. 5.30 has a length 2ℓ and an axial rigidity of AE. It is fixed at its left end. The left half between $0 \le x \le \ell$ undergoes a temperature change $\Delta T = T_0$; the right half between $\ell \le x \le 2\ell$ undergoes a temperature change $\Delta T = 0$. Assume a shape function of

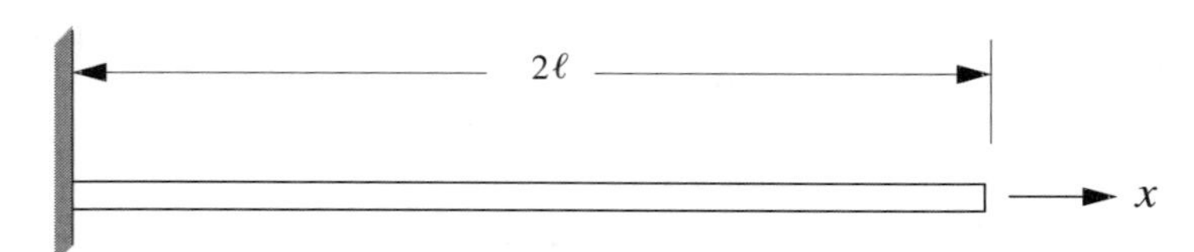

Figure 5.30 Problem 5.23.

$$u(x) = \frac{ax}{2\ell}$$

where a is an unknown parameter to be determined. By stationary potential energy, determine the value of a.

5.24 The simply supported beam in Fig. 5.31 has a flexural rigidity EI and

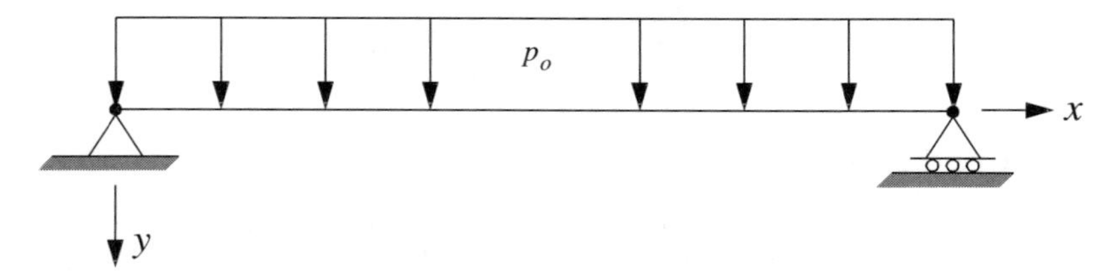

Figure 5.31 Problem 5.24.

is uniformly loaded with load intensity p_0. With an assumed deflection curve defined as

$$v(x) = ax(\ell - x)$$

determine by stationary potential energy the unknown parameter a.

5.25 The simply supported beam in Fig. 5.32 has a flexural rigidity EI and a concentrated load P at C as shown. With an assumed deflection curve defined as

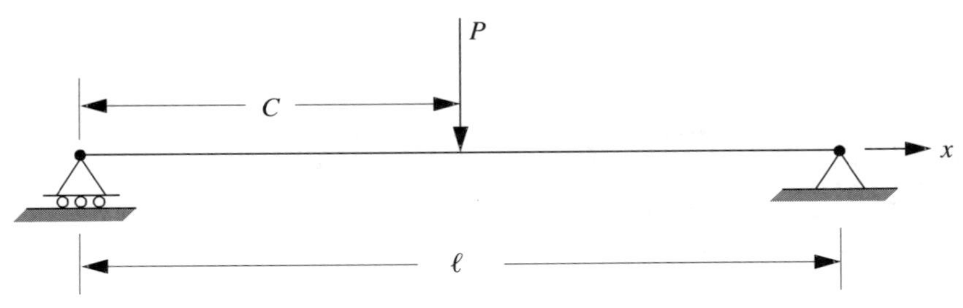

Figure 5.32 Problem 5.25.

$$v(x) = a \sin \frac{\pi x}{\ell} + b \sin \frac{2\pi x}{\ell}$$

determine by stationary potential energy the unknown parameters a and b.

5.26 Solve problem 4.33, as shown in Fig. 4.48, by using the Engesser-Crotti theorem.

5.27 Solve problem 4.34, as shown in Fig. 4.49, by using the Engesser-Crotti theorem.

CHAPTER 6

SOME STATIC AND DYNAMIC STABILITY CONCEPTS

6.1 LINEAR-STABILITY ANALYSIS

In this chapter, we examine some of the effects of geometric nonlinearities and nonconservative forces, as well as the effects of structural stiffness behavior and its relationship to structural stability. In most cases, structural instability is a phenomenon to be avoided; in other cases, however, it provides a useful function, as in the case of the "tin can" effect of keys on calculators.

As an example of geometric nonlinearity leading to instability, consider the beam shown in Fig. 6.1. The compressive load P always remains horizontal but is offset by a distance e. The beam straight length is ℓ, but the moments are measured in the slightly deformed position.

The moment relation is still given as

$$EIv'' = -M \tag{6.1}$$

From the Fig. 6.1, for small angles of rotation, the moment M is given as

$$M = -P[e + v(\ell) - v(x)]$$

Then, Eq. (6.1) becomes

$$EIv'' = P[e + v(\ell) - v(x)]$$

or on rearranging and defining,

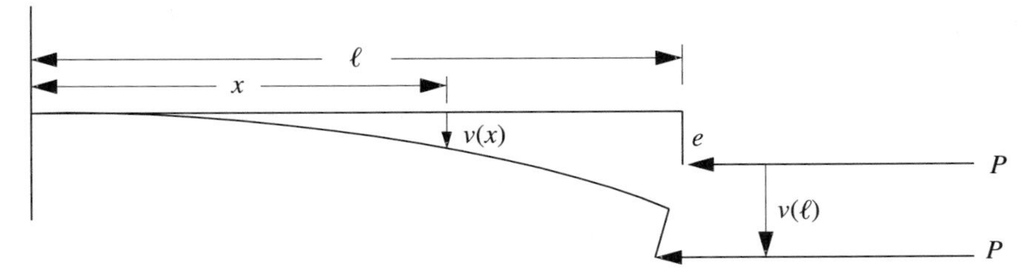

Figure 6.1 Eccentrically loaded deformed beam.

$$k^2 = \frac{P}{EI}$$

it becomes

$$v'' + k^2 v = k^2[e + v(\ell)] \tag{6.2}$$

with the following kinematic boundary conditions:

$$v(0) = v'(0) = 0$$

Observe that $v(\ell)$ is the unknown tip deflection. The general solution is given as

$$v = A \cos kx + B \sin kx + e + v(\ell)$$

or with the given kinematic boundary conditions,

$$v = [e + v(\ell)](1 - \cos kx)$$

The preceding equation can be used to solve for $v(\ell)$ to get

$$v(\ell) = \left(\frac{1}{\cos k\ell} - 1\right) e \tag{6.3}$$

and for any v, we have

$$v = \frac{1 - \cos kx}{\cos k\ell} e$$

We make the important observation that because we have considered the deformed geometry, the displacement v is not proportional to the applied load P.

It is useful to plot the tip deflection as given by Eq. (6.3). Before doing so, note that as $k\ell \rightarrow \pi/2$, the tip deflection becomes unbounded. When $k\ell$ reaches $\pi/2$, we have

$$k = \sqrt{\frac{P}{EI}} = \frac{\pi}{2\ell}$$

or

$$P_{\text{cr}} = \frac{\pi^2 EI}{4\ell^2} \tag{6.4}$$

For convenience in plotting, it is useful to nondimensionalize the tip deflection expression. To do this, define the radius of gyration as $r = \sqrt{I/A}$, where I is the least moment of inertia of the section and A is the cross-sectional area. Then, Eq. (6.3) becomes

$$\frac{v(\ell)}{r} = \left(\frac{1}{\cos k\ell} - 1 \right) \frac{e}{r}$$

A plot of this expression is shown in Fig. 6.2. It is seen from this figure how the deflection becomes more and more asymptotic to the vertical line at $\pi/2$ as the eccentricity becomes smaller and smaller.

If $e = 0$, the differential equation becomes

$$v''(x) + k^2 v(x) = k^2 v(\ell) \tag{6.5}$$

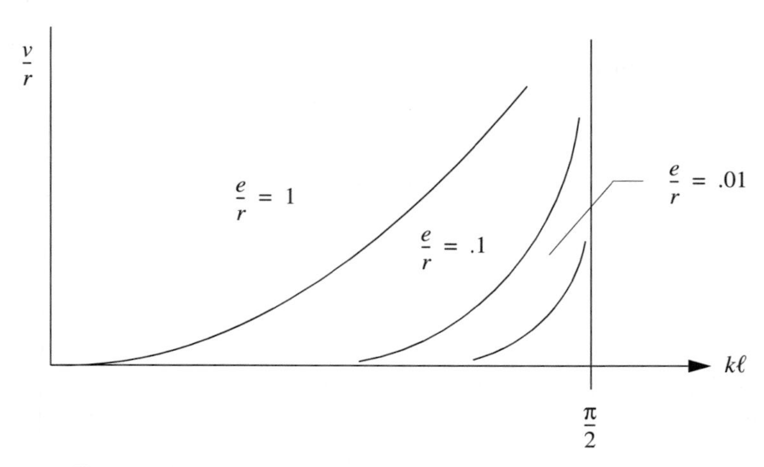

Figure 6.2 Tip deflection of an eccentrically loaded beam.

If we define a new variable $w(x) = v(x) - v(\ell)$, Eq. (6.5) takes the form

$$w'' + \lambda w = 0$$

where $\lambda = k^2$ is considered an unknown parameter. Such an equation, known as an eigenvalue equation, has one important aspect: no forcing function of the form $f(x)$ on the right-hand side. For this reason, an eigenvalue equation therefore predicts the "natural" forms of the structure.

The solution to Eq. (6.5) is

$$v(x) = v(\ell)(1 - \cos kx) \tag{6.6}$$

Note that this solution is in terms of an undetermined tip deflection $v(\ell)$. If we try to determine the tip deflection, we get

$$v(\ell) = v(\ell) - v(\ell)\cos k\ell$$

or

$$v(\ell)\cos k\ell = 0 \tag{6.7}$$

which requires that either $v(\ell) = 0$, in which case there is no deflection, or that $\cos k\ell = 0$. If $\cos k\ell = 0$, we have the following relation:

$$k\ell = (2n - 1)\frac{\pi}{2}, \qquad n = 1, 2, \ldots$$

For $n = 1$, $k\ell$ has the smallest value, and we have as before that

$$P_{\text{cr}} = \frac{\pi^2 EI}{4\ell^2}$$

The $e = 0$ solution is then an eigenvalue or characteristic-value problem, and P_{cr} is the characteristic value or critical load representing the load above which the structure will collapse. For the type of structural stability exhibited by this particular structure, P_{cr} is often called the bifurcation point, as it represents the junction point between two possible equilibrium paths.

For $n = 1$, the deflected shape becomes

$$v(x) = v(\ell)\left(1 - \cos\frac{\pi x}{2\ell}\right) \tag{6.8}$$

This is called the mode shape or eigenvector for the characteristic value ($n = 1$). The tip deflection $v(\ell)$ represents an undetermined amplitude of the deflection. Often, for plotting, it is set to some convenient value such as $v(\ell) = 1$.

When seeking eigenvalues by numerical methods, the problem reduces to what is called the algebraic eigenvalue problem. As a word of caution in working with such problems, the eigenvector associated with the eigenvalue should always be displayed and examined to ensure that the found mode makes engineering sense. For example, in the type of buckling just discussed with an initial tensile load P, the negative eigenvalue (to turn the load around) with the smallest absolute value is the desired solution. For shear buckling of orthotropic plates, the two eigenvalues with the smallest absolute value and opposite sign are usually the desired values. Algebraic eigenvalue solvers often do not adequately determine if any other modes lie between zero and the found mode.

Finally, for the problem just solved it is important to realize that the linearized differential equation loses its validity long before the tip deflection $v(\ell)$ becomes comparable with the length ℓ. Also, inelastic action often takes place before the critical load value is reached. The above results should be considered valid only in the vicinity of the x axis.

6.2 GEOMETRIC MEASURE OF STRAIN

In the discussion of the eccentrically loaded beam of the last section, perhaps the central point was the fact that we needed to look at the geometry in a slightly deformed position. We also might ask why we could start with the second-order differential equation rather than with the fourth-order one to which we are accustomed. Also, we certainly did not use any virtual work or energy principles. To address these questions, let us first develop a measure of strain that includes deformed geometry.

Consider a line element MN of length dS before deformation. After deformation, the element takes on a new orientation and length ds. As a geometric measure of strain, we can define $E_{(MN)}$ as the unit extension of the line element MN as follows:

$$E_{(MN)} = \frac{ds - dS}{dS} \tag{6.9}$$

We can consider this expression to be the extensional strain of a line element initially parallel to the direction MN. We can get an expression for the deformed length in terms of $E_{(MN)}$ as

$$ds = (E_{(MN)} + 1)\, dS \tag{6.10}$$

Consider Eq. (2.33), which measures finite strain in the undeformed system. The expression repeated again is

$$(ds)^2 - (dS)^2 = 2E_{ij}\, dX_i\, dX_j \tag{6.11}$$

where E_{ij} is the Lagrangian strain tensor (the measure of strain relative to the original coordinates), often called Green's strain tensor. Let us expand the left-hand side and use Eq. (6.10) to get

$$\begin{aligned}(ds)^2 - (dS)^2 &= (ds - dS)(ds + dS)\\ &= (ds - dS)(E_{(MN)} + 2)\ dS\\ &= \frac{(ds - dS)}{dS}\ (E_{(MN)} + 2)\ dS^2\\ &= 2\left(E_{(MN)} + \frac{1}{2}\ E^2_{(MN)}\right)\ dS^2\end{aligned}$$

Using this last expression, Eq. (6.11) becomes

$$E_{(MN)} + \frac{1}{2}\ E^2_{(MN)} = E_{ij}\ \frac{dX_i}{dS}\ \frac{dX_j}{dS} \tag{6.12}$$

In this expression, terms such as

$$\frac{dX_i}{dS}$$

represent the direction cosines of the line segment *MN* before deformation with the coordinate axes.

Now consider *MN* to be aligned with the *X* axis. We then have

$$E_{(X)}(1 + \tfrac{1}{2}E_{(X)}) = E_{XX} \tag{6.13}$$

where E_{XX} is defined as

$$E_{XX} = \frac{\partial u}{\partial X} + \frac{1}{2}\left[\left(\frac{\partial u}{\partial X}\right)^2 + \left(\frac{\partial v}{\partial X}\right)^2 + \left(\frac{\partial w}{\partial X}\right)^2\right] \tag{6.14}$$

If the strains are considered small with respect to one, then

$$E_{(X)}(1 + \tfrac{1}{2}E_{(X)}) \approx E_{(X)}$$

and we can write

$$E_{(X)} = \frac{\partial u}{\partial X} + \frac{1}{2}\left[\left(\frac{\partial v}{\partial X}\right)^2 + \left(\frac{\partial w}{\partial X}\right)^2\right] \tag{6.15}$$

Next, consider the rotation of a line element initially parallel to the *X* axis

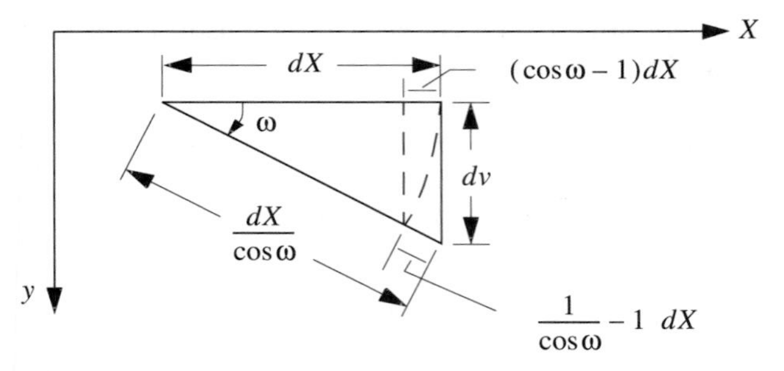

Figure 6.3 Line element rotation without X-direction motion.

through an angle ω about the Z axis in such a fashion that there is no X-direction motion. This rotation is shown in Fig. 6.3, from which we see that

$$\frac{du}{dX} = (\cos\,\omega - 1)$$

and

$$\frac{dv}{dX} = \tan\,\omega$$

Using the results from Eq. (6.15), we get

$$E_{(X)} = (\cos\,\omega - 1) + \tfrac{1}{2}\tan^2\,\omega$$

If we now consider the rotation to be small relative to one, the above expression becomes

$$E_{(X)} \approx \frac{1}{2}\,\omega^2 \approx \frac{1}{2}\left(\frac{dv}{dX}\right)^2$$

Since we have reduced the problem to small strain with small yet significant rotation, let us again drop the distinction between the X_i and x_i coordinates.

We can then define the extensional strain from a small rotation as

$$\epsilon = \tfrac{1}{2}(v')^2 \tag{6.16}$$

and for the beam shown in Fig. 6.1, we can write the total strain as

$$\epsilon = u_0' - yv'' + \tfrac{1}{2}(v')^2 \tag{6.17}$$

The virtual strain becomes

$$\bar{\epsilon} = \bar{u}_0' - y\bar{v}'' + v'\bar{v}' \tag{6.18}$$

where we have treated δ as a differential operator. The stress becomes

$$\sigma = E[u_0' - yv'' + \tfrac{1}{2}(v')^2] \tag{6.19}$$

If we resolve the load P into a moment and a force acting through the axis of the beam, the external virtual work can be written as

$$\delta W_e = -P\bar{u}_0(\ell) + Pe\bar{v}'(\ell)$$

The internal virtual work can be defined as

$$\delta U = \int \sigma\bar{\epsilon}\, dV$$

The last equation requires a bit of explanation. In deriving Eq. (3.5), we stated that we could use small virtual strains; now, however, we are also including a small but nonlinear strain term. The important point is that in Eq. (3.5), the small strains were defined in the updated or deformed geometry. In our current case, even though we have assumed small strain and small rotation and, for convenience, dropped the distinction between X and x axes, the nonlinear strain involved is measured in the undeformed coordinates. Reference [54] shows that, in general, when referring virtual work to the undeformed geometry, we should write

$$\delta U = \int \tilde{\mathbf{T}}_{ij}\, \delta E_{ij}\, dV$$

Since we are assuming small strain and small rotation,

$$\tilde{\mathbf{T}}_{ij} = \sigma_{ij}$$

and δE_{ij} has been reduced to Eq. (6.18).

For the beam, the principle of virtual work now becomes

$$\int E\left[u_0' - yv'' + \frac{1}{2}(v')^2\right](\bar{u}_0' - y\bar{v}'' + v'\bar{v}')\, dA\ dx = -P\bar{u}_0(\ell) + Pe\bar{v}'(\ell) \tag{6.20}$$

To solve Eq. (6.20), consider first the case

$$\bar{u}_0 \neq 0, \qquad \bar{v} = 0$$

Then, the virtual equation becomes

$$\int EA\left[u_0' + \frac{1}{2}(v')^2\right]\bar{u}_0'\,dx = -P\bar{u}_0(\ell) \tag{6.21}$$

Integration by parts yields

$$EA\left[u_0' + \frac{1}{2}(v')^2\right]\bar{u}_0\bigg|_0^{\ell} - \int EA(u_0'' + v'v'')\bar{u}_0\,dx = -P\bar{u}_0(\ell)$$

From this relationship, we get the following differential equation:

$$EA(u_0'' + v'v'') = 0 \tag{6.22}$$

with the kinematic boundary condition of

$$u_0(0) = 0 \tag{6.23}$$

and the natural boundary condition of

$$EA[u_0' + \tfrac{1}{2}(v')^2]|_{\ell} = -P \tag{6.24}$$

The last equation represents the equilibrium relationship between the internal axial load (call it N) and the applied load at the tip of the beam. Therefore, we now define N as

$$N = EA[u_0' + \tfrac{1}{2}(v')^2] \tag{6.25}$$

Taking the derivative of N, we get

$$N' = EA(u_0'' + v'v'') \tag{6.26}$$

that from Eq. (6.22) (a differential equation) yields the following result:

$$N' = 0 \tag{6.27}$$

Also, from Eq. (6.24) and the natural boundary condition, we can in general write

$$N = -P \tag{6.28}$$

for all x.

Consider again Eq. (6.20), but this time use

$$\overline{u}_0 = 0, \qquad \overline{v} \neq 0$$

We then get the following relationship:

$$\int E\left[u_0' - yv'' + \frac{1}{2}(v')^2\right](-y\overline{v}'' + v'\overline{v}')\, dA\ dx = Pe\overline{v}'(\ell) \tag{6.29}$$

Using Eq. (6.25), we can write

$$\int E\left(\frac{N}{EA} - yv''\right)(-y\overline{v}'' + v'\overline{v}')\, dA\ dx = Pe\overline{v}'(\ell) \tag{6.30}$$

which, after integrating on dA, yields

$$\int (Nv')\overline{v}'\, dx + \int EIv''\overline{v}''\, dx = Pe\overline{v}'(\ell) \tag{6.31}$$

Integrating by parts yields

$$(Nv')\overline{v}\,|_0^\ell - \int (Nv')'\overline{v}\, dx$$

$$+ EIv''\overline{v}'\,|_0^\ell - EIv'''\overline{v}\,|_0^\ell + \int EIv^{iv}\overline{v}\, dx = Pe\overline{v}'(\ell) \tag{6.32}$$

which yields the following differential equation:

$$EIv^{iv} - (Nv')' = 0 \tag{6.33}$$

Using Eqs. (6.27) and (6.28), this expression can be written as

$$EIv^{iv} + Pv'' = 0 \tag{6.34}$$

Define k as

$$k^2 = \frac{P}{EI}$$

and obtain the final form of the differential equation as

$$v^{iv} + k^2 v'' = 0 \tag{6.35}$$

with kinematic boundary conditions

$$v(0) = 0 \tag{6.36}$$

and

$$v'(0) = 0 \tag{6.37}$$

The natural boundary conditions become

$$v''(\ell) = k^2 e \tag{6.38}$$

and

$$v'''(\ell) + k^2 v'(\ell) = 0 \tag{6.39}$$

The general solution is

$$v = c_1 \cos kx + c_2 \sin kx + c_3 x + c_4 \tag{6.40}$$

Applying the boundary conditions results in

$$c_3 = 0, \qquad c_2 = 0, \qquad c_4 = -c_1$$

and

$$c_1 = -\frac{e}{\cos k\ell}$$

which again yields

$$v = \frac{1 - \cos kx}{\cos k\ell}$$

and

$$v(\ell) = \left(\frac{1}{\cos k\ell} - 1 \right) e$$

Note that the above fourth-order differential equation can, in this case, be reduced to a second-order equation. Integrate Eq. (6.35) twice to get

$$v''' + k^2 v' = b_3$$

and

$$v'' + k^2 v = b_3 x + b_2$$

and use Eqs. (6.38) and (6.39) to get

$$b_3 = 0, \qquad b_2 = k^2[e + v(\ell)]$$

We have

$$v'' + k^2 v = k^2[e + v(\ell)]$$

which is Eq. (6.2).

If $e = 0$, the differential equation remains the same, but the natural boundary condition as given by Eq. (6.38) changes to

$$v''(\ell) = 0 \tag{6.41}$$

Applying again the general solution given by Eq. (6.40) and the kinematic and natural boundary equations, we get the following system of linear and homogeneous equations:

$$\begin{bmatrix} 1 & 0 & 0 & 1 \\ 0 & k & 1 & 0 \\ -k^2 \cos k\ell & -k^2 \sin k\ell & 0 & 0 \\ 0 & 0 & k^3 & 0 \end{bmatrix} \begin{Bmatrix} c_1 \\ c_2 \\ c_3 \\ c_4 \end{Bmatrix} = \begin{Bmatrix} 0 \\ 0 \\ 0 \\ 0 \end{Bmatrix}$$

This is an underdetermined system with infinitely many solutions that are determined by equating the determinant to zero. We may expand the determinant with respect to the last column (the last row would be just as easy) to obtain the result

$$\cos k\ell = 0$$

leading again to the solution

$$k\ell = (2n - 1)\,\frac{\pi}{2}, \qquad n = 1, 2, \ldots$$

For $n = 1$, $k\ell$ has the smallest value, and, as before, we have

$$P_{\mathrm{cr}} = \frac{\pi^2 EI}{4\ell^2}$$

For $n = 1$ and $k = \pi/2\ell$, by direct substitution into the general solution we again get the following relationships:

$$c_3 = 0, \qquad c_2 = 0, \qquad c_4 = -c_1$$

Evaluating the general solution at $v(\ell)$ yields $c_4 = v(\ell)$, the result

$$v(x) = v(\ell)\left(1 - \cos\frac{\pi x}{2\ell}\right)$$

as previously determined by Eq. (6.8).

6.3 A BEAM WITH INITIAL CURVATURE REVISITED

We can write Eq. (6.11) as

$$\frac{1}{2}\left[\left(\frac{ds}{dS}\right)^2 - 1\right] dS^2 = E_{ij}\, dX_i\, dX_j$$

and use Eq. (6.12) to obtain

$$E_{(MN)} + \frac{1}{2} E_{(MN)}^2 = \frac{1}{2}\left[\left(\frac{ds}{dS}\right)^2 - 1\right]$$

which, on assuming that the strain is small when relative to one, becomes

$$\epsilon \approx \frac{1}{2}\left[\left(\frac{ds}{dS}\right)^2 - 1\right] \tag{6.42}$$

where ϵ is the extensional strain. For the case of the curved beam, consider again Fig. 3.26 and define

$$dS^2 = |\,d\vec{R}_0|^2$$

and

$$ds^2 = |\,d\vec{R}\,|^2$$

From the geometry of the figure, we have the following:

$$\begin{aligned}\frac{d\vec{R}_0}{dS} &= \frac{d\vec{R}_0}{ds}\,\frac{ds}{dS}\\ &= \frac{1}{1-y/R}\,\frac{d\vec{R}_0}{ds}\\ &= \frac{1}{1-y/R}\,\frac{d}{ds}\,(\vec{r}_0 + y\vec{n}_0 + z\vec{k})\\ &= \vec{t}_0\end{aligned}$$

where dS and ds have taken on the meaning described in Fig. 3.27, where we have again used the Frenet formulas and where the scalar R is the radius of curvature at ds.

Also, we have

$$\begin{aligned}\frac{d\vec{R}}{dS} &= \frac{d\vec{R}}{ds}\,\frac{ds}{dS}\\ &= \frac{1}{1-y/R}\,\frac{d\vec{R}}{ds}\\ &= \frac{1}{1-y/R}\,\frac{d}{ds}\,[\vec{r}_0 + U_0\vec{t}_0 + (y+V)\vec{n}_0 + (z+W)\vec{k}]\\ &= \left(1 + U' - \frac{y}{R} - \frac{V}{R}\right)\vec{t}_0 + \left(V' + \frac{U}{R}\right)\vec{n}_0 + W'\vec{k}\end{aligned}$$

By using Eq. (3.104) as well as expanding and assuming (as before) that $V' = v'$ and $W' = w'$, we get

$$\begin{aligned}\frac{d\vec{R}}{dS} = &\left[1 + \frac{1}{1-y/R}\left(u_0' - \frac{v}{R}\right) - \frac{y}{1-y/R}\,\beta' - \frac{z}{1-y/R}\,w''\right]\vec{t}_0\\ &+ \left(\beta - \frac{z}{1-y/R}\,\frac{w'}{R}\right)\vec{n}_0 + \frac{1}{1-y/R}\,w'\vec{k}\end{aligned}$$

where β is defined in Fig. 6.4 as

$$\beta = \left(\frac{u_0}{R} + v'\right) \tag{6.43}$$

and β' is the curvature in the y–s plane.

Using Eq. (3.106a), we can write

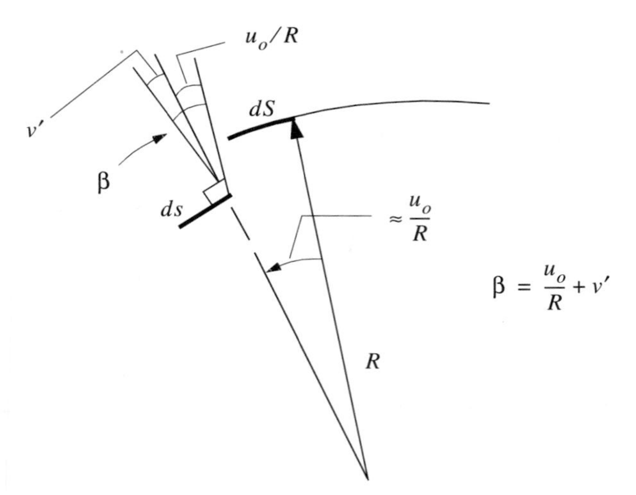

Figure 6.4 Curvature in the y–s plane.

$$\frac{1}{1 - y/R} w' \vec{k} = \left(w' + \frac{y}{1 - y/R} \frac{w'}{R} \right) \vec{k}$$

that for the small-strain assumption can be written as

$$\frac{1}{1 - y/R} w' \vec{k} \approx w' \vec{k}$$

By the same small-strain argument:

$$\left(\beta - \frac{z}{1 - y/R} \frac{w'}{R} \right) \vec{n}_0 = \beta \vec{n}_0$$

Then, $|d\vec{R}|^2$ becomes

$$|d\vec{R}|^2 = \left\{ \left[\left(u_0' - \frac{v}{R} \right) \frac{1}{1 - y/R} + \left(1 - \beta' \frac{y}{1 - y/R} - w'' \frac{z}{1 - y/R} \right) \right]^2 + \beta^2 + (w')^2 \right\} dS^2$$

Using the above expression in Eq. (6.42), we finally get

$$\epsilon = \left(u_0' - \frac{v}{R}\right) + \frac{1}{2}\beta^2 + \frac{1}{2}(w')^2 - \frac{y}{1 - y/R}\beta' - \frac{z}{1 - y/R}w'' \tag{6.44}$$

where we have neglected curvature terms involving $(\beta')^2$, $\beta' w''$, and $(w'')^2$ along with terms $(u_0' - v/R)/2$ when compared with one.

If we consider bending only in the y–s plane and assume that the depth dimensions are small compared with the radius of curvature R, the above equation reduces to

$$\epsilon = \left(u_0' - \frac{v}{R}\right) + \frac{1}{2}\beta^2 - y\beta' \tag{6.45}$$

We select Eq. (3.14) with temperature neglected as the constitutive equation. We may then write the expression for stress as follows:

$$\sigma = E\left[\left(u_0' - \frac{v}{R}\right) + \frac{1}{2}\beta^2 - y\beta'\right] \tag{6.46}$$

As before, we can now use virtual work to determine the stress σ in terms of the stress resultants and second moments of area. Consider the virtual work of the axial force $\int_A \sigma \, dA$ from the virtual displacement $\overline{U}$. This is expressed as

$$\int_A \sigma \, dA(\overline{u}_0 - y\overline{v}') = \int_A E\left[\left(u_0' - \frac{v}{R}\right) + \frac{1}{2}\beta^2 - y\beta'\right](\overline{u}_0 - y\overline{v}') \, dA$$

Collecting the arguments of $\overline{u}_0$ and $\overline{v}'$ and using Eq. (3.61), we get

$$N = EA\left[\left(u_0' - \frac{v}{R}\right) + \frac{1}{2}\beta^2\right] \tag{6.47}$$

and

$$M = -EI\beta' = -EI\left(\frac{u_0'}{R} + v''\right) \tag{6.48}$$

This last expression should be compared with the $v'' + v/R^2$ term of Eq. (3.110), which gives the result

$$M = -EI\left(\frac{v}{R^2} + v''\right)$$

Thus there appears to be two expressions for curvature. These expressions are equivalent because they represent different parametric statements of curvature. The curvature expression in Eq. (3.110) comes from a polar coordinate representation of curvature given by

$$\frac{1}{R_1} = \frac{R^2 - R\ddot{R} + 2\dot{R}^2}{(R^2 + \dot{R}^2)^{3/2}}$$

where R is the radius, θ is the polar angle, and $\dot{R} = dR/d\theta$. The curvature expression in Eq. (6.48) comes from the parametric representation of curvature given by

$$\frac{1}{R_1} = \frac{|\dot{x}\ddot{y} - \dot{y}\ddot{x}|}{[\dot{x}^2 + \dot{y}^2]^{3/2}}$$

where x, y are the coordinates of the point on the curve as functions of R and θ. Figure 6.5 compares the two expressions for curvature.

Using instantaneous polar coordinates, we have from Fig. 6.5(a)

$$R_1 = R - v - dR + u_0\, d\theta \approx R - v$$

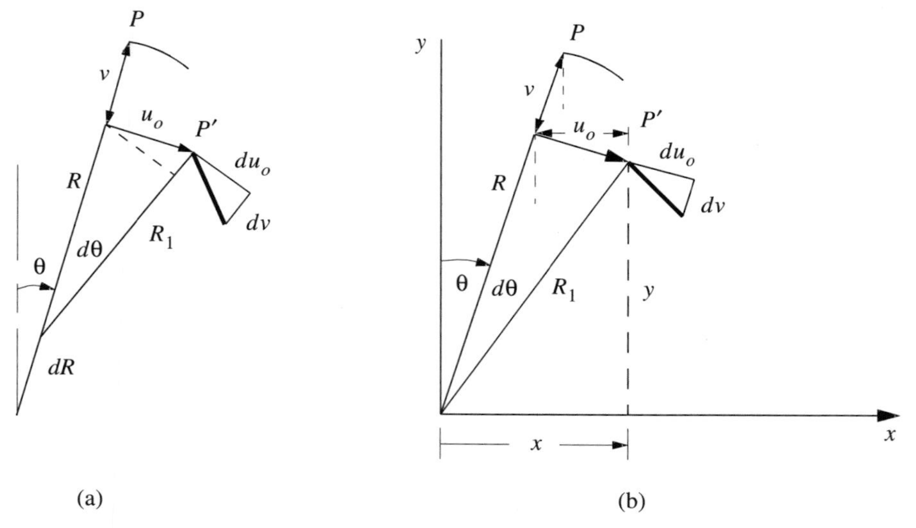

Figure 6.5 Curvature comparisons.

We can then write for the instantaneous radius of curvature

$$\frac{1}{R_1} = \frac{(R-v)^2 - (R-v)(\ddot{R}-\ddot{v}) + 2(\dot{R}-\dot{v})^2}{[(R-v)^2 + (\dot{R}-\dot{v})^2]^{3/2}}$$

If we linearize by neglecting $\dot{R}$ and $\ddot{R}$ and by considering $\dot{v} \ll R$, we get

$$\frac{1}{R_1} = \frac{(R-v)^2 + (R-v)\ddot{v}}{(R-v)^3}$$

Then, the change in curvature after subtracting $1/R$ from both sides becomes

$$\begin{aligned}\frac{1}{R_1} - \frac{1}{R} &= \frac{(R-v)^2 + (R-v)\ddot{v}}{(R-v)^3} - \frac{1}{R} \\ &= \frac{v}{R(R-v)} + \frac{\ddot{v}}{(R-v)^2} \\ &= \frac{(R-v)v + R\ddot{v}}{R^3(1 - 2v/R + v^2/R)}\end{aligned}$$

or

$$\frac{1}{R_1} - \frac{1}{R} = \frac{v + \ddot{v}}{R^2} = \frac{v}{R^2} + v''$$

Using initial Cartesian coordinates, we have the following from Fig. 6.5(b):

$$\begin{aligned}x &= (R-v)\sin\theta + u_0\cos\theta \\ y &= (R-v)\cos\theta - u_0\sin\theta\end{aligned}$$

By computing the appropriate derivatives, neglecting all square- and cross-product terms in v and u_0, and considering $2(v - u_0')/R \ll 1$, we get

$$\frac{1}{R_1} = \frac{1}{R} + \frac{\dot{u}_0 + \ddot{v}}{R^2}$$

or

$$\frac{1}{R_1} - \frac{1}{R} = \frac{u_0'}{R} + v''$$

Thus, for terms involving curvature we have the following relationship:

$$u_0' = \frac{v}{R} \tag{6.49}$$

This is a transformation between parametric forms of curvature only; it cannot be applied indiscriminately. For example, if we neglect β^2 and $(w')^2$ terms in Eq. (6.44), we can arrive again at Eq. (3.105). To do so, multiply the v/R term by $(1 - y/R)/(1 - y/R)$ and expand to get

$$\epsilon = u_0' - \frac{yv''}{1 - y/R} - \frac{zw''}{1 - y/R} - \frac{v/R}{(1 - y/R)} + \left(\frac{v}{R^2} - \frac{u_0'}{R} \right) \frac{y}{1 - y/R}$$

The last term in the above expression is zero by Eq. (6.49).

Defining the virtual strain as

$$\overline{\epsilon} = \left(\overline{u}_0' - \frac{\overline{v}}{R} \right) + \beta\overline{\beta} - y\overline{\beta}' \tag{6.50}$$

the virtual work for the beam shown in Fig. 3.28 now considered a thin ring in the xY plane is

$$\begin{aligned} \int E\left(\frac{N}{AE} + y\,\frac{M}{EI} \right) \left[\left(\overline{u}_0' - \frac{\overline{v}}{R} \right) + \beta\overline{\beta} - y\overline{\beta}' \right] dA\; ds \\ = \int (p_s\overline{u}_0 + p_y\overline{v})\, ds - P_A\overline{u}_0(A) + P_B\overline{u}_0(B) \\ - V_{Ay}\overline{v}(A) + V_{By}\overline{v}(B) + M_{Az}\overline{v}'(A) - M_{Bz}\overline{v}'(B) \end{aligned}$$

These expressions yield two nonlinear differential expressions:

$$\begin{aligned} -N' + \frac{N}{R}\left(\frac{u_0}{R} + v' \right) + \frac{1}{R}\,M' &= p_s \\ -\frac{N}{R} - \left(\frac{u_0}{R} + v' \right)' - M'' &= p_y \end{aligned} \tag{6.51}$$

with the following kinematic boundary conditions or natural boundary conditions:

$$
\begin{aligned}
u_0(B) = 0 &\quad \text{or} \quad \left(N - \frac{M}{R}\right)_B = P_B \\
u_0(A) = 0 &\quad \text{or} \quad \left(N - \frac{M}{R}\right)_A = P_A \\
v'(B) = 0 &\quad \text{or} \quad -M\,|_B = -M_{Bz} \\
v'(A) = 0 &\quad \text{or} \quad M\,|_A = M_{Az} \\
v(B) = 0 &\quad \text{or} \quad \left[N\left(\frac{u_0}{R} + v'\right) + M'\right]_B = V_{By} \\
v(A) = 0 &\quad \text{or} \quad -\left[N\left(\frac{u_0}{R} + v'\right) + M'\right]_A = -V_{Ay}
\end{aligned}
$$

Reference [55] discusses in detail the solution of the expressions in Eq. (6.51).

6.4 THIN WALLED OPEN BEAMS REVISITED

Following the previous results of this chapter, we take the extensional strain on a generic point of the middle surface of a thin walled open section to be of the form

$$
u_0 + \tfrac{1}{2}\,(v_M')^2 + \tfrac{1}{2}\,(w_M')^2
$$

or on using Eq. (3.144),

$$
u_0 + \tfrac{1}{2}\,[v' - (z - a_z)\phi']^2 + \tfrac{1}{2}\,[w' + (y - a_y)\phi']^2
$$

The strain then becomes

$$
\begin{aligned}
\epsilon = u_0 &+ \tfrac{1}{2}\,[v' - (z - a_z)\phi']^2 + \tfrac{1}{2}\,[w' + (y - a_y)\phi']^2 \\
&- yv'' - zw'' - \omega\phi''
\end{aligned} \tag{6.52}
$$

The virtual strain becomes

$$
\begin{aligned}
\overline{\epsilon} = \overline{u}_0 &- y\overline{v}'' - z\overline{w}'' - \omega\overline{\phi}'' \\
&+ [v' - (z - a_z)\phi'][\overline{v}' - (z - a_z)\overline{\phi}'] \\
&+ [w' - (y - a_y)\phi'][\overline{w}' + (y - a_y)\overline{\phi}']
\end{aligned} \tag{6.53}
$$

We also define the internal axial load as

$$N = EA\{u_0 + \tfrac{1}{2}[v' - (z - a_z)\phi']^2 + \tfrac{1}{2}[w' + (y - a_y)\phi']^2\} \tag{6.54}$$

and the stress as

$$\sigma = E\epsilon \tag{6.55}$$

The internal virtual work then becomes

$$\delta U = \int \sigma\bar{\epsilon}\, dV + \int \tau_{SV}\bar{\gamma}_{SV}\, dV \tag{6.56}$$

Equation (6.56) demonstrates both the best and the worst feature of the virtual work principle. The best feature is represented by the fact that the additional virtual strain terms automatically project the stress σ components arising from the deformed position onto the undeformed axes. The worst feature is that *the result will be an inconsistent set of differential equations.*

When dealing with beams, it is usual to neglect the effect of transverse shear except for short, stubby beams; however, when dealing with thin walled beams in the deformed position, the transverse shear also contributes to the torsional stability. The first indication of this occurs when expanding Eq. (6.56). Terms of the form $M_y'\phi'$ occur. The M_y' term represents the transverse shear V_z. Additionally, terms of the form $M_y''v$ also appear. If there is no transverse applied load p_z, then $M_y'' = 0$, but if there is an applied transverse load, then $M_y'' = -p_z$. This indicates that the change in position of the transverse loads also contributes to additional torsional effects.

To develop the differential equations, we essentially consider the beam shown in Fig. 3.39. As we continue the development, we put restrictions on the applied loads to simplify the results. The first restriction is that the axial end loads P always maintain their original orientation and that there are no p_x loads. Also, the edge tractions S_R and S_L are assumed to be zero.

We also assume that the y, z axes are the principal axes passing through the line of centroids. The pole is the principal pole located at the shear center, and the sectorial centroid has been located so that $S_\omega = 0$.

Applying the principle of virtual work and considering the case

$$\bar{u}_0 \neq 0; \qquad \bar{v} = \bar{w} = \bar{\phi} = 0$$

we now get the differential equation

$$N' = 0 \tag{6.57}$$

which, using the associated natural boundary conditions, implies

$$N = -P \tag{6.58}$$

Next, considering the case

$$\overline{v} \neq 0; \qquad \overline{u}_0 = \overline{w} = \overline{\phi} = 0$$

we get the differential equation

$$EI_z v^{iv} - [N(v' + a_z\phi')]' + (M_y\phi')' = p_y + \tilde{p}_y \tag{6.59}$$

where we will add $\tilde{p}_y$ later by considering appropriate shear projections.
Next, considering the case

$$\overline{w} \neq 0; \qquad \overline{u}_0 = \overline{v} = \overline{\phi} = 0$$

we get the differential equation

$$EI_y w^{iv} - [N(w' - a_y\phi')]' - (M_z\phi')' = p_z + \tilde{p}_z \tag{6.60}$$

where we will add $\tilde{p}_z$ later by considering appropriate shear projections.
Next, considering the case

$$\overline{\phi} \neq 0; \qquad \overline{u}_0 = \overline{v} = \overline{w} = 0$$

we get the differential equation

$$\begin{aligned} EI_\omega\phi^{iv} &- JG\phi'' \\ &- [N(a_z v' - a_y w')]' \\ &- [(\beta_N N + \beta_z M_z + \beta_y M_y + \beta_\omega M_\omega)\phi']' \\ &+ (M_y v')' - (M_z w')' \\ &= m_x + \tilde{m}_x \end{aligned} \tag{6.61}$$

where we will add $\tilde{m}_x$ later by considering appropriate shear projections.
In the above equation, we have defined the following properties:

$$H_y = \frac{1}{I_y} \int z(y^2 + z^2)\, dA$$

$$H_z = \frac{1}{I_z} \int y(y^2 + z^2)\, dA$$

$$\beta_N = \frac{1_y + I_z}{A} + a_y^2 + a_z^2$$

$$\beta_y = H_y - 2a_z$$

$$\beta_z = H_z - 2a_y$$

$$\beta_\omega = \frac{1}{I_\omega} \int \omega(y^2 + z^2)\, dA$$

If we now consider the beam to have its supports and transverse loading applied through the line of shear centers, and the end rigidly capped so that the applied axial load cannot cause warping, then $M_\omega = 0$. Consequently, the $\beta_\omega M_\omega$ can be dropped from the differential equation.

To compute the effects of shear, we follow the development given in ref. [35] by first considering Fig. 6.6(a).

This figure shows the differential element $dx\ ds$ with shear τ acting on its curved faces after the element has been rotated a small amount ϕ. If $\vec{n}$ is the normal to the midsurface (see Fig. 3.32), this rotation will rotate the shear stress so that it has a projection along a line parallel to the normal $\vec{n}$.

Let q_t represent the projected shear. We can then write

$$dF_t = q_t\, dx\, ds = (\tau\, d\phi + d\tau\, \phi)t\, ds$$

where we have neglected higher-order terms. Then, we define q_t as

$$q_t = t\tau \frac{d\phi}{dx} + t \frac{d\tau}{dx} \phi = t \frac{d}{dx} (\tau\phi)$$

The projected shear shown in Fig. 6.6(b) will cause transverse forces parallel to the O–y and O–z axes as follows:

$$dF_y = -q_t \cos \psi\, dx\, ds = -q_t \frac{dz}{dx}\, dx\, ds = q_{ty}\, dx\, ds$$

and

$$dF_z = -q_t \sin \psi\, dx\, ds = +q_t \frac{dy}{ds}\, dx\, ds = q_{tz}\, dx\, ds$$

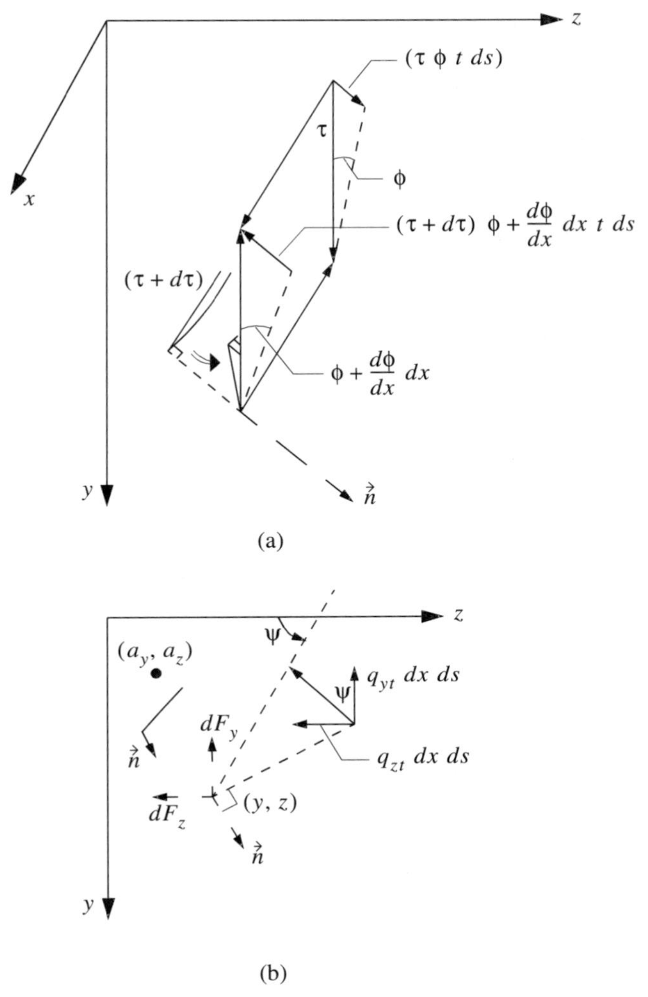

Figure 6.6 Transverse shear projection.

where ψ is the angle betwen the tangent to the midplane and the O–z axis. We can now define the transverse loads after using Eqs. (3.217) and (3.218) as

$$
\begin{aligned}
\tilde{p}_y &= \int q_{ty}\, ds \\
&= -\int q_t \frac{dz}{ds}\, ds
\end{aligned}
$$

$$= -\int t \frac{d}{dx} (\tau\phi)\, dz$$

$$= -\frac{d}{dx} \int \left(-\frac{M_z'}{I_z} \frac{Q_z}{t} - \frac{M_y'}{I_y} \frac{Q_y}{t} - \frac{M_\omega'}{I_\omega} \frac{S_\omega}{t} \right) \phi t\, dz$$

$$= -(M_y'\phi)' \tag{6.62}$$

and

$$\tilde{p}_z = \int q_{ty}\, ds$$

$$= \int q_t \frac{dy}{ds}\, ds$$

$$= \int t \frac{d}{dx} (\tau\phi)\, dy$$

$$= \frac{d}{dx} \int \left(-\frac{M_z'}{I_z} \frac{Q_z}{t} - \frac{M_y'}{I_y} \frac{Q_y}{t} - \frac{M_\omega'}{I_\omega} \frac{S_\omega}{t} \right) \phi t\, dy$$

$$= (M_z'\phi)' \tag{6.63}$$

where we have used the identities obtained by integration by parts

$$\int Q_y\, dy = 0; \qquad \int Q_y\, dz = -I_y$$

$$\int Q_z\, dy = -I_z; \qquad \int Q_z\, dz = 0 \tag{6.64}$$

and the fact that we have chosen $S_\omega = 0$ by virtue of choosing the sectorial centroid. The final results of Eqs. (6.62) and (6.63) represent the tilde terms in Eqs. (6.59) and (6.60), respectively.

The projected shear shown in Fig. 6.6(b) will also cause a twisting moment m_1 about the shear center located at a_y, a_z. The moment is given as

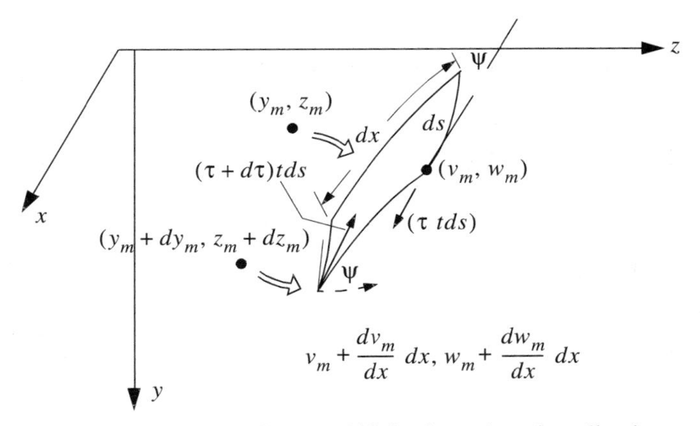

Figure 6.7 Torsion from a shift in the point of application.

$$dm_1 = (|q_{ty}|dx\,dx)(z - a_z) - (|q_{tz}|dx\,ds)(y - a_y)$$

$$m_1 = \int [|q_{ty}|(z - a_z) - |q_{tz}|(y - a_y)]\,ds$$

$$= \int \left[t\,\frac{d}{dx}\,(\tau\phi)\cos\psi(z - a_z) - t\,\frac{d}{dx}\,(\tau\phi)\sin\psi(y - a_y)\right] ds \qquad (6.65)$$

We now consider the effect of a change in the point of application of the shear force on the curved faces of the section and see that this also contributes to a torsion term m_2. Consider Fig. 6.7, from which we can write

$$dm_2 = \left[(\tau + d\tau)\cos\psi\left(v_M + \frac{dv_M}{dx}\,dx\right) - \tau v_M\right] t\,ds$$

$$+ \left[(\tau + d\tau)\sin\psi\left(w_M + \frac{dw_M}{dx}\,dx\right) - \tau w_M\right] t\,ds$$

If we expand and neglect higher-order terms and divide by dx, we get

$$m_2 = \frac{d}{dx}\,p\int [(\tau v_M)\cos\psi + (\tau w_M)\sin\psi]t\,ds$$

$$= \frac{d}{dx}\int \{\tau\cos\psi[v - (z - a_z)\phi] + \tau\sin\psi[w - (y - a_y)\phi]\}t\,ds$$

$$
\begin{aligned}
&= \frac{d}{dx}\left(v \int \tau \cos \psi t \, ds + w \int \tau \sin \psi t \, ds\right) \\
&\quad + \int \left[-t \frac{d}{dx}(\tau\phi) \cos \psi (z - a_z) + t \frac{d}{dx}(\tau\phi) \sin \psi (y - a_y)\right] ds \\
&= \frac{d}{dx}\left(v \int \tau \frac{dz}{ds} t \, ds - w \int \tau \frac{dy}{ds} t \, ds\right) \\
&\quad + \int \left[-t \frac{d}{dx}(\tau\phi) \cos \psi (z - a_z) + t \frac{d}{dx}(\tau\phi) \sin \psi (y - a_y)\right] ds \\
&= (vM'_y)' - (wM'_z)' \\
&\quad + \int \left[-t \frac{d}{dx}(\tau\phi) \cos \psi (z - a_z) + t \frac{d}{dx}(\tau\phi) \sin \psi (y - a_y)\right] ds
\end{aligned} \tag{6.66}
$$

where we have again used Eq. (3.217) and the identities given by the expressions in Eq. (6.64).

Next, we consider the contribution to torsion of the displaced transverse loads p_y and p_z. The applied transverse loads distribute themselves over the surface of a differential element, as shown in Fig. 6.8. When the point of application of the elemental strip moves, the load distribution causes an additional torsion to occur of the form

$$dm_3 = (q_z v_M - q_y w_M) \, dx \, ds$$

or

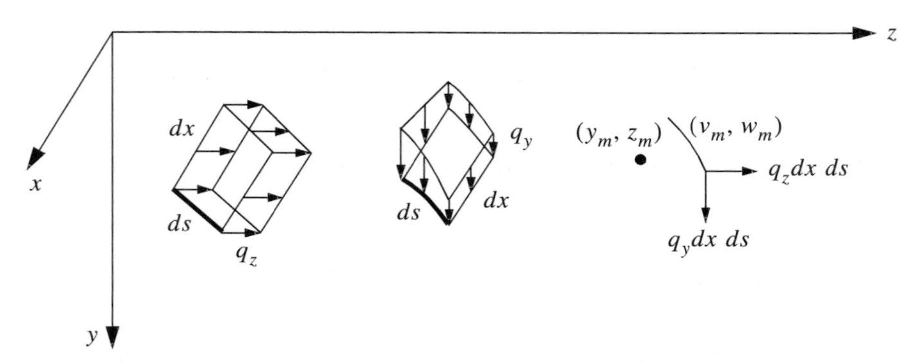

Figure 6.8 Transverse load distribution.

$$
\begin{aligned}
m_3 &= \int (q_z v_M - q_y w_M)\, ds \\
&= \int \{q_z[v - (z - a_z)\phi] - q_y[w + (y - a_y)\phi]\}\, ds \\
&= v \int q_z\, ds - w \int q_y\, ds \\
&\quad - \phi \int q_z(z - a_z)\, ds - \phi \int q_y(y - a_y)\, ds
\end{aligned}
$$

The transverse load per strip dx is given as

$$
p_y = \int q_y\, ds; \qquad p_z = \int q_z\, ds
$$

If we use the equilibrium relationships for an elementary transverse strip,

$$
\begin{aligned}
EI_z v'' &= -M_z'; \qquad & EI_y w'' &= -M_y \\
EI_z v^{iv} &= p_y; \qquad & EI_y w^{iv} &= p_z
\end{aligned}
$$

we get

$$
-wp_y = wM_z''; \qquad vp_z = -vM_y''
$$

If e_y and e_z are the actual points of application of the transverse loads, we can via the first moment of a distribution concept (e.g., the centroid) define

$$
\int yq_y\, ds = e_y p_y; \qquad \int zq_z\, ds = e_z p_z
$$

Then, m_3 becomes

$$
m_3 = -vM_y'' + wM_z'' - \phi[p_y(e_y - a_y) + p_z(e_z - a_z)] \tag{6.67}
$$

We may now define

$$
\tilde{m}_x = m_1 + m_2 + m_3
$$

and note that m_1 cancels the last term of m_2. Then, we get

$$
\begin{aligned}
\tilde{m}_x &= v'M_y' - w'M_z' \\
&\quad - \phi[p_y(e_y - a_y) + p_z(e_z - a_z)]
\end{aligned} \tag{6.68}
$$

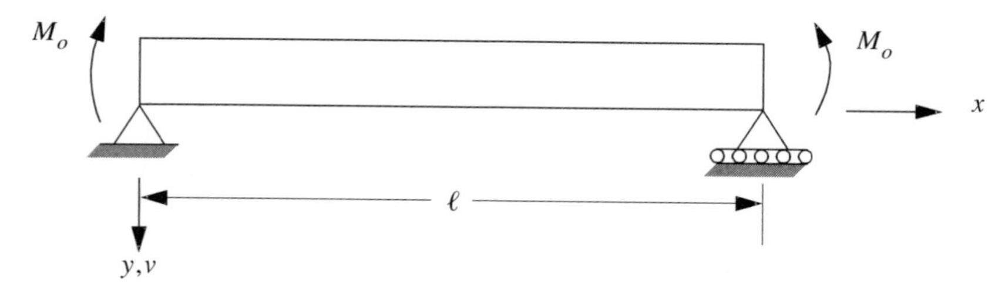

Figure 6.9 Thin walled beam under an applied end moment.

which represents the tilde term of Eq. (6.61).

The differential equations for the beam, on bringing the tilde terms to the left-hand side, now become

$$
\begin{aligned}
EI_z v^{iv} &- [N(v' + a_z\phi')]' + (M_y\phi)'' = p_y \\
EI_y w^{iv} &- [N(w' - a_y\phi')]' - (M_z\phi)'' = p_z \\
EI_\omega \phi^{iv} &- JG\phi'' - [N(a_z v' - a_y w')]' \\
&- [(\beta_N N + \beta_z M_z + \beta_y M_y)\phi']' \\
&+ \phi[p_y(e_y - a_y) + p_z(e_z - a_z)] \\
&\qquad + M_y v'' - M_z w'' = m_x
\end{aligned}
\tag{6.69}
$$

Example 6.1 To see the meaning of these equations, consider a doubly symmetric "simply" supported beam under applied end moments M_0 in the x–y plane as shown in Fig. 6.9. By simply supported, we mean

$$
\begin{aligned}
v(0) &= 0; & v(\ell) &= 0 \\
v''(0) &= -\frac{M_0}{EI_z}; & v''(\ell) &= -\frac{M_0}{EI_z} \\
w(0) &= 0; & w(\ell) &= 0 \\
w''(0) &= 0; & w''(\ell) &= 0 \\
\phi(0) &= 0; & \phi(\ell) &= 0 \\
\phi''(0) &= 0; & \phi''(\ell) &= 0
\end{aligned}
$$

The relationship on ϕ'' implies that even though the supported ends are not allowed to twist ($\phi = 0$ at ends), they are free to warp and hence do not develop a bimoment M_ω. Since the section is doubly symmetric $\beta_z = 0$, the differential equations become

$$
\begin{aligned}
EI_z v^{iv} &= 0 \\
EI_y w^{iv} - M_0\phi'' &= 0 \\
EI_\omega \phi^{iv} - JG\phi'' - M_0 w'' &= 0
\end{aligned}
$$

The first differential equation yields

$$v = \frac{M_0 \ell^2}{2EI_z} \left[\frac{x}{\ell} - \left(\frac{x}{\ell} \right)^2 \right]$$

which is the standard small-deflection theory for a simply supported beam under equal end moments.

Having two integrations on the second differential equation yields

$$w'' = \frac{M_0}{EI_y} \phi$$

Substituting this result into the last differential equation yields

$$\phi^{iv} - k_1 \phi'' - k_2 \phi = 0$$

where

$$k_1 = \frac{JG}{EI_\omega}; \qquad k_2 = \frac{M_0^2}{E^2 I_\omega I_y}$$

If we assume a solution of the form $e^{\lambda x}$, we get the expression

$$\lambda^4 - k_1 \lambda^2 - k_2 = 0$$

with roots

$$\lambda_{I,II}^2 = \frac{k_1 \pm \sqrt{k_1^2 + 4k_2}}{2}$$

or

$$\lambda_I^2 = \frac{k_1 + \sqrt{k_1^2 + 4k_2}}{2}$$

and

$$\lambda_{II}^2 = -\frac{-k_1 + \sqrt{k_1^2 + 4k_2}}{2}$$

Thus we have the four roots

$$\lambda_{1,2} = \pm\sqrt{\frac{k_1 + \sqrt{k_1^2 + 4k_2}}{2}} = \pm\alpha_1$$

and

$$\lambda_{3,4} = \pm i\sqrt{\frac{-k_1 + \sqrt{k_1^2 + 4k_2}}{2}} = \pm i\alpha_2$$

Then, the solution for ϕ becomes

$$\phi = A\cosh\,\alpha_1 x + B\sinh\,\alpha_1 x + C\sin\,\alpha_2 x + D\cos\,\alpha_2 x$$

The solution of this equation yields a set of homogeneous algebraic equations with the following determinant:

$$\begin{vmatrix} 1 & 0 & 0 & 1 \\ \alpha_1^2 & 0 & 0 & -\alpha_2^2 \\ \cosh\,\alpha_1\ell & \sinh\,\alpha_1\ell & \sin\,\alpha_2\ell & \cos\,\alpha_2\ell \\ \alpha_1^2\cosh\,\alpha_1\ell & \alpha_1^2\sinh\,\alpha_1\ell & -\alpha_2^2\sin\,\alpha_2\ell & -\alpha_2^2\cos\,\alpha_2\ell \end{vmatrix}$$

If the above is expanded by elements of the first row, we get

$$(\alpha_1^2 + \alpha_2^2)^2\sinh\,\alpha_1\ell\,\sin\,\alpha_2\ell = 0$$

Since α_1 and α_2 are positive real numbers, the above can hold only for

$$\sin\,\alpha_2\ell = 0$$

or

$$\alpha_2 = \frac{n\pi}{\ell} = \sqrt{\frac{-k_1 + \sqrt{k_1^2 + 4k_2}}{2}}$$

Solving for k_2, we get

$$k_2 = \frac{n^2\pi^2}{\ell^2}\left(\frac{n^2\pi^2}{\ell^2} + k_1\right)$$

or replacing k_1 and k_2 with their actual values, we get for the critical load

$$(M_0)_{\text{cr}} = \frac{n\pi}{\ell}\sqrt{JGEI_y\left(\frac{n^2\pi^2}{\ell^2}\frac{EI_\omega}{JG}+1\right)}$$

On substituting $\sin\alpha_2\ell = 0$ into the algebraic equation, we get

$$A = B = D = 0$$

with C undetermined. Thus we get the mode shapes

$$\phi = C\sin\frac{n\pi x}{\ell}$$

and

$$w = -C\frac{M_0\ell^2}{n^2\pi^2 EI_y}\sin\frac{n\pi x}{\ell}$$

Thus the behavior of the beam is such that up to the critical load, we have simple bending in the x–y plane; after a critical value for M_0 is reached, we have in addition a twisting of the beam accompanying a lateral w displacement.

6.5 SOME STABILITY CONCEPTS

The above examples have shown two approaches for stability analysis: namely, the equilibrium approach and the virtual work approach. Two other approaches we will explore shortly are the energy and dynamic approaches. When the loads are velocity-independent and can be derived from a potential function, these methods are equivalent.

Obviously, if loading is a function of time, then the dynamic approach is needed. If, however, the loading is stationary and velocity-independent but nonconservative, the dynamic approach is the only consistently valid approach. If the loading is velocity-independent and conservative, the energy approach is equally valid and is usually easier to use. The equilibrium approach in this case is valid for bifurcation-type stability but is usually not useful for snapthrough-type instability. Virtual work is valid for all types of instability because, as we shall shortly see, it can he extended to dynamics.

The type of loading on a structure largely determines the stability behavior. Reference [56] classifies loads into two prime categories and four subcategories, depending on the whether the loads are explicit or implicit functions of time t and whether they are functions of velocity $\dot{u}_i$ as well as functions of displacement u_i. Thus we have the following:

1. Instationary $P(u_i, \dot{u}_i, t)$—for example, $P = P_0\cos\omega t$
2. Stationary $P(u_i, \dot{u}_i)$
   a. Velocity-dependent $P(u_i, \dot{u}_i)$

(1) Gyroscopic: $W = 0$ (Coriolis forces)

(2) Dissipative: $\bar{d}W < 0$ (damping forces)

b. Velocity-independent $P(u_i)$

(1) Circulatory: $U_e \neq -W_e$ (follower forces)

(2) Noncirculatory: $U_e = -W_e$ (gravity)

Gyroscopic and noncirculatory loads represent conservative systems, whereas instationary, dissipative, and circulatory loads represent nonconservative systems.

The three examples we have thus far examined represent noncirculatory systems. In the next section, we introduce some energy criteria concepts and discuss how they relate to stability, as well as use the criteria to examine the concept of equilibrium paths more closely for noncirculatory systems and see how the structural stiffness concepts enter into the stability concept. We then generalize virtual work to include dynamic effects and study both circulatory and instationary systems.

6.6 ENERGY CRITERIA OF STABILITY

Consider a structure with the n generalized coordinates

$$\Delta_1, \Delta_2, \ldots, \Delta_n$$

These coordinates represent the equilibrium configuration of the structure. Now, let this equilibrium configuration be given a small virtual displacement without violating the kinematic boundary conditions. The structure will move into a new configuration. The total potential energy of the original system is

$$\Pi_{(\text{actual})} = \Pi(\Delta_1, \Delta_2, \ldots, \Delta_n)$$

and that of the new configuration is

$$\Pi_{(\text{admissible})} = \Pi(\Delta_1 + \delta\Delta_1, \Delta_2 + \delta\Delta_2, \ldots, \Delta_n + \delta\Delta_n)$$

The change in potential energy is given as

$$\Delta\Pi = \Pi_{(\text{admissible})} - \Pi_{(\text{actual})}$$

Let

$$\Pi_i = \frac{\partial \Pi}{\partial \Delta_i}$$

and

$$\Pi_{ij} = \frac{\partial^2 \Pi}{\partial \Delta_i \partial \Delta_j}$$

Then, we can write the Taylor's series

$$\Delta\Pi = \sum_{i=1}^{n} \Pi_i \, \delta\Delta_i + \frac{1}{2!} \sum_{i=1}^{n} \sum_{j=1}^{n} \Pi_{ij} \, \delta\Delta_i \, \delta\Delta_j + (\text{higher-order terms}) \tag{6.70}$$

This can be written as

$$\Delta\Pi = \delta\Pi + \delta^2\Pi + \cdots$$

Typical terms of $\delta\Pi$ are

$$\delta\Pi = \frac{\partial \Pi}{\partial \Delta_1} \, \delta\Delta_1 + \cdots + \frac{\partial \Pi}{\partial \Delta_i} \, \delta\Delta_i + \cdots + \frac{\partial \Pi}{\partial \Delta_n} \, \delta\Delta_n$$

Typical terms of $\delta^2\Pi$ are

$$\delta^2\Pi = \frac{1}{2} \left(\frac{\partial^2 \Pi}{\partial \Delta_1^2} \, \delta\Delta_1^2 + \cdots + \frac{\partial^2 \Pi}{\partial \Delta_i \partial \Delta_j} \, \delta\Delta_i \, \delta\Delta_j + \cdots + \frac{\partial^2 \Pi}{\partial \Delta_j \partial \Delta_i} \, \delta\Delta_j \, \delta\Delta_i \right.$$
$$\left. + \cdots + \frac{\partial^2 \Pi}{\partial \Delta_n^2} \, \delta\Delta_n^2 \right)$$

This expression is quadratic in the variables $\delta\Delta_i$ and represents a quadratic form.

The principle of stationary potential energy requires that at equilibrium $\delta\Pi = 0$, the sign of $\Delta\Pi$ be controlled by $\delta^2\Pi$. Consider a two-dimensional case where $\Pi(\Delta_1, \Delta_2)$ is a function of two variables. We then define the five following quadratic forms:

- $\delta^2\Pi > 0$: Positive definite $\Rightarrow \delta\Delta_1^2 + \delta\Delta_2^2$. This is required for stable equilibrium; Π is a minimum, and the frequency of vibration is real.
- $\delta^2\Pi \geq 0$: Positive semidefinite $\Rightarrow (\delta\Delta_1 + \delta\Delta_2)^2$. Higher-order terms are needed to evaluate stability.
- $\delta^2\Pi < 0$: Negative definite $\Rightarrow -(\delta\Delta_1^2 + \delta\Delta_2^2)$. The equilibrium is unstable; Π is a maximum, and the frequency of vibration is imaginary.
- $\delta^2\Pi \leq 0$: Negative semidefinite $\Rightarrow -(\delta\Delta_1 + \delta\Delta_2)^2$.
- $-\infty < \delta^2\Pi < \infty$: Indefinite $\Rightarrow \delta\Delta_1\delta\Delta_2$.

If the structure depends upon only a single coordinate Δ, we have on dropping the numerical coefficient

$$\delta^2\Pi = \frac{d^2\Pi}{d\Delta^2}\,\delta\Delta^2$$

Higher-even-order variations may be written as

$$\delta^{2n}\Pi = \frac{d^{2n}\Pi}{d\Delta^{2n}}\,\delta\Delta^{2n}$$

and higher-odd-order variations become

$$\delta^{2n-1}\Pi = \frac{d^{2n-1}\Pi}{d\Delta^{2n-1}}\,\delta\Delta^{2n-1}$$

with the following requirement for equilibrium:

$$\frac{d\Pi}{d\Delta} = 0$$

Then, we conclude that if the first nonvanishing derivative is of an even order, Π is a minimum and the equilibrium is stable if its value is positive. If its value is negative, the equilibrium is unstable, and if the first nonvanishing derivative is of an odd order, the equilibrium is unstable (a horizontal inflection point).

If a structure is stable, $\delta^2\Pi > 0$. Then,

$$\Pi_{(\text{admissible})} = \Pi_{(\text{actual})} + \delta^2\Pi$$

or

$$\Pi_{(\text{admissible})} > \Pi_{(\text{actual})}$$

which is the principle of minimum total potential energy stating,

Among all admissible displacement functions, the actual displacements make the total potential energy an absolute minimum.

The concept of the section variation $\delta^2\Pi$ has important theoretical applications. Nevertheless, in a practical sense it would be too expensive to compute for many modern structural models that involve ten to hundreds of thousands of degrees of freedom. In these cases, structural instability is usually signaled by sudden abrupt changes in deflection or negative terms that occur during the decomposition of the structural stiffness matrix.

6.7 STIFFNESS

The concepts of stability cannot be separated from the concept of stiffness. On loading, some structures, such as automobile tires during inflation, become stiffer; other structures, such as long, straight columns, exhibit buckling and lose stiffness, but on further loading again increase their stiffness. Structures such as rigid, jointed frames exhibit buckling and stiffness loss that continues to the point of collapse. Shallow shells under pressure loading lose stiffness and then suddenly undergo dynamic snap to new equilibrium positions where, under continued loading, their stiffness can increase.

Consider a simply supported beam of length ℓ and flexure rigidity EI with a concentrated load P at midspan. Its midspan deflection is

$$v(\ell/2) = \frac{P\ell^3}{48EI}$$

or

$$\frac{48EI}{\ell^3}\, v(\ell/2) = P$$

and we can define the scalar stiffness as

$$k_e = \frac{48EI}{\ell^3}$$

This idealization of the stiffness into a scalar stiffness then allows the beam structure to be treated as a simple spring-mass system. This can be a convenient representation of an actual structure from which important dynamic characteristics can be obtained.

The stiffness k_e is associated with the rigidity EI of the material. Stiffness can arise, however, from the geometry of deformation, which is different from material stiffness.

Consider Fig. 6.10. It represents a perfectly flexible cello string of length $2L$ stretched under tension T between two fixed points. If the string is then "plucked" at its centerpoint by a force P so that the resulting displacement is small, the force N in the string in the deformed position will still be T. Then, equilibrium in the slightly deformed position is represented by

$$P = 2T \sin\phi = 2T\,\frac{v}{L} = \frac{2T}{L}\,v$$

where we have assumed that $\tan\phi = \sin\phi$ and v is the deflection at the center of the string. Therefore, we can define a stiffness k_T as

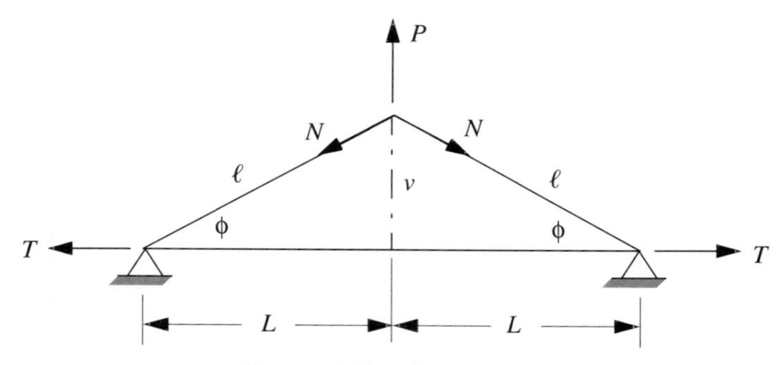

Figure 6.10 Cello string.

$$k_T = \frac{2T}{L}$$

Thus the state of initial stress (here characterized by T) induces mechanical properties that depend upon the magnitude of the stress and are quite distinct from those associated with the rigidity of the material.

Next, consider the cello string under a large elastic deformation. The following geometry holds true:

$$\tan\phi = \frac{v}{L}; \qquad \cos\phi = \frac{L}{\ell}; \qquad \sin\phi = \frac{v}{\ell}$$

where ℓ is the deformed length of the string. Then, equilibrium is given by the relationship

$$P = 2N \sin\phi$$

where the force N in the string is related to the current level of strain.

To avoid complications with incremental stress, which must be rotated during deformation, we consider the pretensioning of the string to be caused by an initial strain $\epsilon_0 = T/AE$. In a numerical code, such a strain could be induced by thermal strain that would follow automatically the rotation of the string. We can write the stress as

$$\sigma = E(\epsilon - \epsilon_0)$$

and N, the total instantaneous force in the sting, is

$$N = A\sigma = AE(\epsilon - \epsilon_0)$$

We now need to choose a definition for ϵ. The geometry of this problem is relatively simple and allows us to get closed-formed expressions for the strain measure. Therefore, we digress a bit and examine several alternative definitions of strain to which we have alluded throughout the text.

The "engineering" Lagrangian strain can be written as

$$\epsilon_l = \frac{\ell}{L} - 1$$
$$= \frac{1 - \cos\phi}{\cos\phi}$$
$$= \sqrt{1 + \left(\frac{v}{L}\right)^2} - 1$$

with a virtual strain of

$$\overline{\epsilon}_l = \frac{\sin\phi}{L}\,\overline{v}$$

The "engineering" Eulerian strain can be written as

$$\epsilon_e = 1 - \frac{L}{\ell}$$
$$= 1 - \cos\phi$$
$$= 1 - \sqrt{1 - \left(\frac{v}{\ell}\right)^2}$$

with a virtual strain of

$$\overline{\epsilon}_e = \frac{\sin\phi\cos^2\phi}{L}\,\overline{v}$$

The logarithmic strain can be written as

$$\epsilon_{\ln} = \ln\frac{\ell}{L}$$

with a virtual strain of

$$\overline{\epsilon}_{\ln} = \frac{\sin\phi}{\ell}\,\overline{v} = \frac{\sin\phi\cos\phi}{L}\,\overline{v}$$

The Green strain would be

$$\epsilon_g = \epsilon_l \left(1 + \frac{1}{2} \epsilon_l \right) = \frac{\ell^2 - L^2}{2L^2}; \qquad \overline{\epsilon}_g = \frac{\tan \phi}{L} \overline{v}$$

and the Almansi strain would be

$$\epsilon_a = \frac{\ell^2 - L^2}{2\ell^2}; \qquad \overline{\epsilon}_a = \frac{\sin \phi \cos^3 \phi}{L} \overline{v}$$

In general-purpose finite element codes, the form chosen depends on the application. For hyperelastic (rubber) applications, the Lagrangian form written in terms of the strain energy as a function of strain invariants is used. For large-displacement theory, an updated Lagrangian form is often used. For large plastic strain with flow, a logarithmic strain is used. For fluid-and-crash dynamics, Eulerian forms are used. For the problem at hand, to remain consistent we will use the Lagrangian form.

For the string in Fig. 6.10, we can define an instantaneous or tangent stiffness as

$$\begin{aligned} K_t &= \frac{dP}{dv} \\ &= \frac{dP}{d\phi} \frac{d\phi}{dv} \\ &= \frac{\cos^2 \phi}{L} \frac{dP}{d\phi} \\ &= \frac{\cos^2 \phi}{L} \left(2 \frac{dN}{d\phi} \sin \phi + 2N \cos \phi \right) \end{aligned}$$

and using the Lagrangian strain

$$\epsilon = \frac{1 - \cos \phi}{\cos \phi}$$

we have

$$\frac{dN}{d\phi} = \frac{AE \sin \phi}{\cos^2 \phi}$$

and the instantaneous stiffness becomes

$$K_t = \frac{\cos^2 \phi}{L} \left(2AE \frac{\sin \phi}{\cos^2 \phi} \sin \phi + 2N \cos \phi \right)$$

$$= \frac{2AE}{L} \sin^2 \phi + \frac{2N}{\ell} \cos^2 \phi$$

$$= K_m + K_d$$

where K_m is the conventional material stiffness matrix or, in our current case, the elastic stiffness, and K_d is the differential (or incremental or geometric) stiffness matrix. The $K_m\overline{v}$ is composed of the virtual elongation $\overline{v} \sin \phi$ ($\overline{u}$ zero by symmetry), as well as the virtual internal force $(AE/L)\overline{v} \sin \phi$ and its vertical projection $(AE/L)\overline{v} \sin \phi \sin \phi$. The $K_d\overline{v}$ is composed of the $\overline{v} \cos \phi$ displacement normal to the current position of the string, as well as the virtual force $(2N/\ell)\overline{v} \cos \phi$ and its vertical projection $(2N/\ell)\overline{v} \cos \phi \cos \phi$. Note that in the initial straight prestrained state, the string has zero K_t stiffness and is thus inherently unstable in this position.

Observe that K_t also arises directly from the equilibrium equation by taking the variation in equilibrium and considering that N is a function of the current value of strain. Hence:

$$\overline{P} = 2\overline{N} \sin \phi + 2N\overline{\phi} \cos \phi$$

$$= \left(\frac{2AE}{L} \sin^2 \phi + \frac{2N}{\ell} \cos^2 \phi \right) \overline{v}$$

The foregoing equation is nonlinear on the right-hand side in both the geometry, characterized by the angle ϕ, and the internal load N. It is usually solved by the Newton's method. Here, an incremental value of load is assumed and the geometry and internal loads are iterated until equilibrium is reached with the applied increment $\overline{P}$, at which time an additional load increment is applied and the procedure repeated.

Often, an attempt is made to expedite the iteration process by finding an incremental secant stiffness K_s rather than the tangent stiffness K_t. Here again the concept of scalar stiffness is invaluable.

Figure 6.11 represents a structure where the applied load is denoted by P and the generalized internal element loads are denoted by F. At any time in the iteration process, there may be an imbalance between the applied load and the internal load measured by $R = P - F$. Rather than continuously computing the tangent stiffness K_t, it is desired to compute a secant stiffness K_s. From the figure we have

$$K_s = \frac{R_{i-1} - R_i}{\Delta u}$$

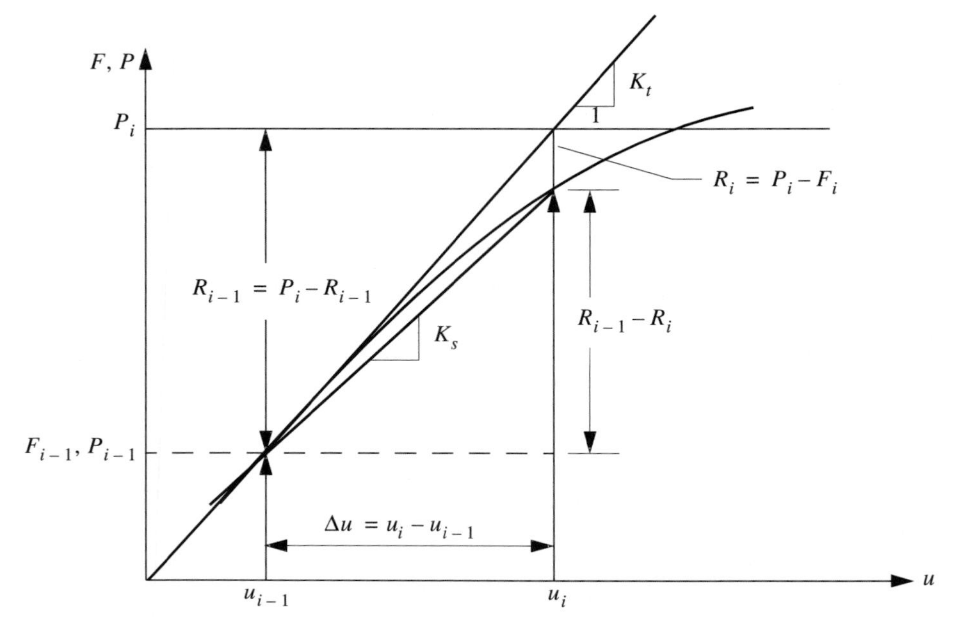

Figure 6.11 Secant stiffness.

where the subscript $i - 1$ represents the previous iteration and i the current iteration, and where $\Delta u = u_i - u_{i-1}$. The expression for K_s can be rearranged as

$$K_s = \frac{R_{i-1}}{\Delta u} - \frac{R_i}{\Delta u}$$
$$= K_t - K_{\text{ubf}}$$

where $K_{\text{ubf}} = R_i/\Delta u$ is a spring in the direction of the unbalanced force R_i. This concept can be generalized to an n-dimensional case as follows: We wish to determine the secant stiffness

$$[K_s]\{\Delta u\} = \{R_{i-1} - R_i\}$$

Form the matrix

$$[K_t + a\{w\}\lfloor w \rfloor]\{\Delta u\} = \{R_{i-1} - R_i\}$$

where

$$\{w\}\lfloor w \rfloor$$

is called a rank 1 matrix. It has the property that if it were to diagonalize, only one term would be nonzero. Since $[K_t]\{\Delta u\} = \{R_{i-1}\}$, we have

$$[a\{w\}\lfloor w \rfloor]\{\Delta u\} = \{-R_i\}$$

Let

$$\{w\} = \{-R_i\}$$

Then

$$a\{w\}\lfloor w \rfloor\{\Delta u\} = \{-R_i\}a\lfloor w \rfloor\{\Delta u\} = \{-R_i\}$$

only if

$$a\lfloor w \rfloor\{\Delta u\} = 1$$

or

$$a = \frac{-1}{\lfloor R_i \rfloor\{\Delta u\}}$$

Then

$$[K_s] = [K_t] - \frac{\{R_i\}\lfloor R_i \rfloor}{\lfloor R_i \rfloor\{\Delta u\}}$$

and

$$[K_{\text{ubf}}] = \frac{\{R_i\}\lfloor R_i \rfloor}{\lfloor R_i \rfloor\{\Delta u\}}$$

is a generalized spring in the direction of the unbalanced force.

This technique can be further generalized. For example, form

$$[K_t + a\{v\}\lfloor v \rfloor + b\{w\}\lfloor w \rfloor]\{\Delta u\} = \{R_{i-1} - R_i\}$$

This is called a rank 2 update. If we assume

$$\{v\} = \{R_{i-1} - R_i\}$$

and

$$\{w\} = [K_t]\{\Delta u\}$$

we can determine a and b and arrive at the two generalized springs

$$[K_{\text{ubf}1}] = \frac{\{R_{i-1} - R_i\}\lfloor R_{i-1} - R_i \rfloor}{\lfloor R_{i-1} - R_i \rfloor\{\Delta u\}}$$

which represents a spring in the direction of the difference in unbalanced force between two iterations, and

$$[K_{\text{ubf}2}] = \frac{\{R_{i-1}\}\lfloor R_{i-1} \rfloor}{\lfloor R_{i-1} \rfloor\{\Delta u\}}$$

which represents a spring in the direction of the unbalanced force from the previous iteration. These two springs define the Broyden-Fletcher-Goldfarb-Shanno (BFGS) method. The secant stiffness then becomes

$$[K_s] = [K_t + K_{\text{ubf}1} - K_{\text{ubf}2}]$$

Notice that some springs are applied so that they increase the structural flexibility in their direction.

The equivalent scalar spring and related displacement can be found. For example, for the $K_{\text{ubf}2}$ term, we can determine its equivalent scalar value as

$$k_{\text{ubf}2} = \frac{\lfloor R_{i-1} \rfloor\{R_{i-1}\}}{\lfloor R_{i-1} \rfloor\{\Delta u\}}$$

with the corresponding displacement

$$u_{\text{ubf}2} = -\frac{R_{i-1}}{\sqrt{\lfloor R_{i-1} \rfloor\{R_{i-1}\}}}$$

6.8 STIFFENING AND UNSTIFFENING MODELS

To gain a better understanding for the effect of stiffness on a model, we now investigate several rigid-link models. These types of models are useful because they are simple yet ideally suited for study using exact- and intermediate-solution techniques, as demonstrated in great detail in refs. [56] and [57].

Figure 6.12 represents two rigid links of equal length ℓ. The support at A is pinned. A load P is applied at the pin connection between the two links, and at the connection is a rotational spring C. The support at B can move horizontally. A translational spring K at B represents the support stiffness. The system is pretensioned by a horizontal load $T \geq 0$. The links are initially horizontal. The deformed geometry yields

$$\Delta_P = \ell \sin\theta \quad \text{and} \quad \Delta_B = 2\ell(1 - \cos\theta)$$

The total potential energy is

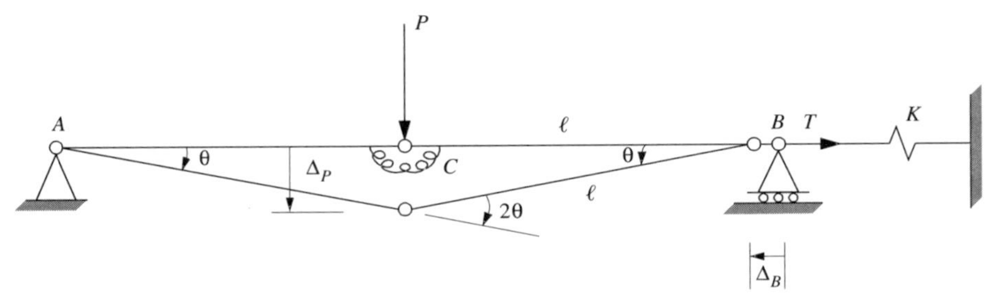

Figure 6.12 A stiffening structure.

$$\Pi = \tfrac{1}{2}C(2\theta)^2 + \tfrac{1}{2}K[2\ell(1 - \cos\theta)]^2 + T[2\ell(1 - \cos\theta)] - P\ell\sin\theta$$

Using θ as the independent variable, the first derivative of potential becomes

$$\frac{d\Pi}{d\theta} = 4C[\theta + k\sin\theta(1 - \cos\theta) + t\sin\theta - p\cos\theta] \tag{6.71}$$

where we have factored $4C$ out and defined the nondimensional quantities

$$p = \frac{P\ell}{4C}; \qquad k = \frac{K\ell^2}{C}; \qquad t = \frac{T\ell}{2C}$$

Equilibrium requires that $d\Pi/d\theta = 0$; hence, we can solve for the nondimensional load p to get

$$p = \frac{1}{\cos\theta}[\theta + k\sin\theta(1 - \cos\theta) + t\sin\theta] \tag{6.72}$$

Let us now define a nondimensional deflection in the direction of the load as

$$v = \frac{\Delta_P}{\ell}$$

Now, we can define a scalar stiffness S as

$$S = \frac{dp}{dv} = \frac{dp}{d\theta}\frac{d\theta}{dv} = \frac{dp/d\theta}{dv/d\theta}$$

or

$$S = \frac{1}{\cos^3\theta}[\cos\theta + \theta\sin\theta + k(1 - \cos^3\theta) + t] \tag{6.73}$$

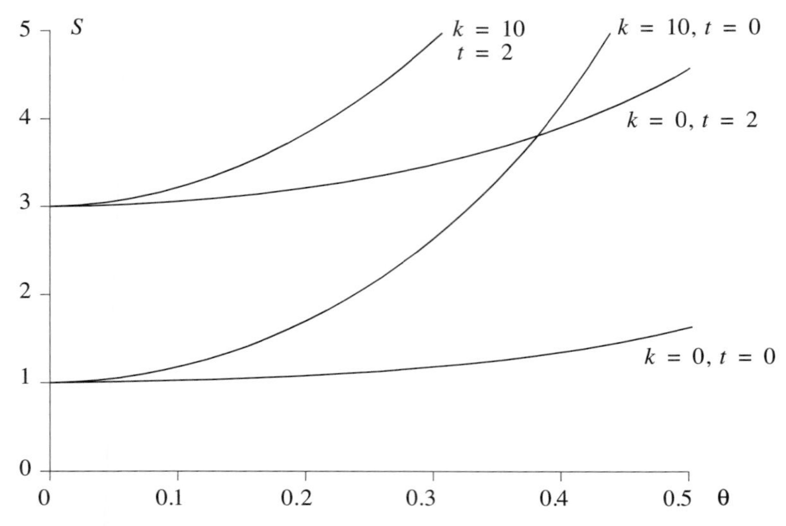

Figure 6.13 Stiffness-deflection curve for various prestress values.

Figure 6.13 shows the results of Eq. (6.73) for various values of t and k. We see that with both the support stiffness $k = 0$ and the prestress $t = 0$, the structure has a slight increase in stiffness with load. A finite value of support stiffness $k = 10$ and $t = 0$ causes the structure to have a large increase of stiffness with load but no change in the initial stiffness. The prestress $t = 2$, on the other hand, has a large effect on the initial stiffness.

To check for the stability of the solution, we look at $d^2\Pi/d\theta^2$. Taking the derivative of Eq. (6.71), we get

$$\frac{d^2\Pi}{d\theta^2} = 4C[1 + k(\cos\theta - \cos 2\theta) + t\cos\theta + p\sin\theta]$$

This equation must now be evaluated on the equilibrium path as determined by Eq. (6.72). Substituting for p, we get

$$\begin{aligned}\frac{d^2\Pi}{d\theta^2} &= 4C\{1 + k(\cos\theta - \cos 2\theta) + t\cos\theta \\ &\quad + \tan\theta[\theta + k(1 - \cos\theta) + t\sin\theta]\}\end{aligned}$$

For $0 \le \theta < \pi/2$ and $t \ge 0$, $d^2\Pi/d\theta^2 > 0$, and the solution is always stable.

The next example shows unstiffening behavior and demonstrates that a structure is sensitive to whether a dead load or an enforced motion is used to initiate deformation.

Consider again the two rigid links of equal length ℓ of Fig. 6.12, now shown in Fig. 6.14 with an initial position characterized by the angle α. The support

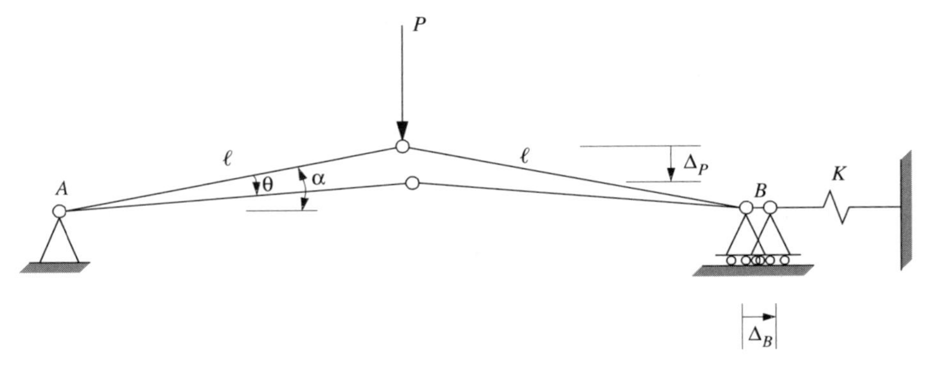

Figure 6.14 An unstiffening structure.

at A is pinned. A load P is applied at the pin connection between the two links. The support at B can move horizontally. A translational spring K at B represents the support stiffness.

The deformed geometry yields

$$\Delta_P = \ell[\sin\alpha - \sin(\alpha - \theta)] \quad \text{and} \quad \Delta_B = 2\ell[\cos(\alpha - \theta) - \cos\alpha]$$

The total potential energy is

$$\Pi = \tfrac{1}{2}K\{2\ell[\cos(\alpha - \theta) - \cos\alpha]\}^2 - P\ell[\sin\alpha - \sin(\alpha - \theta)]$$

Using θ as the independent variable, the first derivative of potential becomes

$$\frac{d\Pi}{d\theta} = 4K\ell^2\{\sin(\alpha - \theta)[\cos(\alpha - \theta) - \cos\alpha] - p\cos(\alpha - \theta)\} \tag{6.74}$$

where we have factored $4K\ell^2$ out and defined the nondimensional load

$$p = \frac{P}{4K\ell}$$

Equilibrium requires that $d\Pi/d\theta = 0$; hence, we can solve for the nondimensional load p to get

$$p = \tan(\alpha - \theta)[\cos(\alpha - \theta) - \cos\alpha] \tag{6.75}$$

To check for the stability of the solution, we look at $d^2\Pi/d\theta^2$. Taking the derivative of Eq. (6.74), we get

$$\frac{d^2\Pi}{d\theta^2} = 4K\ell^2\{\sin^2(\alpha - \theta) - \cos(\alpha - \theta)[\cos(\alpha - \theta) - \cos\alpha] - p\sin(\alpha - \theta)\}$$

This equation must now be evaluated on the equilibrium path as determined by Eq. (6.75). Substituting for p, we get

$$\frac{d^2\Pi}{d\theta^2} = 4K\ell^2 \frac{\cos\alpha - \cos^3(\alpha - \theta)}{\cos(\alpha - \theta)}$$

Not allowing the structure to fold over itself, $|\alpha - \theta| < \pi/2$. The denominator is, therefore, always positive, and the sign of $\delta^2\Pi$ is determined by the numerator $\cos\alpha - \cos^3(\alpha - \theta)$. Thus we can identify three conditions:

$$\frac{d^2\Pi}{d\theta^2} = \begin{cases} > 0 & \text{stable when } \cos^3(\alpha - \theta) < \cos\alpha \\ = 0 & \text{critical when } \cos^3(\alpha - \theta) = \cos\alpha \\ < 0 & \text{unstable when } \cos^3(\alpha - \theta) > \cos\alpha \end{cases}$$

Let us now define a nondimensional deflection in the direction of the load as

$$v = \frac{\Delta_P}{\ell} = \sin\alpha - \sin(\alpha - \theta)$$

We again define a scalar stiffness S as

$$S = \frac{dp}{dv} = \frac{dp}{d\theta}\frac{d\theta}{dv} = \frac{dp/d\theta}{dv/d\theta}$$

or

$$S = \frac{\cos\alpha - \cos^3(\alpha - \theta)}{\cos^3(\alpha - \theta)} \tag{6.76}$$

Observe that when the numerator of this expression is zero, S is zero. Comparing the numerator for a zero value of S with the critical value of $d^2\Pi/d\theta^2$, we see that the critical equilibrium corresponds to the zero value of stiffness. The value of θ at critical is

$$\theta = \alpha \pm \cos^{-1}(\cos\alpha)^{1/3}$$

A plot of Eq. (6.75) is shown in Fig. 6.15 using a value of $\alpha = \pi/6$. The type of instability shown represents a single equilibrium path with a portion of the path (shown as dotted) as unstable. At point A, the structure experiences a loss of stability and will move (snap) dynamically to point C. If the structure is loaded to E, it will continue to increase in stiffness. If the structure is unloaded at C, the stiffness will decrease until at B it snaps to D.

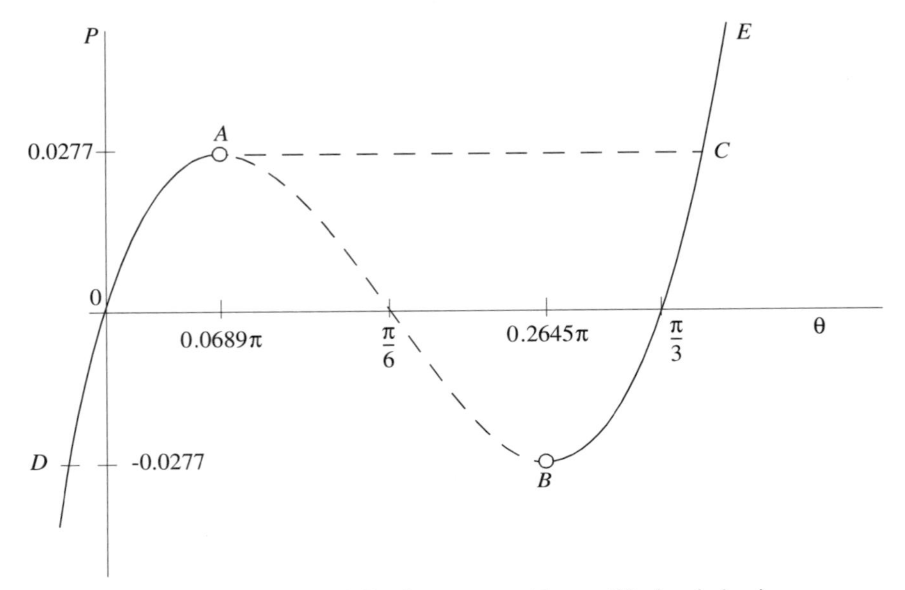

Figure 6.15 Load-deflection curve with unstiffening behavior.

A plot of Eq. (6.76) for the above value of α is shown in Fig. 6.16. We see that at the point A of instability, the stiffness goes negative and remains so until point B is reached on the p–θ curve.

There is an important variation to this problem. As just described, a physical load P is applied at the junction of the two rigid links. If an enforced displacement is applied rather than a physical load, then the behavior will be somewhat different. Consider again the structure shown in Fig. 6.14 but with the load P replaced with a loading jack, as shown in Fig. 6.17.

Figure 6.17 shows the two rigid links of equal length ℓ with an initial position characterized by the angle α. The support at A is pinned. The support at B can move horizontally. A translational spring K_B at B represents the support stiffness. At the pin connection between the two links is a jack J rigidly supported at one end and connected to a spring K_J at the other end, which represents the stiffness of the jack against vertical motion. The lower end of the spring K_J connects to the joining pin at C of the two rigid links.

As the jack screw turns, the end of the jack pressing against the upper end of the spring K_J produces an enforced displacement Δ_J at that location. The lower end of the spring K_J and its attached link connection move through a displacement Δ_C. This relative motion between the ends of the spring K_J produces an internal force

$$P = K_J(\Delta_J - \Delta_C) \tag{6.77}$$

that is used to define the scalar stiffness. We assume that the loading device as

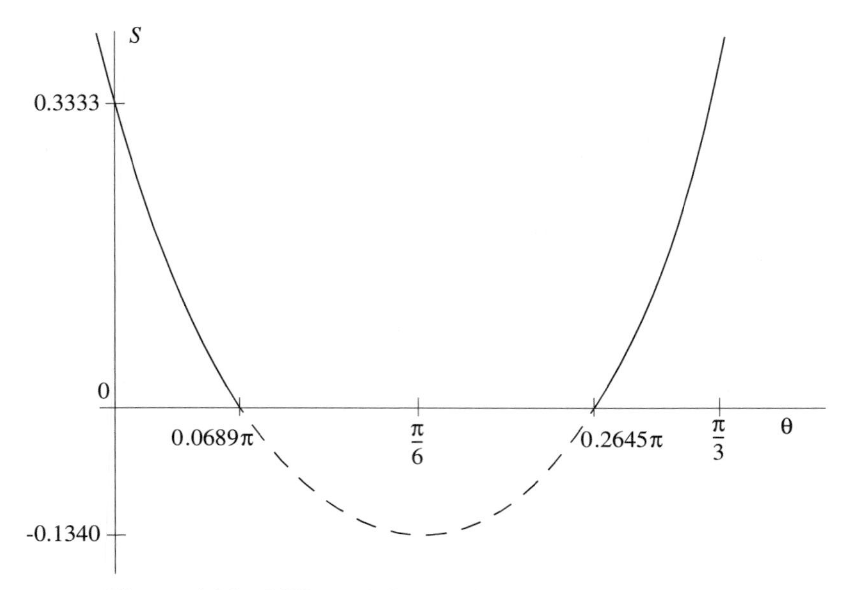

Figure 6.16 Stiffness deflection for an unstiffening model.

represented by the jack does not itself undergo stability problems. In an actual structure, this would be accomplished by pretensioning the loading device. Such a device is built and analyzed in detail in ref. [57].

The deformed geometry yields

$$\Delta C = \ell[\sin\alpha - \sin(\alpha - \theta)] \quad \text{and} \quad \Delta_B = 2\ell[\cos(\alpha - \theta) - \cos\alpha]$$

The total potential energy is

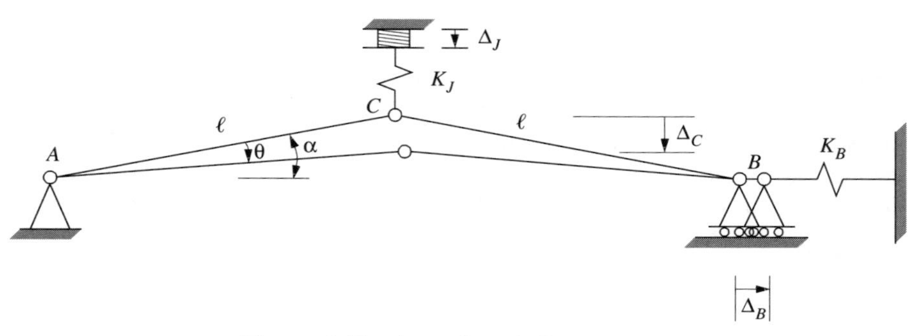

Figure 6.17 An enforced displacement.

$$\begin{aligned}\Pi &= \tfrac{1}{2}K_B\Delta_B^2 + \tfrac{1}{2}K_J(\Delta J - \Delta C)^2\\ &= 2K_B\ell^2[\cos(\alpha-\theta) - \cos\alpha]^2\\ &\quad + \tfrac{1}{2}K_J\{\Delta_J - \ell[\sin\alpha - \sin(\alpha-\theta)]\}^2\end{aligned}$$

Using θ as the independent variable, the first derivative of potential becomes

$$\begin{aligned}\frac{d\Pi}{d\theta} &= 4K_B\ell^2[\cos(\alpha-\theta)\cos\alpha]\sin(\alpha-\theta)\\ &\quad - K_J\ell\{\Delta_J - \ell[\sin\alpha - \sin(\alpha-\theta)]\}\cos(\alpha-\theta)\\ &= 4K_B\ell^2[\cos(\alpha-\theta) - \cos\alpha]\sin(\alpha-\theta)\\ &\quad - K_J\ell[\Delta_J - \Delta_C]\cos(\alpha-\theta)\end{aligned}$$

where we have used the identity

$$\Delta_J - \Delta_C = \Delta_J - \ell[\sin\alpha - \sin(\alpha-\theta)]$$

Then, the first derivative of potential becomes

$$\frac{d\Pi}{d\theta} = 4K_B\ell^2\{[\cos(\alpha-\theta) - \cos\alpha]\sin(\alpha-\theta) - p\cos(\alpha-\theta)\} \tag{6.78}$$

where we have factored $4K\ell^2$ out and defined the nondimensional load

$$p = \frac{P}{4K_B\ell}$$

after first using Eq. (6.77).

Equilibrium requires that $d\Pi/d\theta = 0$; hence, we can solve for the nondimensional load p to get

$$p = \tan(\alpha-\theta)[\cos(\alpha-\theta) - \cos\alpha] \tag{6.79}$$

which is identical to Eq. (6.75).

To check for the stability of the solution, we look at $d^2\Pi/d\theta^2$. Taking the derivative of Eq. (6.78), we get

$$\begin{aligned}\frac{d^2\Pi}{d\theta^2} &= 4K_B\ell^2\{\sin^2(\alpha-\theta) - \cos(\alpha-\theta)[\cos(\alpha-\theta) - \cos\alpha]\\ &\quad - p\sin(\alpha-\theta) + k\cos^2(\alpha-\theta)\}\end{aligned}$$

where the nondimensional stiffness k is defined as

$$k = \frac{K_J}{4K_B}$$

This equation must now be evaluated on the equilibrium path as determined by Eq. (6.79). Substituting for p, we get

$$\frac{d^2\Pi}{d\theta^2} = 4K_B\ell^2 \, \frac{\cos(\alpha - \cos^3(\alpha - \theta) + k\cos^3(\alpha - \theta)}{\cos(\alpha - \theta)}$$

Not allowing the structure to foldover itself, $|\alpha - \theta| < \pi/2$. The denominator is, therefore, always positive, and the sign of $\delta^2\Pi$ is determined by the numerator $\cos\alpha - \cos^3(\alpha - \theta) + k\cos^3(\alpha - \theta)$.

For the case $k < 1$, we can identify three conditions:

$$\frac{d^2\Pi}{d\theta^2} = \begin{cases} > 0 & \text{stable when } \cos^3(\alpha - \theta) < \cos\alpha/(1 - k) \\ = 0 & \text{critical when } \cos^3(\alpha - \theta) = \cos\alpha/(1 - k) \\ < 0 & \text{unstable when } \cos^3(\alpha - \theta) > \cos\alpha/(1 - k) \end{cases}$$

For the case $k > 1$, we can identify three conditions:

$$\frac{d^2\Pi}{d\theta^2} = \begin{cases} > 0 & \text{stable when } \cos^3(\alpha - \theta) > \cos\alpha/(1 - k) \\ = 0 & \text{critical when } \cos^3(\alpha - \theta) = \cos\alpha/(1 - k) \\ < 0 & \text{unstable when } \cos^3(\alpha - \theta) < \cos\alpha/(1 - k) \end{cases}$$

For both of the above cases, the critical value of θ is

$$\theta = \alpha \pm \cos^{-1}\left(\frac{\cos\alpha}{1 - k}\right)^{1/3}$$

Let us now define a nondimensional deflection in the direction of the load as

$$v = \frac{\Delta_C}{\ell} = \sin\alpha - \sin(\alpha - \theta)$$

We again define a scalar stiffness S as

$$S = \frac{dp}{dv} = \frac{dp}{d\theta}\,\frac{d\theta}{dv} = \frac{dp/d\theta}{dv/d\theta}$$

or

$$S = \frac{\cos\alpha - \cos^3(\alpha - \theta)}{\cos^3(\alpha - \theta)} \tag{6.80}$$

which is identical to Eq. (6.76).

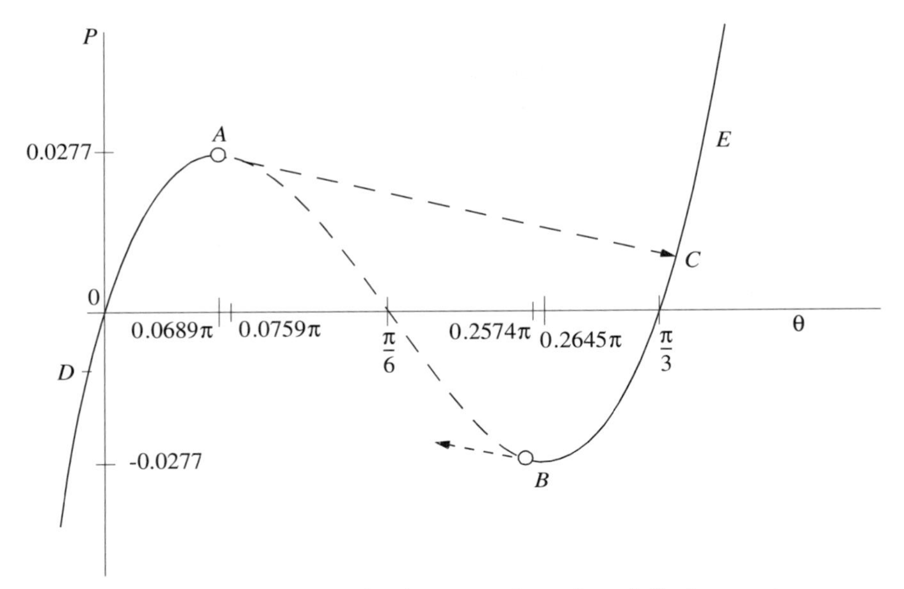

Figure 6.18 Load-deflection curve for enforced displacement.

Observe that when the numerator of this expression is zero, S is zero. Comparing the numerator for a zero value of S with the critical value of $d^2\Pi/d\theta^2$, we see that the critical equilibrium corresponds to a *negative* value of stiffness.

A plot of Eq. (6.79) is shown in Fig. 6.18 for a value of $\alpha = \pi/6$ and $k = 0.02$. The type of instaility again represents a single equilibrium path with a portion of the path (shown as dotted) as unstable. At point A, the structure experiences a loss of stability and will move (snap) dynamically to point C. If the structure is loaded to E, it will continue to increase in stiffness. If the structure is unloaded at C, the stiffness will decrease until at B it snaps to D.

A plot of Eq. (6.80) for the above values of α and k is shown in Fig. 6.19. We see that at the point A of instability, the stiffness is negative with a value of $S = -0.02$. The reason for this is that the jack-loading device has become part of the system. Thus, at

$$\theta_{\text{critical}} = \begin{cases} 0.0759\pi \\ 0.2574\pi \end{cases}$$

the whole system including the jack is at the point of instability.

6.9 BIFURCATION ANALYSIS

In the preceding section, the load deflection curve was characterized by a single equilibrium path with stable and unstable regions. In this section, we look at

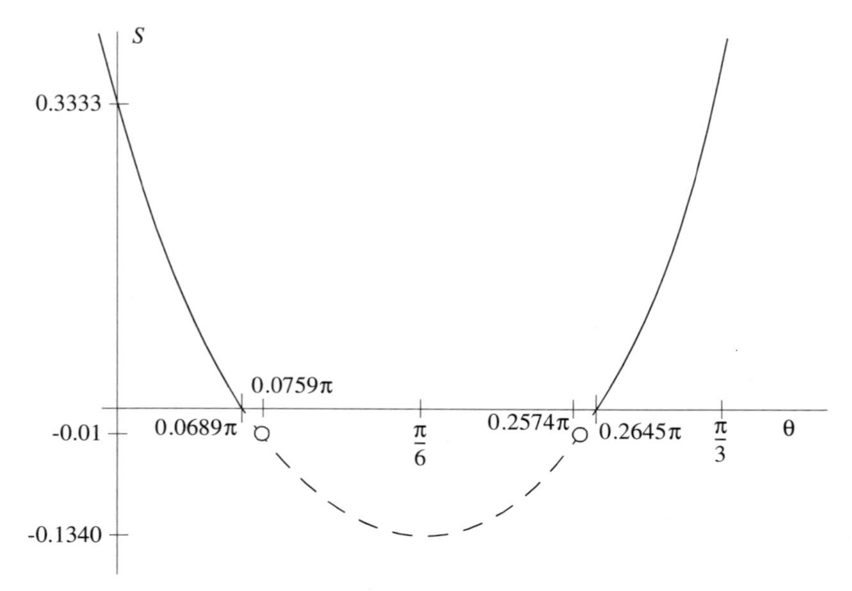

Figure 6.19 Stiffness-deflection curve for enforced displacement.

the case where multiple equilibrium paths are possible. Some of these paths can be stable; others, unstable.

Figure 6.20 shows a rigid bar of length ℓ with a rotational spring C at the hinge. The applied load P always remains vertical.

The deformed geometry yields

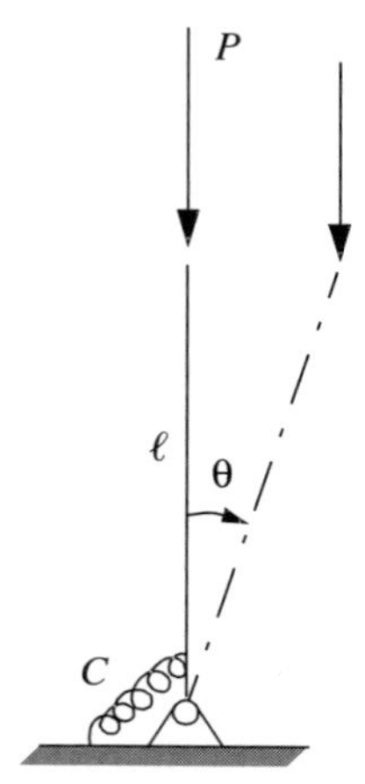

Figure 6.20 Rigid bar with a rotational spring.

$$\Delta_P = \ell(1 - \cos\theta)$$

The total potential energy is

$$\begin{aligned}\Pi &= \tfrac{1}{2}C\theta^2 - P\Delta_P \\ &= \tfrac{1}{2}C\theta^2 - P\ell(1 - \cos\theta)\end{aligned}$$

Using θ as the independent variable, the first derivative of potential becomes

$$\frac{d\Pi}{d\theta} = C(\theta - p\sin\theta) \tag{6.81}$$

where we have factored C out and defined the nondimensional load

$$p = \frac{P\ell}{C}$$

Equilibrium requires that $d\Pi/d\theta = 0$; hence, we can solve and arrive at two possible solutions:

$$\theta = 0 \quad \text{or} \quad p = \frac{\theta}{\sin\theta} \tag{6.82}$$

To check for the stability of the solution, we look at $d^2\Pi/d\theta^2$. Taking the derivative of Eq. (6.81), we get

$$\frac{d^2\Pi}{d\theta^2} = C(1 - p\cos\theta)$$

This equation must now be evaluated on each of the equilibrium paths as determined by Eq. (6.82). First substituting for $\theta = 0$, we get

$$\frac{d^2\Pi}{d\theta^2} = C(1-p)\begin{cases} > 0 & \text{stable when } p < 1 \\ = 0 & \text{critical when } p = 1 \\ < 0 & \text{unstable when } p > 1 \end{cases}$$

Next, substituting for p, we get

$$\frac{d^2\Pi}{d\theta^2} = C\left(1 - \frac{\theta}{\tan\theta}\right)\begin{cases} > 0 & \text{stable when } \theta < \tan\theta \\ = 0 & \text{critical when } \theta = \tan\theta \\ < 0 & \text{unstable when } \theta > \tan\theta \end{cases}$$

The critical value occurs at $p = 1$, so the critical load is

$$P_{\text{cr}} = \frac{C}{\ell} \tag{6.83}$$

Evaluating the third and fourth derivatives of potential at $\theta = 0$ and $p = 1$ gives

$$\frac{d^3\Pi}{d\theta^3} = 0$$

and

$$\frac{d^4\Pi}{d\theta^4} = C$$

Thus, the structure is stable at $p = 1$.

Let us now define a nondimensional deflection in the direction of the load as

$$v = \frac{\Delta_P}{\ell} = 1 - \cos\theta$$

We again define a scalar stiffness S as

$$S = \frac{dp}{dv} = \frac{dp}{d\theta}\,\frac{d\theta}{dv} = \frac{dp/d\theta}{dv/d\theta}$$

or

$$S = \frac{\sin\theta - \theta\cos\theta}{\sin^3\theta} \tag{6.84}$$

Figure 6.21 shows the equilibrium paths. The path at $\theta = 0$ is stable up to $p = 1$. Beyond this point, it is an unstable equilibrium path; any imperfection or "nudge" cause the structure to move to the $\theta/\sin\theta$ path that, for the structure shown, represents a stable equilibrium path. The point $p = 1$ is the bifurcation point, as it represents the junction between two possible paths.

Figure 6.22 represents the stiffness of the structure. When $p < 1$, the structure has essentially infinite stiffness (only because of the rigid-rod assumption). At $p = 1$, the stiffness of the structure abruptly drops to a value of $1/3$; however, as further deflection takes place, the stiffness of the structure increases.

6.10 IMPERFECTION ANALYSIS

In the first section of this chapter, we looked at the deflection of a beam subject to an eccentricity in the load. In this section, we analyze the example of the

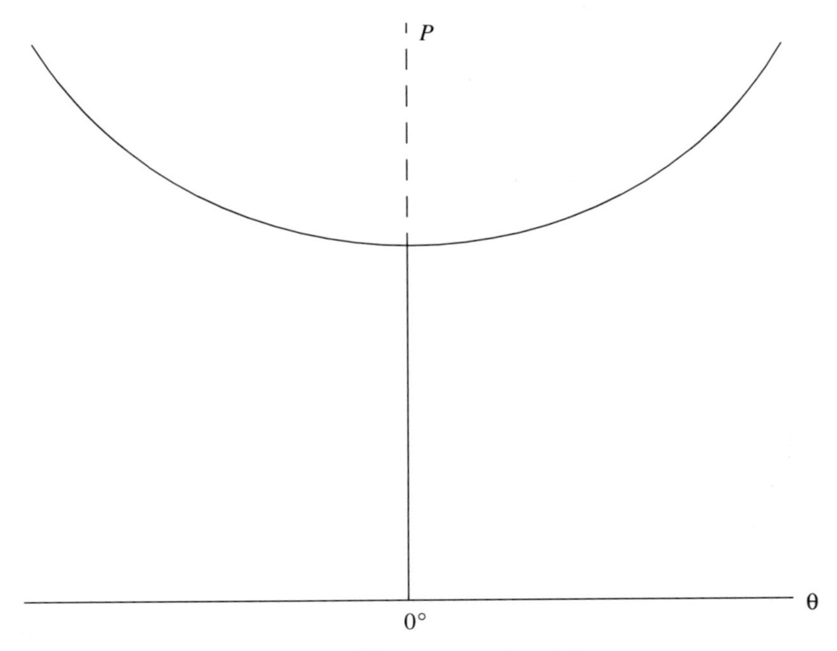

Figure 6.21 Equilibrium paths for bifurcation.

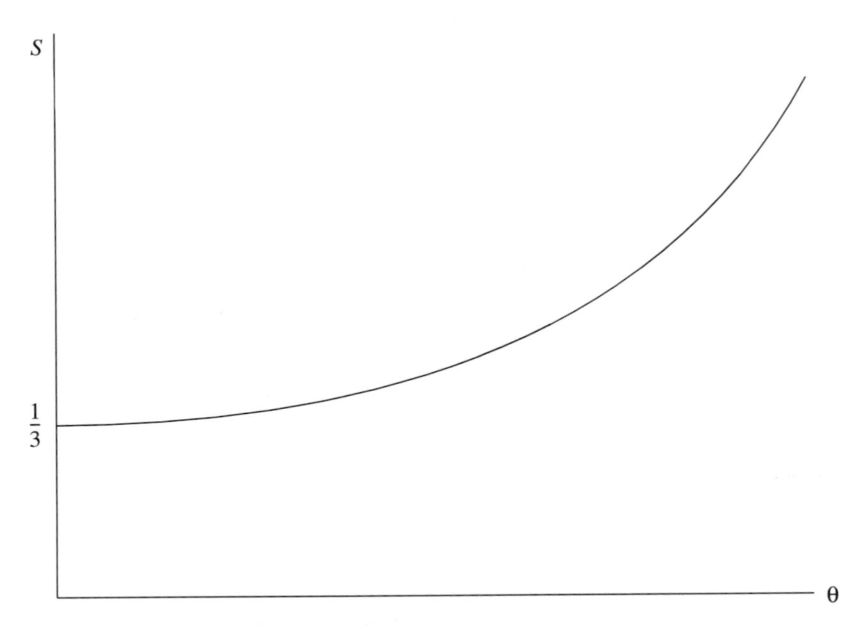

Figure 6.22 Stiffness-deflection for bifurcation.

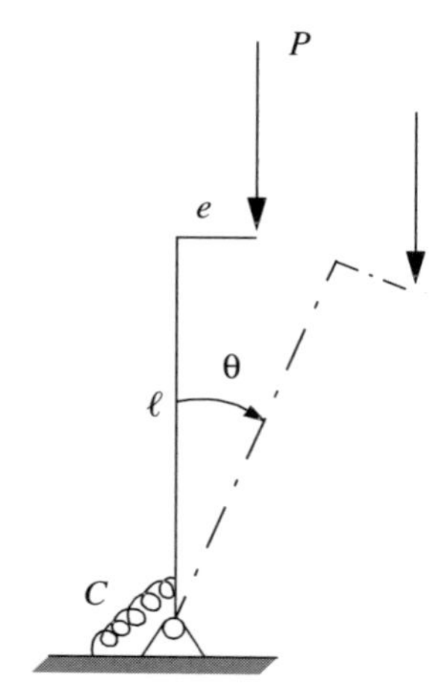

Figure 6.23 Rigid bar with a load eccentricity.

last section subject to a load eccentricity. Imperfection in load application and in geometries can have a significant effect on the critical load.

Figure 6.23 shows the same geometry of Fig. 6.20, except that the load application point is subject to the eccentricity e as shown.

The deformed geometry yields

$$\begin{aligned}\Delta_P &= \ell(1 - \cos\theta) + e\sin\theta \\ &= \ell\left(1 - \cos\theta + \frac{e}{\ell}\sin\theta\right)\end{aligned}$$

The total potential energy is

$$\Pi = \frac{1}{2}C\theta^2 - P\ell\left(1 - \cos\theta + \frac{e}{\ell}\sin\theta\right)$$

Using θ as the independent variable, the first derivative of potential becomes

$$\frac{d\Pi}{d\theta} = C\left[\theta - p\left(\sin\theta + \frac{e}{\ell}\cos\theta\right)\right] \tag{6.85}$$

where we have factored C out and defined the nondimensional load

$$p = \frac{P\ell}{C}$$

Equilibrium requires that $d\Pi/d\theta = 0$; hence, we can solve for p and arrive at the solution

$$p = \frac{\theta}{\cos\theta\left(\tan\theta + \dfrac{e}{\ell}\right)} \tag{6.86}$$

To check for the stability of the solution, we look at $d^2\Pi/d\theta^2$. Taking the derivative of Eq. (6.85), we get

$$\frac{d^2\Pi}{d\theta^2} = C\left[1 - p\cos\theta\left(1 - \frac{e}{\ell}\tan\theta\right)\right]$$

This equation must now be evaluated on each equilibrium path as determined by Eq. (6.86). Thus, we get

$$\frac{d^2\Pi}{d\theta^2} = \frac{C}{\left(\tan\theta + \dfrac{e}{\ell}\right)}\left[\tan\theta + \frac{e}{\ell} - \theta\left(1 - \frac{e}{\ell}\tan\theta\right)\right]$$

Therefore, $d^2\Pi/d\theta^2 > 0$ for all $0 \le \theta \le \pi/2$.

Let us now define a nondimensional deflection in the direction of the load as

$$v = \frac{\Delta_P}{\ell} = 1 - \cos\theta + \frac{e}{\ell}\sin\theta$$

We again define a scalar stiffness S as

$$S = \frac{dp}{dv} = \frac{dp}{d\theta}\,\frac{d\theta}{dv} = \frac{dp/d\theta}{dv/d\theta}$$

or

$$S = \frac{\tan\theta + \dfrac{e}{\ell} - \theta\left(1 - \dfrac{e}{\ell}\tan\theta\right)}{\cos^2\theta\left(\tan\theta + \dfrac{e}{\ell}\right)^3} \tag{6.87}$$

Figure 6.24 shows the equilibrium path compared with those of Fig. 6.21. The upper branch of the load deflection curve that exhibits both an unstable region and a stable region is not physically realizable, although an iterative procedure gone astray could conceivably reach such a branch. The most important observation is that the stable path no longer exhibits a bifurcation point.

Figure 6.25 represents the stiffness of the structure compared with the stiffness shown in Fig. 6.22.

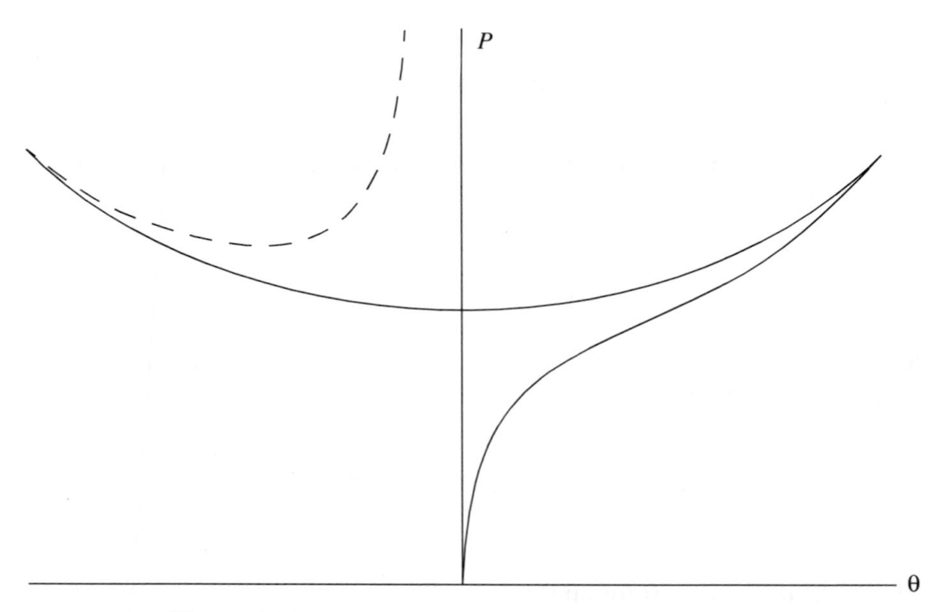

Figure 6.24 Equilibrium paths for imperfection model.

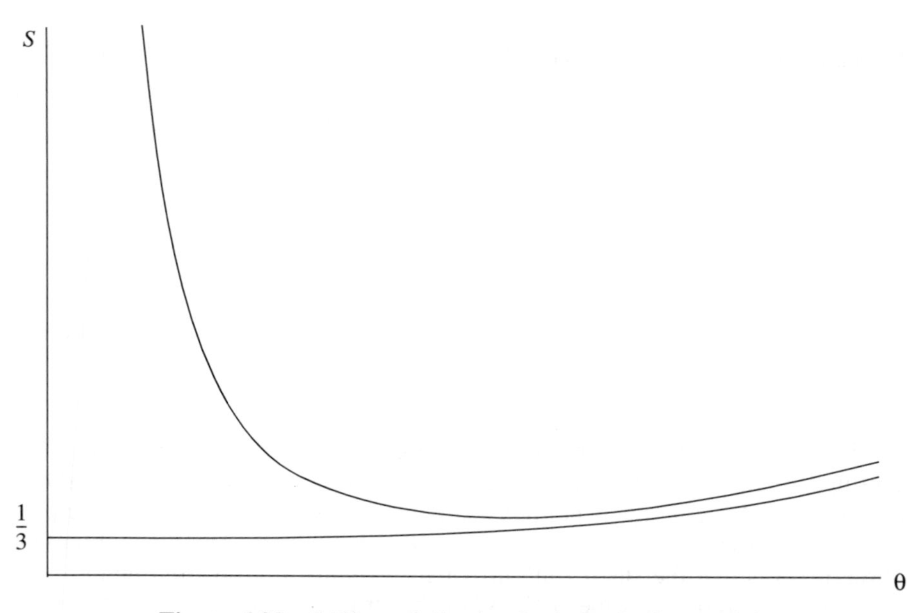

Figure 6.25 Stiffness-deflection for imperfection model.

6.11 CIRCULATORY DYNAMIC STABILITY

So far, we have described the stability of structural systems using static equilibrium or the principle of stationary potential energy. If the structural loading is a function of time, it is reasonable to expect that it would be necessary to study the system dynamically. However, there are even static loading conditions that require the dynamic approach, an example of which is a circulatory system where the static loading in nonconservative. Under such a system, the static energy approach in unreliable and may not exhibit buckling when the physical structure does indeed buckle.

Consider the stability of a circulatory system as shown in Fig. 6.26, which represents an idealized column composed of two rigid rods each of length ℓ. Since we desire to treat the system dynamically, we consider the system mass distribution to be represented by concentrated masses m and $2m$ as shown. The two linear-torsional springs, each with a spring constant C, represent the column's flexural rigidity. The unit vectors $\vec{t}_1$, $\vec{t}_2$, $\vec{n}_1$, and $\vec{n}_2$ are fixed to the rods as shown. Unit vectors $\vec{k}$ are defined as $\vec{t}_i \times \vec{n}_i$, $i = 1, 2$. The chosen generalized coordinates are θ_1 and θ_2, both measured from a fixed vertical line.

The applied load P is a follower load and is applied so that it always remains parallel to the top link. Thus $\vec{P} = -P\vec{t}_2$. The load P is nonconservative, which can be demonstrated as follows:

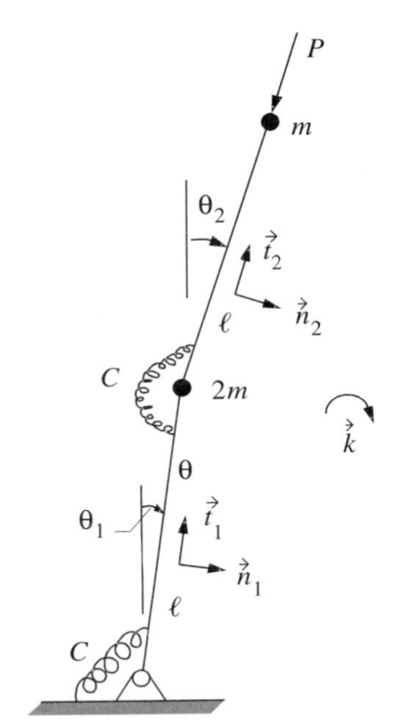

Figure 6.26 A circulatory model.

1. A way to obtain the configuration shown in Fig. 6.26 is to first hold $\theta_2 = 0$ while rotating θ_1 into its final position. The work done will be $W = P\ell(1 - \cos\theta_1)$. Then, rotate θ_2 into its final position. No work will be done because P always acts tangent to the top link.
2. Another way to obtain the configuration shown in Fig. 6.26 is to first hold $\theta_1 = \theta_2$ while rotating θ_1 into its final position. Then, rotate the top rod $\theta_2 - \theta_1$ into its final position. The total work done is $W = 0$.

Since the work done is path-dependent, it is by definition nonconservative.

To further study the system, we need to obtain the equations of motion. In structural mechanics (and especially when dealing with multibodies), the equations of motion are usually obtained by use of one of three methods: Lagrange's equations [58], Kane's equations [4] and [59][1] or the Gibbs-Appell equations [59]. The first two methods are the most common. From a computational mechanics' point of view, Kane's equations have at least two advantages: (1) they are obtained directly from the kinematics and hence are easily computerized, and (2) they may be consistently linearized prior to obtaining the full set of nonlnear equations.

Consider only the first two methods, Lagrange's equation for our example is most conveniently written as

$$\frac{d}{dt}\frac{\partial K}{\partial q_r} = F_r \qquad r = 1 \text{ to } n \tag{6.88}$$

where

K is the kinetic energy
q_r is a generalized coordinate
F_r is a generalized active force
n is the number of degrees of freedom

If the force is obtainable from a potential function U, then from Eq. (5.3) we have

$$F_r = -\frac{\partial U}{\partial q_r}$$

For example, the potential energy of the springs is

$$U = \tfrac{1}{2}C\theta_1^2 + \tfrac{1}{2}C(\theta_2 - \theta_1)^2$$

and

[1]In ref. [4], they are called Lagrange's form of D'Alembert's principle.

$$F_{s\theta_1} = -\frac{\partial U}{\partial \theta_1} = C(\theta_2 - 2\theta_1)$$

$$F_{s\theta_2} = -\frac{\partial U}{\partial \theta_2} = C(\theta_1 - \theta_2) \tag{6.89}$$

To compute the nonconservative generalized force, we can compute its virtual work and get an expression of the form

$$\delta W = F_r \, \delta q_r$$

and by inspection determine the generalized force as the coefficient of the virtual coordinate. Thus

- First, hold $\delta\theta_1 = 0$ and $\delta\theta_2 \neq 0$; thus $\delta W_{\theta_2} = 0$, since the load P is always tangent to the tip rod.
- Second, hold $\delta\theta_1 \neq 0$ and $\delta\theta_2 = 0$ and resolve the load P into two components. The component $P\cos(\theta_2 - \theta_1)$, is always tangent to the bottom rod and does no virtual work for the prescribed virtual displacement; the other component $P\sin(\theta_2 - \theta_1)$ performs the virtual work $\delta W_{\theta_1} = -P\ell\sin(\theta_2 - \theta_1)\,\delta\theta_1$.

We obtain therefore

$$F_{P\theta_1} = -P\ell\sin(\theta_2 - \theta_1)$$
$$F_{P\theta_2} = 0$$

We will see that by using Kane's method, the above expressions are automatic and of natural consequence.

We now determine the velocities of the masses as

$$\vec{v}_{2m} = \ell\dot{\theta}_1\vec{n}_1$$
$$\vec{v}_m = \ell\dot{\theta}_1\vec{n}_1 + \ell\dot{\theta}_2\vec{n}_2$$

where, for example, we have used the relationships

$$\vec{v}_{2m} = \frac{d}{dt}(\ell\vec{t}_1) = \ell\vec{\omega}_1 \times \vec{t}_1$$

and

$$\vec{\omega}_1 = \dot{\theta}_1\vec{k}; \qquad \vec{\omega}_2 = \dot{\theta}_2\vec{k}_2$$

We define the coefficients of the $\dot{\theta}_1$ and $\dot{\theta}_2$ terms in the velocity expressions as the partial velocities and write them as

$$\vec{v}^{\,2m}_{\dot{\theta}_1} = \ell\vec{n}_1; \qquad \vec{v}^{\,m}_{\dot{\theta}_1} = \ell\vec{n}_1$$
$$\vec{v}^{\,2m}_{\dot{\theta}_2} = 0; \qquad \vec{v}^{\,m}_{\dot{\theta}_2} = \ell\vec{n}_2 \tag{6.90}$$

Partial angular velocities can be defined in a similar fashion but are not needed for this particular example.

The following dot products are useful and are obtained from Fig. 6.26:

•	$\vec{n}_1$	$\vec{t}_1$	$\vec{n}_2$	$\vec{t}_2$
$\vec{n}_1$	1	0	$\cos(\theta_2 - \theta_1)$	$\sin(\theta_2 - \theta_1)$
$\vec{n}_2$	$\cos(\theta_2 - \theta_1)$	$-\sin(\theta_2 - \theta_1)$	1	0

For Lagrange's equations, we need the velocity squared, which are

$$v_{2m}^2 = \vec{v}_{2m} \cdot \vec{v}_{2m} = \ell^2\dot{\theta}_1^2$$
$$v_m^2 = \vec{v}_m \cdot \vec{v}_m = \ell^2\dot{\theta}_1^2 + 2\ell^2\dot{\theta}_1\dot{\theta}_2\cos(\theta_2 - \theta_1) + \ell^2\dot{\theta}_2^2$$

The resulting kinetic energy is

$$K = \tfrac{3}{2}m\ell^2\dot{\theta}_1^2 + m\ell^2\dot{\theta}_1\dot{\theta}_2\cos(\theta_2 - \theta_1) + \tfrac{1}{2}m\ell^2\dot{\theta}_2^2$$

If we now take the three partial derivatives indicated in Lagrange's equations for each of the two generalized coordinates and combine the terms as indicated, we obtain the equations of motion. This is, however, a lot of work, so we will obtain the final equations by using Kane's method.

The accelerations of the two masses are obtained as

$$\vec{a}_{2m} = \ell\ddot{\theta}_1\vec{n}_1 - \ell\dot{\theta}_1^2\vec{t}_1$$
$$\vec{a}_m = \ell\ddot{\theta}_1\vec{n}_1 - \ell\dot{\theta}_1^2\vec{t}_1 + \ell\ddot{\theta}_2\vec{n}_2 - \ell\dot{\theta}_2^2\vec{t}_2$$

Kane's equations are

$$F_r + F_r^* = 0 \qquad r = 1 \text{ to } n \tag{6.91}$$

and F_r are the generalized active forces defined as

$$F_r = \sum_{i=1}^{N} \vec{V}_r^{M_i} \cdot \vec{F}_i \qquad r = 1 \text{ to } n$$

and F_r^* are the generalized inertia forces defined as

$$F_r^* = \sum_{i=1}^{N} \vec{V}_r^{M_i} \cdot \vec{F}_i^* \qquad r = 1 \text{ to } n$$

where

N is the number of mass points
$\vec{V}_r$ are the partial speeds, such as those defined in Eq. (6.90)
$\vec{F}_r$ are the active loadings, such as the springs and the load P
$\vec{F}_i^* = -M_i\vec{a}_i$ is the inertia force on the ith mass

Then, the contribution of the nonconservative load P to the generalized forces is computed as

$$\begin{aligned}
F_{P\theta_1} &= \vec{v}_{\dot{\theta}_1}^{2m} \cdot \vec{F}_{2m} + \vec{v}_{\dot{\theta}_1}^{m} \cdot \vec{F}_m \\
&= \ell\vec{n}_1 \cdot 0 + \ell\vec{n}_1 \cdot (-P\vec{t}_2) \\
&= -P\ell \sin(\theta_2 - \theta_1) \\
F_{P\theta_2} &= \vec{v}_{\dot{\theta}_2}^{2m} \cdot \vec{F}_{2m} + \vec{v}_{\dot{\theta}_2}^{m} \cdot \vec{F}_m \\
&= 0 \cdot 0 + \ell\vec{n}_2 \cdot (-P\vec{t}_2) \\
&= 0
\end{aligned} \tag{6.92}$$

In a similar fashion, the generalized inertia forces are

$$\begin{aligned}
-F_{\theta_1}^* &= 3m\ell^2\ddot{\theta}_1 + m\ell^2\ddot{\theta}_2\cos(\theta_2 - \theta_1) - m\ell^2\dot{\theta}_2^2\sin(\theta_2 - \theta_1) \\
-F_{\theta_2}^* &= m\ell^2\ddot{\theta}_1\cos(\theta_2 - \theta_1) + m\ell^2\dot{\theta}_1^2\sin(\theta_2 - \theta_1) + m\ell^2\ddot{\theta}_2
\end{aligned}$$

Collecting all the generalized forces, including Eq. (6.89), we get the following equations of motion:

$$\begin{aligned}
&3m\ell^2\ddot{\theta}_1 + m\ell^2\ddot{\theta}_2\cos(\theta_2 - \theta_1) - m\ell^2\dot{\theta}_2^2\sin(\theta_2 - \theta_1) \\
&\qquad + P\ell\sin(\theta_2 - \theta_1) - C(\theta_2 - 2\theta_1) = 0 \\
&m\ell^2\ddot{\theta}_1\cos(\theta_2 - \theta_1) + m\ell^2\dot{\theta}_1^2\sin(\theta_2 - \theta_1) + m\ell^2\ddot{\theta}_2 \\
&\qquad + C(\theta_2 - \theta_1) = 0
\end{aligned} \tag{6.93}$$

The linear equations of motion are

$$\begin{aligned}
3m\ell^2\ddot{\theta}_1 + m\ell^2\ddot{\theta}_2 + P\ell(\theta_2 - \theta_1) - C(\theta_2 - 2\theta_1) &= 0 \\
m\ell^2\ddot{\theta}_1 + m\ell^2\ddot{\theta}_2 + C(\theta_2 - \theta_1) &= 0
\end{aligned} \tag{6.94}$$

Assume a solution of the form

$$\begin{Bmatrix} \theta_1 \\ \theta_2 \end{Bmatrix} = \begin{Bmatrix} \Phi_1 \\ \Phi_2 \end{Bmatrix} e^{\omega t}$$

The equations of motion can be written as

$$\begin{bmatrix} 3m\ell^2\omega^2 + 2C - P\ell & m\ell^2\omega^2 - C + P\ell \\ m\ell^2\omega^2 - C & m\ell^2\omega^2 + C \end{bmatrix} \begin{Bmatrix} \Phi_1 \\ \Phi_2 \end{Bmatrix} = \begin{Bmatrix} 0 \\ 0 \end{Bmatrix} \tag{6.95}$$

If we factor C from the agove equation and define

$$\Omega^2 = \frac{m\ell^2\omega^2}{C}$$

and

$$p = \frac{P\ell}{C}$$

we finally get

$$\begin{bmatrix} 3\Omega^2 + 2 - p & \Omega^2 - 1 + p \\ \Omega^2 - 1 & \Omega^2 + 1 \end{bmatrix} \begin{Bmatrix} \Phi_1 \\ \Phi_2 \end{Bmatrix} = \begin{Bmatrix} 0 \\ 0 \end{Bmatrix} \tag{6.96}$$

Nontrivial solutions of the above equation occur when its determinant is zero, or

$$2\Omega^4 + (7 - 2p)\Omega^2 + 1 = 0 \tag{6.97}$$

with roots

$$\Omega^2_{I,II} = \frac{-7 - 2p) \pm \sqrt{(7 - 2p)^2 - 8}}{4} \tag{6.98}$$

Whether the roots are real-positive, real-negative, or conjugate-complex depends on $(7 - 2p)$. Figure 6.27 plots

$$b = 7 - 2p$$

versus the discriminat

$$\Delta = (7 - 2p)^2 - 8$$

The following should be noted:

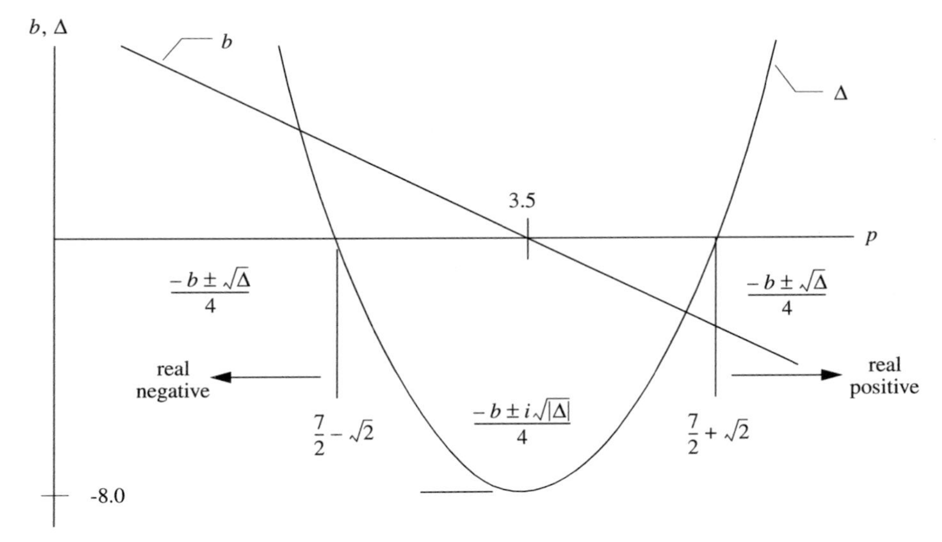

Figure 6.27 Roots for circulatory model.

$$
\begin{aligned}
|b| > \sqrt{\Delta}; &\qquad b^2 > 8 \\
7 - 2p > 0; &\qquad p < 3.5 \\
7 - 2p < 0; &\qquad p > 3.5
\end{aligned}
$$

The discriminant is zero when

$$p_{1,2} = \tfrac{7}{2} \pm \sqrt{2} = 2.086, \;\; 4.914$$

Therefore:

$$p \leq 2.086; \qquad \Omega^2 \text{ is real-negative} \tag{6.99a}$$

$$2.086 < p < 4.914; \qquad \Omega^2 \text{ is conjugate-complex} \tag{6.99b}$$

$$p \leq 4.914; \qquad \Omega^2 \text{ is real-positive} \tag{6.99c}$$

The differential equation we are dealing with has solutions of the form $e^{\lambda t}$. The roots of this solution occur in six possible forms:

1. $-A \pm iB$; $y = Ce^{-At}\cos(\lambda t + \phi)$ damped stable.
2. $-A$; $y = Ce^{-At}$ overdamped stable.
3. $\pm iB$; $y = C\cos(\lambda t + \phi$ undamped stable.
4. 0; $y = C_1$ neutrally stable.

5. $A \pm iB$; $y = Ce^{At}\cos(\lambda t + \phi)$ "flutter" unstable.
6. A; $y = Ce^{At}$ "divergence" unstable.

Thus, for values of p less then the root given by Eq. (6.99a), all of the roots take the form

$$\Omega_{1,2} = \pm iB$$

and the system is stable. For values of p between the roots given by Eq. (6.99b), the roots[2] take the form

$$\Omega_{1,2} = A \pm iB$$

and the system is dynamically unstable, exhibiting a "flutter-like" condition. For values of p greater than the root given by Eq. (6.99c), at least some of the roots take the form

$$\Omega_1 = A$$

and the system is statically unstable, exhibiting divergence. Therefore, the critical load is given as

$$P_{\text{cr}} = 2.086\,\frac{C}{\ell}$$

6.12 INSTATIONARY DYNAMIC STABILITY

An instationary system contains loads explicitly dependent on time. Their differential equations of motion are characterized by time-dependent coefficients. This type of instability is important in the field of rotating machinery, particularly in situations of lateral instability caused by pulsating torque, longitudinal compression, or asymmetric bearing characteristics.

Consider the rigid rod shown in Fig. 6.28, which represents an idealized column composed of a rigid rod of length ℓ with a concentrted mass m and a linear-torsional spring with constant C representing the column's flexural rigidity. The unit vectors $\vec{t}$ and $\vec{n}$ are fixed to the rod as shown. A unit vector $\vec{k}$ is defined as $\vec{n} \times \vec{t}$. The chosen generalized coordinate is θ, measured from a fixed vertical line.

The applied load follows the harmonic law $P = P_0 + P_1 \cos \omega t$ and always remains in the vertical. The resulting equation of motion is called a para-

[2]The complex surd is given as follows:

$$\sqrt{A \pm iB} = \sqrt{\frac{\sqrt{A^2 + B^2} + A}{2}} \pm i\sqrt{\frac{\sqrt{A^2 + B^2} - A}{2}}$$

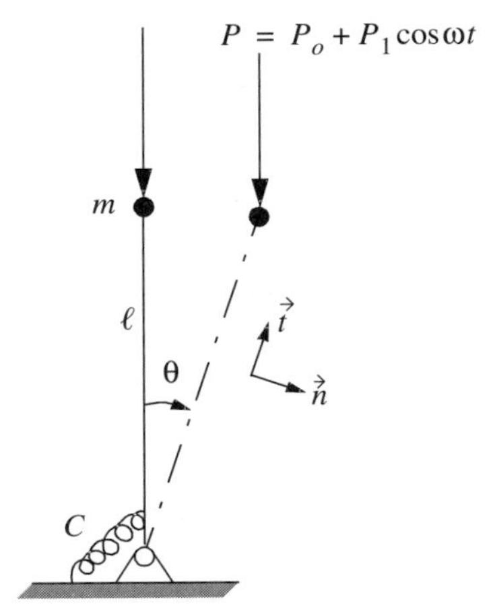

Figure 6.28 Rigid rod with a lumped mass under a periodic load.

metrically forced oscillation.

The velocity of the mass is

$$\vec{v} = \ell\dot{\theta}\vec{n}$$

and its acceleration is

$$\vec{a} = \ell\ddot{\theta}\vec{n} - \ell\dot{\theta}^2\vec{t}$$

Then, the inertia force is

$$\begin{aligned}\vec{F}^* &= -m\vec{a} \\ &= -m(\ell\ddot{\theta}\vec{n} - \ell\dot{\theta}^2\vec{t})\end{aligned}$$

As in the preceding section, we define the generalize inertia force as

$$F^*_{m\theta} = -m\ell\vec{n} \cdot (\ell\ddot{\theta}\vec{n} - \ell\dot{\theta}^2\vec{t}) \tag{6.100}$$

where we have again defined the coefficient of the $\dot{\theta}$ as the partial velocity written as

$$\vec{v}_{\dot{\theta}} = \ell\vec{n}$$

The potential energy of the spring is

$$U = \tfrac{1}{2}C\theta^2$$

and its generalized force is

$$F_{s\theta} = -\frac{\partial U}{\partial \theta} = -C\theta \tag{6.101}$$

The applied load is

$$\begin{aligned} P &= -(P_0 + P_1 \cos \omega t)\vec{j} \\ &= -(P_0 + P_1 \cos \omega t)(\cos \theta \vec{t} - \sin \theta \vec{n}) \end{aligned}$$

and its generalized force is

$$\begin{aligned} F_{P\theta} &= -(P_0 + P_1 \cos \omega t)(\cos \theta \vec{t} - \sin \theta \vec{n}) \cdot \ell \vec{n} \\ &= \ell (P_0 + P_1 \cos \omega t) \sin \theta \end{aligned} \tag{6.102}$$

Kane's equations are

$$F_{s\theta} + F_{P\theta} + F^{*}_{m\theta} = 0$$

or

$$\begin{aligned} m\ell^2 \ddot{\theta} + C\theta - \ell(P_0 + P_1 \cos \omega t) \sin \theta &= 0 \\ \ddot{\theta} + \frac{C}{m\ell^2}\,\theta - \frac{1}{m\ell}\,(P_0 + P_1 \cos \omega t) \sin \theta &= 0 \end{aligned} \tag{6.103}$$

For small θ, this becomes

$$\ddot{\theta} + \frac{1}{m\ell^2}\,(C - P_0 \ell - P_1 \ell \cos \omega t)\theta = 0 \tag{6.104}$$

If we define the nondimensional time parameter τ as

$$2\tau = \omega t$$

and note that

$$\frac{d^2\theta}{d\tau^2} = \frac{4}{\omega^2}\,\ddot{\theta}$$

and also define the two parameters

$$\begin{aligned} a &= \frac{4}{m\omega^2 \ell}\left(\frac{C}{\ell} - P_0\right) \\ q &= \frac{2P_1}{m\omega^2 \ell} \end{aligned} \tag{6.105}$$

we get the linear homogeneous differential equation with variable coefficients

$$\frac{d^2\theta}{d\tau^2} + (a - 2q\cos 2\tau)\theta = 0 \tag{6.106}$$

This is the Mathieu equation in standard form.

To obtain solutions that are periodic in 2π, assume a function of the form

$$\theta = \cos n\tau + qc_1(\tau) + q^2c_2(\tau) + \cdots$$

and

$$a = n^2 + \alpha_1 q + \alpha^2 q^2 + \cdots$$

The solutions obtained are in the form of Mathieu functions. These are cosine-elliptic functions of class $ce_{2n}(\tau, q)$ or $ce_{2n+1}(\tau, q)$. If sin $n\tau$ is used instead, the Mathieu functions obtained are in the form of sine-elliptic functions of class $se_{2n+1}(\tau, q)$ or $se_{2n+2}(\tau, q)$. For a given q, there is a value for a called the *characteristic number* of the Mathieu function. The Mathieu functions and their related characteristic numbers are tabulated in ref. [60].

Figure 6.29 represents the stability regions of (a, q) plane. The figure is symmetric about the a axis, and in its white area, all solutions of the Mathieu equation are stable, whereas in its cross-hatched regions, the solutions are unstable. For $q = 0$, the stability boundaries occur at $a = r^2 (r = 0, 2, 3, \ldots)$. Reference [61] gives a concise discussion of the stability boundaries[3] of the Mathieu equation.

The stability diagram of Fig. 6.29 completely eliminates the necessity of any operations to solve the Mathieu equation. It is sufficient to write the Mathieu equation and determine the parameters a and q. The diagram immediately answers the question of whether the system does or does not have stability.

From the expression in Eq. (6.105), we see that both parameters a and q decrease in proportion to an increase in forcing frequency ω. Since the ratio of both parameters remains constant, as ω increases points on the stability diagram are determined by the line $q = ka$. Thus k is

$$k = \frac{P_1}{2(P_{\text{cr}} - P_0)}$$

where we have used Eq. (6.83) to define the critical static buckling load.

For a given value of P_1, k depends on the difference $P_{\text{cr}} - P_0$. The more the static load P_0 approaches P_{cr}, the steeper the slope and the wider the instability regions. When $P_0 = P_{\text{cr}}$, then $a = 0$, and $q = ka$ coincides with the ordinate axis of the stability diagram. Then, the system remains stable if $0 < |q| < 1$. This will happen if

[3]In Ref. [61], the term cos τ is used rather than cos 2τ; hence, the roots occur at $a = r^2/4$.

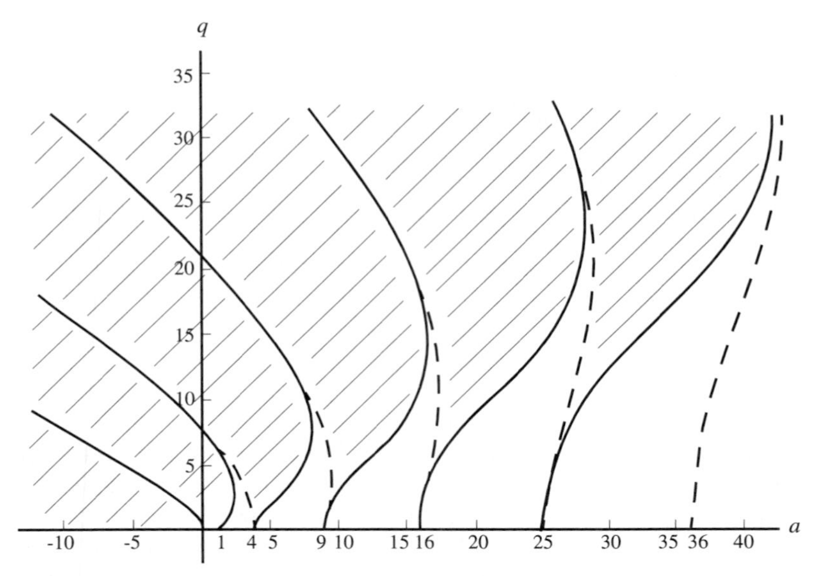

Figure 6.29 Stability boundaries of solutions to the Mathieu equation.

$$\omega > \sqrt{\frac{2P_1}{m\ell}}$$

In the case of $P_0 > P_{cr}$, the line $q = ka$ is located in the second quadrant of the stability diagram. Again, for appropriately selected ranges of frequency ω, the system may be made stable.

The structure discussed in this example has a stable static secondary equilibrium path. However, under instationary loading the relationship between the applied static load and the static critical load only control the slope of the $q = ka$ line. The driving frequency ω determines what regions of the line are in stable regions of the stability diagram.

An important consideration begins to emerge from the above discussion, namely, if a structure's secondary static equilibrium path is unstable, under certain conditions of oscillations the unstable static system can be made stable by the addition of oscillations.

PROBLEMS

6.1 The total potential energy of a certain system is

$$\Pi = \frac{x^2}{5} - \frac{a}{4}x^4 - \frac{2a^2}{3}x^3 + a^3$$

where a is a constant. Determine all of the equilibrium configurations of this system and indicate which ones are stable and which ones are unstable for the constant a, either positive or negative.

6.2 In Fig. 6.30, the frictionless pin bears on the rigid, semicircular ring of radius a that is supported on rigid rollers. If the body moves horizontally a distance s, use large-displacement analysis and do the following:

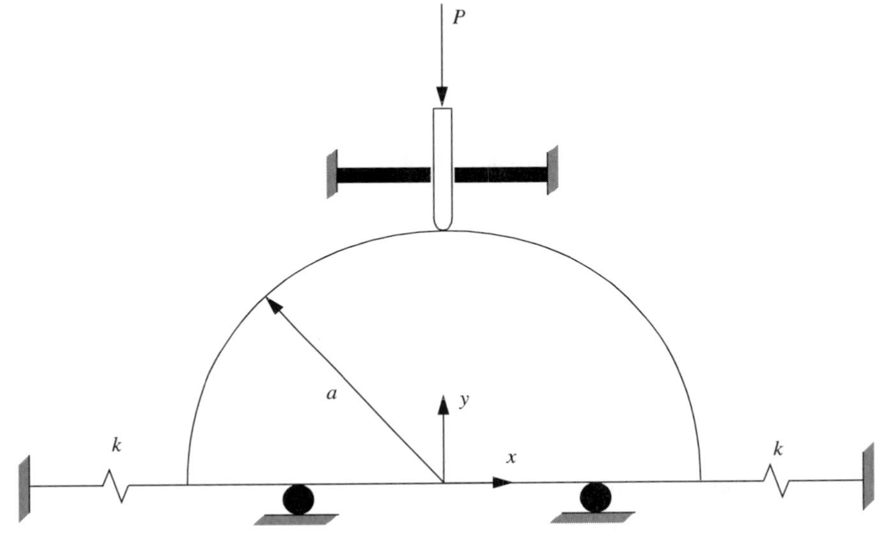

Figure 6.30 Problem 6.2.

(a) Determine the potential energy of the system.

(b) Determine the equilibrium paths. [Define the nondimensional load $p = P/(ka)$.]

(c) Determine the critical load P_{cr}.

(d) Determine if the paths are stable.

(e) Sketch the paths.

(f) Determine the stiffness $S = dp/d\delta$, where δ is a nondimensional displacement.

(g) Sketch the S–θ curve.

(h) Solve using a linearized theory, and comment on the results.

(i) How would you classify this model?

6.3 In Fig. 6.31, the two rigid links are supported as shown. The links are each of length ℓ. The support at A is pinned and that at B can roll horizontally. The load P is applied at B and always remains horizontal. At the pin connection C joining the two links, there is a rotational spring of modulus C. Before P is applied, the links are horizontal. Using large-displacement analysis,

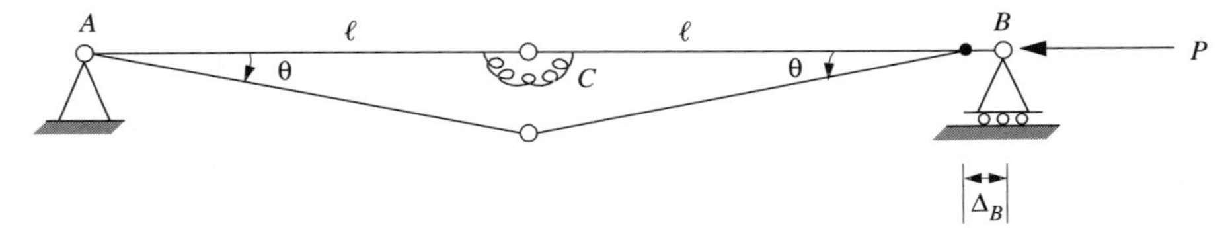

Figure 6.31 Problem 6.3.

(a) Determine the potential energy of the system.

(b) Determine the equilibrium paths. [Define the nondimensional load $p = P\ell/(2C)$.]

(c) Determine the critical load P_{cr}.

(d) Determine if the paths are stable.

(e) Sketch the paths.

(f) Determine the stiffness $S = dp/d\delta$, where δ is a nondimensional displacement.

(g) Sketch the p–δ and S–θ curves.

(h) Solve using a linearized theory, and comment on the results.

(i) How would you classify this model?

6.4 For the enforced displacement shown in Fig. 6.17, assume that the jack spring K_J is first pretensioned by hanging a weight W at point C. Determine the following:

(a) The total potential energy of the system.

(b) The force F in the jack spring K_J.

(c) By virtual work, the nondimensional load

$$p = \frac{P}{4K_B\ell} = [\cos(\alpha - \theta) - \cos\alpha]\tan(\alpha - \theta)$$

6.5 In Fig. 6.32, the bar is rigid and the hinge is frictionless. The spring K always remains horizontal and the load P always remains vertical. Using large displacement analysis, do the following:

(a) Determine the potential energy of the system.

(b) Determine the equilibrium paths. [Define the nondimensional load $p = P/(K\ell)$.]

(c) Determine the critical load P_{cr}.

(d) Determine if the paths are stable.

(e) Sketch the paths.

(f) Determine the stiffness $S = dp/d\delta$, where δ is a nondimensional displacement.

(g) Sketch the S–θ curves.

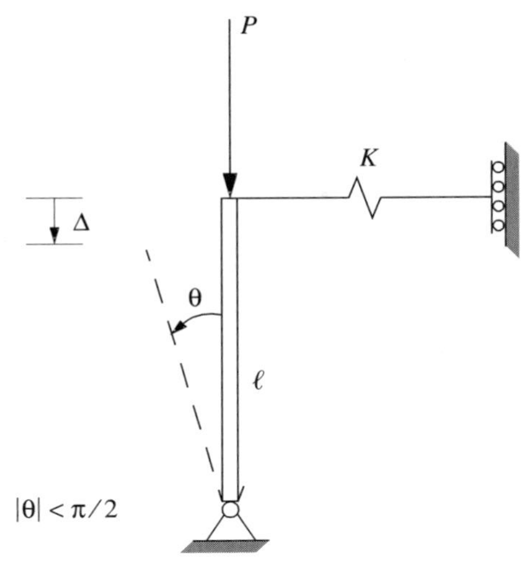

Figure 6.32 Problem 6.5.

6.6 In Fig. 6.33, the bar is rigid and the hinge is frictionless. The spring K always remains horizontal and the load P always remains vertical. Rotatikon at the hinge is resisted by a rotational spring of modulus C. Using large displacement analysis,

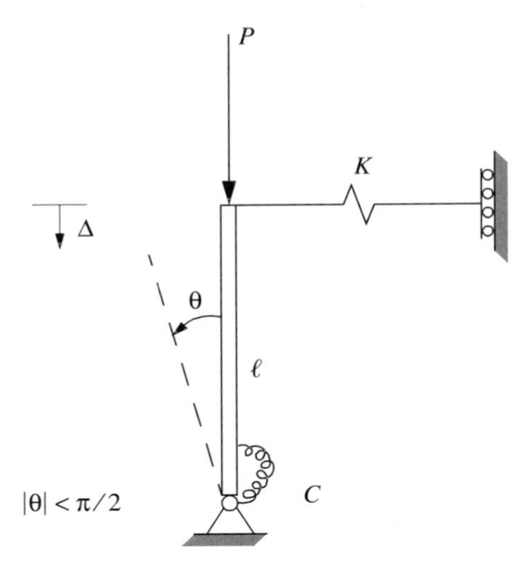

Figure 6.33 Problem 6.6.

(a) Determine the potential energy of the system.

(b) Determine the equilibrium paths. (Define the nondimensional load $p = P\ell/C$ and the nondimensional stiffness $k = K\ell^2/C$.)

(c) Determine the critical load P_{cr}.

(d) Determine if the paths are stable.

(e) Determine the stiffness $S = dp/d\delta$, where δ is a nondimensional displacement.

6.7 In Fig. 6.34, the two rigid horizontal links are supported as shown. The links are each of length ℓ. The support at A is pinned and that at C can roll horizontally. The load P is applied at C and always remains horizontal. At the pin connection B joining the two links, there is a vertical spring of modulus K that is supported in such a way that it always remains above the connection B. Using large-displacement analysis,

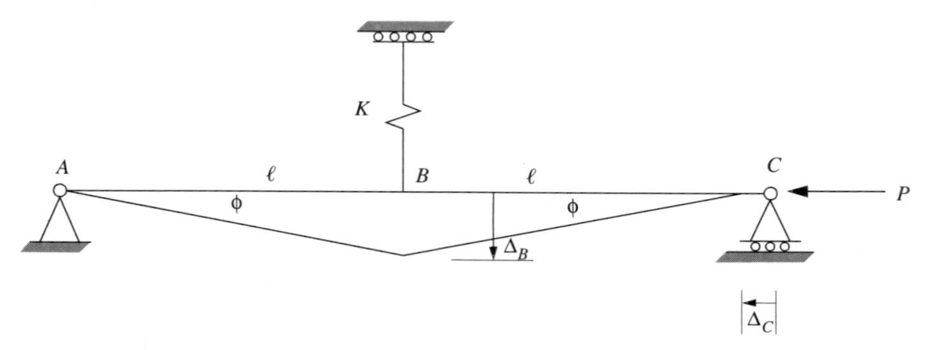

Figure 6.34 Problem 6.7.

(a) Determine the potential energy of the system.

(b) Determine the equilibrium paths. [Define the nondimensional load $p = 2P/(K\ell)$.]

(c) Determine the critical load P_{cr}.

(d) Determine if the paths are stable.

(e) Sketch the paths.

(f) Determine the stiffness $S = dp/d\delta$, where δ is a nondimensional displacement.

(g) Sketch the p–δ and S–δ curves.

6.8 In Fig. 6.35, the rigid bar is originally vertical. Its length is ℓ, and it is hinged at A and supported by a frictionless slider S connected to a spring of modulus k, which is constrained to move horizontally at a distance a above the point A. At end B, a load P, which always remains vertical, is applied as shown. Using large-displacement analysis,

(a) Determine the potential energy of the system.

(b) Determine the equilibrium paths. [Define the nondimensional load $p = P\ell/(ka^2)$.]

(c) Determine the critical load P_{cr}.

(d) Determine if the paths are stable.

(e) Sketch the paths.

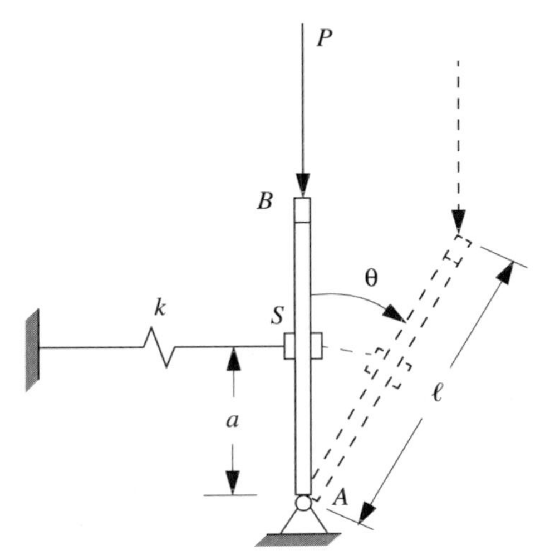

Figure 6.35 Problem 6.8.

(f) Determine the stiffness $S = dp/d\delta$, where δ is a nondimensional displacement.

(g) Sketch the S–δ curves.

6.9 In Fig. 6.36, the rigid bar is originally at an angle θ_0 as shown. Its length is ℓ and it is hinged at A. The hinge is restrained with a rotational spring of modulus C. At θ_0, the rotational stretch in the spring C is zero. At end B, a load P, which always remains vertical, is applied as shown. Using large-displacement analysis for $\theta_1 \leq \theta \leq \pi/2$,

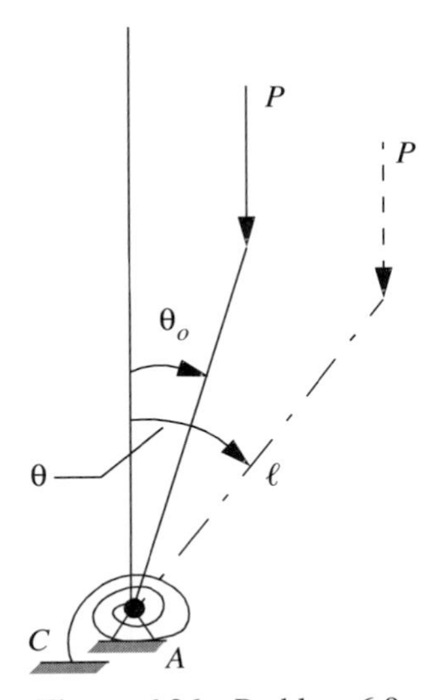

Figure 6.36 Problem 6.9.

(a) Determine the potential energy of the system.

(b) Determine the equilibrium paths. (Define the nondimensional load $p = P\ell/C$.)

(c) Determine if the path is stable.

(d) Sketch the path and compare with the case $\theta_0 = 0$.

(e) Determine the stiffness $S = dp/d\delta$, where δ is a nondimensional displacement.

(f) Sketch the S–δ curve and compare with the case $\theta_0 = 0$.

6.10 In Fig. 6.37, the rigid bar is originally vertical. Its length is ℓ, and it is hinged at A and supported at C by a spring of modulus K constrained to move horizontally, as its left end is pushed by a force Q along a frictionless incline to make an angle β with the vertical. At end B, a load P, which always remains vertical, is applied as shown. Using large-displacement analysis,

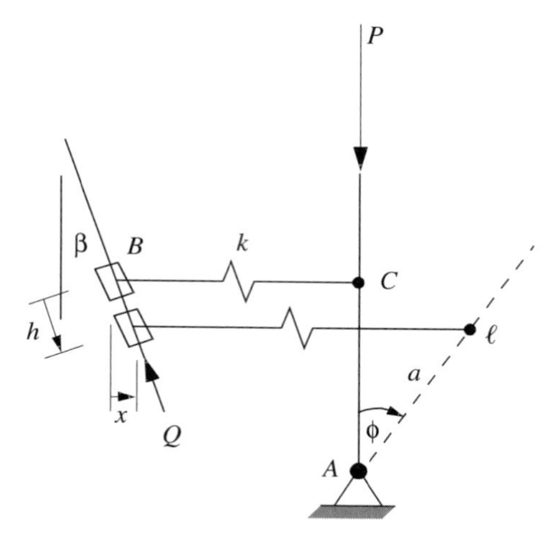

Figure 6.37 Problem 6.10.

(a) Determine the potential energy of the system.

(b) Determine the equilibrium paths. [Define the nondimensional load $p = P\ell/(ks^2)$.]

(c) Determine if the paths are stable.

(d) Sketch the paths.

(e) Determine the stiffness $S = dp/d\delta$, where δ is a nondimensional displacement.

(f) Sketch the S–δ curve.

6.11 In Fig. 6.38, the rigid bar is originally vertical. Its length is ℓ, and it is hinged at A and supported at C at a distance a from the hinge C by a spring k that always moves horizontally. At its end, a load P, which always remains vertical, is applied at an eccentricity e as shown. Using large-displacement analysis,

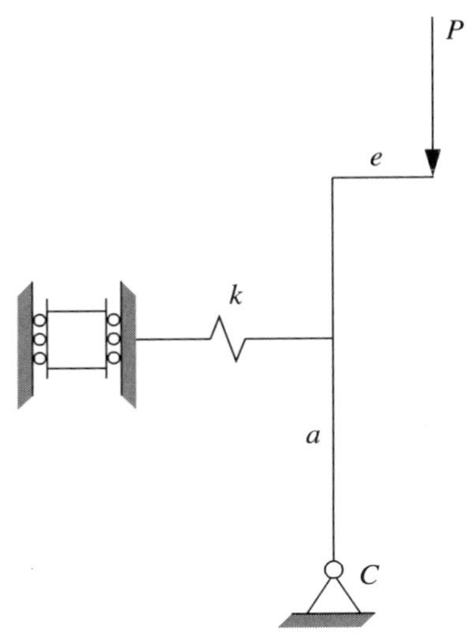

Figure 6.38 Problem 6.11.

(a) Determine the potential energy of the system.

(b) Determine the equilibrium paths. [Define the nondimensional load $p = P\ell/(ka^2)$.]

(c) Determine if the paths are stable.

(d) Determine stiffness $S = dp/d\delta$, where δ is a nondimensional displacement.

(e) Sketch the S–δ curve for $e/\ell = 0.1$ and $e/\ell = 0.001$.

6.12 Derive Eq. (6.93) by using Lagrange's equations.

REFERENCES

[1] S. Timoshenko and J. N. Goodier, *Theory of Elasticity*, McGraw-Hill, New York, 1951.

[2] R. Weinstock, *Calculus of Variations*, McGraw-Hill, New York, 1952.

[3] C. Lanczos, *The Variational Principles of Mechanics*, University of Toronto Press, Toronto, 1949.

[4] T. R. Kane, *Dynamics*, Holt, Rinehart, and Winston, New York, 1968.

[5] J. W. Gibbs, *Vector Analysis*, Dover, New York, 1960, pp. 139–142.

[6] I. Stakgold, *Boundary Value Problems of Mathematical Physics*, Vols. I and II, Macmillan, London, 1967.

[7] J. N. Readdy and M. L. Rasmussen, *Advanced Engineering Analysis*, John Wiley and Sons, New York, 1982, pp. 153–294.

[8] È. Goursat, *A Course in Mathematical Analysis*, Vol. I, *Derivatives and Differentials Definite Integrals Expansion in Series Applications to Geometry*, Dover, New York, 1959, pp. 66–83.

[9] H. B. Callen, *Thermodynamics*, John Wiley and Sons, New York, 1960.

[10] L. Brand, *Vector and Tensor Analysis*, John Wiley and Sons, New York, 1947, pp. 51–55.

[11] A. J. Durelli and V. J. Parks, *More Analysis of Strain*, Prentice-Hall, Englewood Cliffs, NJ, 1970, p. 95.

[12] L. E. Malvern, *Introduction to the Mechanics of a Continuous Medium*, Prentice-Hall, Englewood Cliffs, NJ, 1969, pp. 160–161.

[13] V. V. Novozhilov, *Foundations of the Nonlinear Theory of Elasticity*, Graylock Press, Rochester, NY, 1953, pp. 39–56.

[14] A. C. Eringen, *Nonlinear Theory of Continuous Media*, McGraw-Hill, New York, 1952, pp. 44–47.

[15] I. S. Sokolnikoff, *Mathematical Theory of Elasticity*, 2d Ed., McGraw-Hill, New York, 1956, pp. 25–29.

[16] L. E. Malvern, *Introduction to the Mechanics of a Continuous Medium*, Prentice-Hall, Englewood Cliffs, NJ, 1969, p. 214.

[17] K. Washizu, *Variational Methods in Elasticity and Plasticity*, Pergamon Press, Oxford, 1969, p. 66.

[18] H. B. Callen, *Thermodynamics*, John Wiley and Sons, New York, 1960, pp. 213–237.

[19] L. E. Malvern, *Introduction to the Mechanics of a Continuous Medium*, Prentice-Hall, Englewood Cliffs, NJ, 1969, pp. 243–266.

[20] D. T. Greenwood, *Principles of Dynamics*, Prentice-Hall, Englewood Cliffs, NJ, 1965, pp. 332–336.

[21] A. A. F. van de Ven, *Interaction of Electromagnetic and Elastic Fields in Solids*, Druk Vam Voorschoten, The Netherlands, 1975.

[22] K. Washizu, *Variational Methods in Elasticity and Plasticity*, Pergamon Press, Oxford, 1969, pp. 93–101.

[23] M. A. Biot, *Mechanics of Incremental Deformations*, John Wiley and Sons, New York, 1965.

[24] L. E. Malvern, *Introduction to the Mechanics of a Continuous Medium*, Prentice-Hall, Englewood Cliffs, NJ, 1969, pp. 273–422.

[25] H. L. Langhaar, *Energy Methods in Applied Mechanics*, John Wiley and Sons, New York, 1962, pp. 81–82.

[26] R. H. Gallagher, *Finite Element Analysis Fundamentals*, Prentice-Hall, Englewood Cliffs, NJ, 1975, pp. 218–223.

[27] S. O. Asplund, *Structural Mechanics: Classical and Matrix Methods*, Prentice-Hall, Englewood Cliffs, NJ, 1966, pp. 212–216.

[28] B. A. Boley and J. H. Weiner, *Theory of Thermal Stress*, John Wiley and Sons, New York, 1960, pp. 307–335.

[29] L. Brand, *Vector and Tensor Analysis*, John Wiley and Sons, New York, 1947, pp. 90–94.

[30] F. B. Seely and J. O. Smith, *Advanced Mechanics of Materials*, 2d Ed., John Wiley and Sons, New York, 1952, pp. 137–187.

[31] S. F. Borg and J. J. Gennaro, *Advanced Structural Analysis*, D. Van Nostrand, Princeton, NJ, 1959, pp. 193–200.

[32] B. A. Boley and E. S. Barrekette, "Thermal Stress in Curved Beams," *Journal of the Aero Space Sciences*, Vol. 25, No. 10 (October 1958): pp. 627–630, 643.

[33] B. A. Boley and J. H. Weiner, *Theory of Thermal Stress*, John Wiley and Sons, New York, 1960, pp. 356–359.

[34] R. J. Roark and W. C. Young, *Formulas for Stress and Strain*, 5th Ed., McGraw-Hill, New York, 1975, pp. 92–93, 209–213.

[35] V. Z. Vlasov, *Thin-Walled Elastic Beams*, 2d Ed. revised and augmented 1959, translated from Russian, distributed by Clearinghouse for federal and technical information: TT61-11400, Washington, DC.

[36] T. V. Galambos, *Structural Members and Frames*, T. V. Galambos, St. Louis, MO, 1978, Library of Congress Catalog Card Number: 68-17530.

[37] C. F. Kollbrunner and K. Basler, *Torsion in Structures, An Engineering Approach*, translated from the German edition by E. C. Glauser with Annotations and Appendix by B. G. Johnston, Springer-Verlag, Berlin, 1969.

[38] E. P. Popov, *Introduction to Mechanics of Solids*, Prentice-Hall, Englewood Cliffs, NJ, 1968, pp. 47–53.

[39] Y. B. Zeldovicj, *Higher Mathematics for Beginners and its Application to Physics*, translated from Russian by G. Yankovsky, Mir Publishers, Moscow, 1973.

[40] J. M. Gere and W. Weaver, *Analysis of Framed Structures*, D. Van Nostrand, New York, 1965.

[41] A. E. Taylor, *Advanced Calculus*, Ginn and Company, Boston, 1955, pp. 479–493.

[42] W. T. Koiter, "On the Principle of Stationary Complementary Energy in the Nonlinear Theory of Elasticity," *SIAM Journal Applied Mathematics*, Vol. 25, No. 3 (November 1973): pp. 424–434.

[43] C. T. Wang, *Applied Elasticity*, McGraw-Hill, New York, 1973, pp. 77–105.

[44] H. Weyl, *Symmetry*, lectures given at the Institute for Advanced Study, reprinted in the *World of Mathematics*, Simon and Schuster, New York, Vol. I, 1956, pp. 671–724.

[45] C. E. Fortescue, "Method of Symmetrical Coordinates Applied to the Solution of Polyphase Networks," *American Institute of Electrical Engineers Transactions*, Vol. 37, Part II (1918): pp. 1027–1140.

[46] R. W. Hamming, *Numerical Methods for Scientists and Engineers*, McGraw Hill, New York, 1962, pp. 67–78.

[47] J. Robinson, *Integrated Theory of Finite Element Methods*, John Wiley and Sons, New York, 1973, pp. 15–67.

[48] D. V. Wallerstein, "Thermal Deformation Vector for a Bilinear Temperature Distribution in an Anisotropic Quadrilateral Membrane Element," *International Journal for Numerical Methods in Engineering*, Vol. 9 (1975): pp. 325–336.

[49] A. E. H. Love, *A Treatise on the Mathematical Theory of Elasticity*, Dover, New York, 1944, pp. 11, 92–100.

[50] J. H. Argyris, and S. Kelsey, *Energy Theorems and Structural Analysis*, Butterworth, London, 1960, p. 3.

[51] L. E. Malvern, *Introduction to the Mechanics of a Continuous Medium*, Prentice-Hall, Englewood Cliffs, NJ, 1969, pp. 226–272.

[52] T. J. R. Hughes, *The Finite Element Method*, Prentice-Hall, Englewood Cliffs, NJ, 1987, pp. 2–10.

[53] K. Washizu, *Variational Methods in Elasticity and Plasticity*, Pergamon Press, Oxford, 1969, pp. 31–38, 68–69.

[54] K. Washizu, *Variational Methods in Elasticity and Plasticity*, Pergamon Press, Oxford, 1969, pp. 63–64.

[55] D. O. Bush and B. O. Almroth, *Buckling of Bars, Plates, and Shells*, McGraw-Hill, New York, 1975, pp. 120–141.

[56] H. Ziegler, *Principles of Structural Stability*, Blaisdell Publishing Company, Waltham, MA, 1968.

[57] J. G. Croll and A. C. Walker, *Elements of Structural Stability*, John Wiley and Sons (Halsted Press), New York, 1972.

[58] J. L. Synge and B. A. Griffith, *Principles of Mechanics*, McGraw-Hill, New York, 1959, pp. 411–436.

[59] T. R. Kane and D. A. Levinson, *Dynamics: Theory and Application*, McGraw-Hill, New York, 1985.

[60] M. Abramowitz and I. A. Stegun, Eds., *Handbook of Mathematical Functions with Formulas, Graphs, and Mathematical Tables*, National Bureau of Standards Applied Mathematics, Ser. 55, June 1964 (7th printing: May 1968 with corrections, pp. 722–750).

[61] C. M. Bender and S. A. Orszag, *Advanced Mathematical Methods for Scientists and Engineers*, McGraw-Hill, New York, 1978, pp. 560–566.

INDEX